Manfred Wendisch and Ping Yang
**Theory of Atmospheric Radiative
Transfer**

Manfred Wendisch and Ping Yang

Theory of Atmospheric Radiative Transfer

A Comprehensive Introduction

WILEY-VCH

WILEY-VCH Verlag GmbH & Co. KGaA

The Authors

Prof. Manfred Wendisch
University of Leipzig
Leipzig Institute for Meteorology (LIM)
Leipzig, Germany
m.wendisch@uni-leipzig.de

Prof. Ping Yang
Texas A&M University
Department Atmospheric Science
College Station, USA
pyang@tamu.edu

Cover
The photograph was taken in Pegnitz, Germany
by Stefan Bauer on 15 July 2008. Composed
of five individual images taken with different
exposure times in order to handle the difficult
lighting conditions and the contours of the
landscape and the sun using HDRI (high
dynamic range imaging) software.

All books published by Wiley-VCH are carefully
produced. Nevertheless, authors, editors, and
publisher do not warrant the information
contained in these books, including this book, to
be free of errors. Readers are advised to keep in
mind that statements, data, illustrations,
procedural details or other items may
inadvertently be inaccurate.

Library of Congress Card No.: applied for

British Library Cataloguing-in-Publication Data:
A catalogue record for this book is available
from the British Library.

**Bibliographic information published by the
Deutsche Nationalbibliothek**
The Deutsche Nationalbibliothek lists this
publication in the Deutsche Nationalbibliografie;
detailed bibliographic data are available on the
Internet at http://dnb.d-nb.de.

© 2012 WILEY-VCH Verlag GmbH & Co. KGaA,
Boschstr. 12, 69469 Weinheim, Germany

Typesetting   le-tex publishing services GmbH,
Leipzig
Printing and Binding   Markono Print Media
Pte Ltd, Singapore
Cover Design   Grafik-Design Schulz,
Fußgönheim

Printed in Singapore
Printed on acid-free paper

ISBN Print 978-3-527-40836-8

Contents

Preface

The energy that drives terrestrial weather and climate comes from the Sun. The absorption of solar radiation by the Earth-atmosphere system warms our planet. Solar warming is balanced by the cooling due to the atmospheric and surface emission of thermal infrared radiation. Thus, knowledge about the transfer of solar and terrestrial radiation in the atmosphere is critical to understanding the radiative energy budget and current and future climates. Furthermore, the theory of radiative transfer plays a vital role in atmospheric remote sensing, which is becoming increasingly important in geosciences. For these reasons, atmospheric radiation is now part of the core curriculum in atmospheric sciences.

The theme of the book is to introduce and illustrate the governing equations of atmospheric radiative transfer. Specifically, the book offers a comprehensive but concise survey of all physical processes relevant to single-scattering, multiple-scattering, absorption, and emission of electromagnetic radiation in the atmosphere. Important equations are derived from basic principles, and key results are illustrated graphically. Some basic mathematical tools that are widely used in the discipline are also briefly reviewed.

In the authors' opinion, existing texts on this subject can be categorized into two groups as follows:

Advanced texts The texts in this category are lengthy and in-depth, and therefore beyond the level of even those senior undergraduate students who already have quite extensive training in mathematics and physics. In most European and American universities, atmospheric radiation is taught as a one-semester course, and the scope of these texts is beyond a one-semester course. Some topics covered in many existing atmospheric radiation texts are outdated. For example, band models for calculating the infrared transmission are usually described in detail in many existing books. However, with the development of the correlated k-distribution method that is computationally efficient and accurate, band models are largely obsolete.

General introductory level texts The texts in this category are at a rather basic level and supply little detail. In other words, these books lack depth or are too brief for senior undergraduate or first year graduate students, as most quantitative information is missing.

This book is intended to fill the gap between the aforementioned two types of texts, within the constraint that the material should be appropriate for a one-semester course. Because the book is not aimed at specialists, all mathematical derivations are performed without omitting major intermediate steps. This is important for the readability of the text and will help the students to master the subject in detail.

The recommended prerequisite for the reader is completion of the bachelor in European universities or two years of college-level physics and mathematics courses in the United States (US) system. Although European and US systems are converging due to the introduction of the bachelor and master system in Europe, certain differences still exist. With the authors' teaching experience in Germany (MW) and the US (PY), we hope the book will help to facilitate teaching atmospheric radiation in various international educational systems.

This book is the result of the authors' lecture notes. Needless to say, any errors in the book are the authors' own.

We would like to thank the former and current members of our research groups at the University of Leipzig and Texas A&M University for their dedication and contribution to their assigned research projects. It has been a great inspiration for us to work with students.

Manfred Wendisch is particularly grateful to his main scientific mentor, Jost Heintzenberg. In the past several years, there were many fruitful collaborations on various topics of atmospheric radiation with colleagues around the world which are acknowledged in alphabetic order in certainly an incomplete list: Howard Barker, Darrel Baumgardner, Franz Berger, Birger Bohn, Stephan Borrmann, Jean-Louis Brenguier, Phil Brown, Anthony Bucholtz, Susanne Crewell, Uwe Feister, John Foot, Jürgen Fischer, Paola Formenti, Barbara Früh, Jean-Francois Gayet, Tim Garrett, Hermann Gerber, Xingfa Gu, Andrew Heymsfield, Tadahiro Hayasaka, Jim Haywood, Andreas Herber, Wolfgang von Hoyningen-Huene, Ruprecht Jänicke, Ralph Kahn, Alex Kokhanovsky, Alexei Korolev, Barry Lefer, Zhengqiang Li, Andreas Macke, Bernhard Mayer, Teruyuki Nakajima, Klaus Pfeilsticker, Peter Pilewski, Steven Platnick, Heinrich Quenzel, Erhard Raschke, Jens Redemann, Phil Russell, Maria von Schönermark, Rick Shetter, Clemens Simmer, Walter Strapp, Jonathan Taylor, Thomas Trautmann, Victor Venema, Matthias Wiegner, and Warren Wiscombe.

Ping Yang would like to take this opportunity to sincerely thank several senior (not age-wise) scientists, Kuo-Nan Liou, Gerald North, George Kattawar, Thomas Wilheit, Kenneth Bowman, Warren Wiscombe, Michael King, Thomas Vonder Haar, James Coakley, William L. Smith, and Andrew Heymsfield, for their mentoring and guidance. Ping Yang also wishes to express his special gratitude to Hal Maring, Donald Anderson, Bradley Smull, Rangasayi Halthore and S. Daniel Jacob for their support and encouragement. Ping Yang is very thankful for the fruitful collaborative educational and research efforts with several colleagues at the Department of Atmospheric Sciences, Texas A&M University (in alphabetic order): Sarah Brooks, Andrew Dessler, Shaima Nasiri, and R. Lee Panetta. Ping Yang is also very grateful for the inspiring and pleasant research collaborations with external collaborators: Anthony Baran, Bryan Baum, Helene Chepfer, Xiquan Dong, Oleg Dubovik, Qiang Fu, Bo-Cai Gao, Thomas Greenwald, Yong Han, Andrew Heidinger, N. Christina Hsu, Yongxiang Hu, Hung-Lung (Allen) Huang, Hironobu Iwabuchi, Ralph Kahn, Jhoon Kim, Jun Li, Istvan Laszlo, Tang-Huang Lin, Quanhua Liu, Xiaodong Liu, Xu Liu, Alexander Marshak, Patrick Minnis, Michael Mishchenko, Martin Mlynczak, Steven Platnick, Peter Pilewskie, Jerome Riedi, Byung-Ju Sohn, Wenbo Sun, Si-Chee Tsay, Heli Wei, Fuzhong Weng, Dong L. Wu, Pengwang Zhai, Zhibo Zhang, and Daniel Zhou. During the preparation of this text, Ping Yang's research was supported by the following US agencies: the National Science Foundation (NSF), National Aeronautics and Space Administration (NASA), National Oceanic and Atmospheric Administration (NOAA) and Federal Aviation Administration (FAA).

This book would not have been possible without the selfless help of Frank Werner (Leipzig Institute for Meteorology, LIM, University of Leipzig) whose advanced drawing skills produced most of the figures. We are very grateful to Ms. Mary Gammon (Texas A&M University) who did an extraordinarily good job of editing the manuscript. Our thanks also go to Shouguo Ding, Lei Bi, Yu Xie, Bingqi Yi, Chao Liu, Hyoun-Myoung Cho, Chenxi Wang, Bingqiang Sun, Xin Huang, and Meng Gao (Texas A&M University) for preparing diagrams and proofreading mathematical equations in the manuscript. We would like to express our gratitude to André Ehrlich, Eike Bierwirth, Heike Kalesse, Sebastian Otto, and Evi Jäkel (LIM, University of Leipzig) who have carefully commented on the manuscript and provided problems and solutions, jointly with Sebastian Schmidt (Laboratory for Atmospheric and Space Physics, University of Colorado). Stefan Bauer (LIM, University of Leipzig) supplied the cover page photo.

Last, but not least, we thank our family members for having kindly endured our long preoccupation with writing this book. Amanda Yang participated in the editing process and offered some good suggestions to improve the manuscript.

Leipzig, December 2011
College Station, December 2011

Manfred Wendisch
Ping Yang

1
Introduction

1.1
Brief Survey of Atmospheric Radiation

The budget of electromagnetic (EM[1]) radiation in the Earth-atmosphere system is critical to atmospheric energetics. The major portion of energy exchanged between the Earth and the cosmic background, including the solar system, is EM radiation with wavelengths ranging from the ultraviolet (UV) to the far infrared (far IR, FIR) regimes. The transfer of solar and thermal infrared (TIR) radiation, which involves absorption, scattering, and emission in the Earth-atmosphere system, is the primary influence on the terrestrial climate and has a significant impact on the weather system. Atmospheric radiation is an important branch of modern atmospheric sciences because of the role radiation plays in the atmosphere. A number of excellent texts are fully, or partially, dedicated to the discipline of atmospheric radiative transfer including single- and multiple-scattering and absorption processes (Bohren and Clothiaux, 2006; Bohren and Huffman, 1983; Goody and Yung, 1989; Liou, 2002; Mishchenko et al., 2006; Petty, 2006; Pomraning, 1973; Stephens, 1994; Thomas and Stamnes, 1999; Zdunkowski et al., 2007). Some of these texts assume the reader to be well trained in mathematics and physics, particularly in electrodynamics and optics.

The transfer of EM radiation in the atmosphere is a multiple-scattering process involving absorption by various gases along with scattering and absorption by particulate matter (aerosol particles, liquid water droplets, and ice crystals). Early studies of EM radiative transfer were primarily completed by astrophysicists, including the eminent scientists A.S. Eddington, E.A. Milne, K. Schwarzschild, V.A. Ambarzumian, and S. Chandrasekhar, who were interested in the transfer of EM radiation in stellar and planetary atmospheres. Schuster (1905) used a simple two-stream approximation to discuss the transfer of EM radiation through a foggy atmosphere, and is generally believed to be the first to study the multiple-scattering process. However, the study of radiative transfer in a scattering medium can be traced (Mishchenko, 2008) to the work of Lommel (1887) and Chwolson (1889).

---

1) Acronyms used in this book are written in capital letters and explained in Appendix A.1.

Theory of Atmospheric Radiative Transfer, First Edition. Manfred Wendisch and Ping Yang
© 2012 WILEY-VCH Verlag GmbH & Co. KGaA. Published 2012 by WILEY-VCH Verlag GmbH & Co. KGaA.

The Radiative Transfer Equation (RTE) is the cornerstone of radiative transfer theory. Traditionally, the RTE had been regarded as a phenomenological formula, until the work by Tsang and Kong (2001) and Mishchenko (2008, and references cited therein), who demonstrated the theory of radiative transfer to be a corollary of Maxwell's equations. During the early development stages of the radiative transfer theory, significant effort was dedicated to solving the RTE, subject to the plane-parallel approximation, and the results have been summarized in several classic texts (Chandrasekhar, 1950; Lenoble, 1985; Preisendorfer, 1965; Sobolev, 1975; van de Hulst, 1980).

The RTE for a scattering medium consisting of an ensemble of particles requires an understanding of the interaction between radiation and particulate matter. The Lorenz–Mie theory, first described by Lorenz (1890) with independent subsequent development by Mie (1908) and Debye (1909), was formulated to study the scattering of EM radiation by spherical particles. Logan (1965) surveyed some early, prior to World War II, studies of the scattering of plane EM waves by spheres; however, many atmospheric particles, such as airborne dust aerosol particles and ice crystals, are nonspherical. Within the past fifty years, significant effort (Kokhanovsky, 2006; Mishchenko et al., 2000) has been focused on the description of EM scattering by nonspherical and inhomogeneous particles, resulting in the development of new computational techniques, for example, the T-matrix method (Mishchenko and Travis, 1994b; Waterman, 1965), the finite-difference time domain technique (Yee, 1966), and the discrete dipole approximation (Purcell and Pennypacker, 1973).

The absorption of EM radiation by atmospheric gases has been an active research area with laboratory measurements of various gases documenting spectral line parameters. A long range project, initiated in the 1960s, has compiled and continued updating a High-resolution Transmission Molecular Absorption Database, known as HITRAN (Rothman et al., 1992). Based on HITRAN, the Line-By-Line Radiative Transfer Models (LBLRTMs), for example, see Clough et al. (1992), can accurately compute the transmission of radiation through atmospheric gases, but require substantial computational resources. Various other spectral band models have evolved to increase the computational efficiency of radiative transfer simulation. Other techniques, the k-distribution method (Ambartzumian, 1936), the correlated k-distribution method (Fu and Liou, 1992b; Lacis and Oinas, 1991; Lacis et al., 1979), the principal component-based radiative transfer technique (Liu et al., 2006), and the optimal spectral sampling method (Moncet et al., 2008), have been developed for both atmospheric radiation budget studies and remote sensing applications.

In the plane-parallel approximation, the optical and microphysical properties of a medium are assumed to be horizontally homogeneous. Attempts to solve the RTE, using the plane-parallel approximation, encountered problems with three-dimensional (3D) radiative transfer, for example, in the transfer of radiation within cumulus clouds. A summary of advances in the study of 3D radiative transfer within cloudy atmospheres can be found in a monograph edited by Marshak and Davis (2005).

Both radiation intensity and polarization properties are required for a complete description of EM radiation. A concise introduction to the transfer of polarized radiation is offered in an excellent treatise by Hovenier et al. (2004). The polarization state of radiation has been employed in the astrophysics and optics disciplines (Mishchenko et al., 2010) for many years; however, atmospheric remote sensing techniques based on polarimetric measurements are in their infancy. The benefits associated with the use of polarization measurements in various passive and active atmospheric remote sensing implementations have become increasingly obvious.

Atmospheric radiation is a rapidly evolving field with new methods and algorithms constantly being developed to solve numerous questions about the radiative transfer process in the atmosphere. The focus of this text will not be on the new developments, but rather on the basic definitions of radiometric and polarimetric quantities and the fundamental principles of atmospheric radiation.

1.2
A Broadbrush Picture of the Atmospheric Radiation Budget

In atmospheric sciences, the EM radiation emitted by the Sun is usually considered to lie within a spectral region of 0.2–5 μm, and is normally referred to as solar radiation; whereas, the terrestrial thermal emission of interest lies within the spectral region primarily from ∼ 3 to 100 μm. Both solar radiation and the terrestrial thermal emission have complicated spectral structures, which will be briefly discussed in Section 1.3.

Neglecting the detailed spectral features of the solar spectrum observed at the top of the atmosphere (TOA), Figure 1.1 approximately describes the overall spectral irradiance by the two solid lines calculated by assuming the Sun as a blackbody with a temperature of $T_S = 5800$ K. The two dashed lines in Figure 1.1 represent the terrestrial thermal emission simulated by assuming the effective temperature of the Earth as $T_E = 288$ K without accounting for the absorption of atmospheric gases, such as, carbon dioxide, ozone, and water vapor. The logarithmic scales are applied to the abscissae in Figure 1.1a,b. In Figure 1.1a, the same ordinate scale is applied to both curves; in Figure 1.1b, different scales are applied to the two curves in order to make them appear to be equally manifest. In Figure 1.1b, it is evident that solar radiation is primarily concentrated within a narrow spectral region in comparison with terrestrial thermal emission. The solar and terrestrial spectra are often treated separately in practical applications. As shown in Figure 1.1, the solar spectrum has its maximum within the visible spectral interval of 0.4–0.7 μm and can be perceived by the human eye. The terrestrial thermal emission maximum is located approximately at an infrared (IR) wavelength near 10 μm, which the human eye, without being equipped with a special instrument, cannot see. The locations of the spectral maxima in Figure 1.1 are directly determined by temperature. Stellar radiation has its spectral maximum in the visible region; whereas, the planetary thermal emission has a spectral maximum in the IR region. For this reason, the

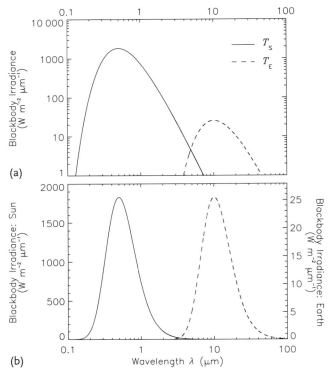

Figure 1.1 (a) Spectra of solar radiation and terrestrial thermal emission simulated by assuming the Sun and the Earth to be black-bodies at temperatures of $T_S = 5800\,$K and $T_E = 288\,$K. For the spectra, the Planck law, see Eq. (3.58), is applied for isotropic radi-ation. For the solar spectrum, a mean Sun–Earth distance of 149.6×10^6 km is assumed. (b) The same as in (a), but the left vertical scale is linear and applied to solar spectrum, while the right scale (linear) is applied to the terrestrial thermal emission.

visible color of a star depends on its temperature, and the perceived color of a planet is not determined by its temperature.

Both the solar radiation and the terrestrial thermal emission are modified by several processes before leaving the atmosphere. These processes are illustrated in Figure 1.2 adapted from Trenberth et al. (2009). From the $341\,\mathrm{W\,m^{-2}}$ of incoming solar radiation, $79\,\mathrm{W\,m^{-2}}$ are reflected back into space by clouds, aerosol particles, and the molecules of atmospheric gases, and $23\,\mathrm{W\,m^{-2}}$ are reflected by the Earth's surface. Thus, a total of $102\,\mathrm{W\,m^{-2}}$, approximately 30% of the solar radiation incident at the TOA, is reflected back into space (left half of Figure 1.2) due to the planetary albedo effect.

Approximately 23%, or $78\,\mathrm{W\,m^{-2}}$, of the solar radiation incident at the TOA is absorbed within the atmosphere. The absorption of solar radiation by atmospheric trace gases, mostly water vapor, contributes to the globally and annually averaged value of $78\,\mathrm{W\,m^{-2}}$. The absorption by aerosol particles and clouds is much less

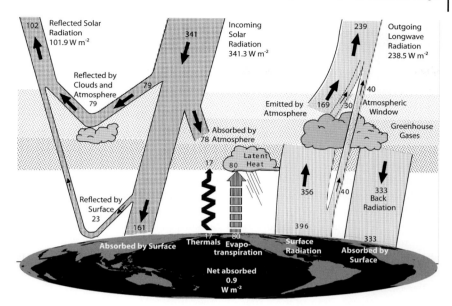

Figure 1.2 The Earth's annual global mean energy budget. The numbers are averaged globally, annually, and wavelength-integrated (broadband-solar: left half of the figure, broadband-thermal: right half) energy flux densities (irradiances) in $W\,m^{-2}$. Adapted from Trenberth et al. (2009), courtesy of K. Trenberth, and reprinted with permission of the American Meteorological Society (AMS) © AMS.

than that by the gaseous atmospheric components. $161\,W\,m^{-2}$ ($\approx 47\%$) are absorbed by the land and oceans.

The solar radiation absorbed within the atmosphere and by the surface is transformed into thermal energy (right half of Figure 1.2). Approximately $396\,W\,m^{-2}$ of the thermal energy is reemitted by the surface, and about $40\,W\,m^{-2}$ penetrates through the atmosphere in the "atmospheric window regions" without being reabsorbed. Of the surface-emitted terrestrial radiation, about $356\,W\,m^{-2}$ are absorbed by clouds and trace gases, particularly by water vapor, and reemitted at lower temperatures. Thus, out of the $356\,W\,m^{-2}$, only $169\,W\,m^{-2}$ from gases, $30\,W\,m^{-2}$ from clouds, and $40\,W\,m^{-2}$ from the surface are reemitted to space. The remaining radiant energy is absorbed by the atmosphere-surface system of the Earth and leads to the greenhouse effect discussed in Section 1.4.

At the TOA, $239\,W\,m^{-2}$ of terrestrial radiation escapes the Earth-atmosphere system, while the same amount effectively enters the system in the solar spectral region [$(239 = 341 - 102)\,W\,m^{-2}$] to maintain the energy equilibrium of the climate system. However, the solar and terrestrial radiation is not in equilibrium at Earth's surface: $(161 - 396 + 333)\,W\,m^{-2} = 98\,W\,m^{-2}$. As illustrated in Figure 1.2, the surface continuously gains EM radiant energy, the atmosphere loses radiant energy, and the deficit is balanced by latent and thermal heat transfer.

1.3
Solar and Terrestrial Thermal Infrared Spectra in a Cloudless Atmosphere

Let us take a more detailed look at the downward and upward radiation in a cloud-free atmosphere by observing typical spectra. Figure 1.3 shows the simulated solar spectra observed at both the TOA and the surface. The simulations were performed with the Moderate Spectral Resolution Atmospheric Transmittance Algorithm and Computer Model (MODTRAN), see Berk et al. (1999). The downward solar radiation reveals typical Fraunhofer absorption lines most obvious at the TOA caused by absorption processes in the solar atmosphere. By the time the solar radiation reaches the surface, the EM radiation has been strongly absorbed at selected wavelengths by atmospheric gases, such as the ozone Hartley band (O_3: < 0.31 μm, see Table 8.3), oxygen (O_2: 0.76 μm; O_2-A band), water vapor (H_2O: 0.72, 0.82, 0.94, 1.1, 1.38, 1.87, 2.7 μm, see Table 8.2), or carbon dioxide (CO_2: 1.4, 2.0, 2.7 μm). The pronounced minimum at the 1.38 μm wavelength is caused by water vapor absorption. This water vapor absorption band is very useful for detecting cirrus clouds located high in the atmosphere (Gao and Kaufman, 1995).

Two examples of the spectral distribution of the upward (outgoing) terrestrial IR radiation, corresponding to the surface temperatures 273 and 288 K, at the TOA are presented in Figure 1.4. In some wavelength bands, a cloud-free atmosphere is far from transparent; however, in the 8–13 μm spectral region, the atmosphere is quite transparent except for the ozone absorption band at 9.6 μm. The spectral observations within this atmospheric window region can be used to retrieve typical properties, for example, the surface temperature from space. The strong CO_2 absorption band centered at 15 μm is another distinct feature shown in Figure 1.4.

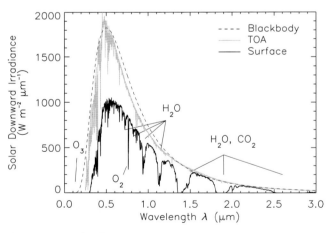

Figure 1.3 Simulated spectral distribution of solar radiation (downward irradiance) at the TOA and at the surface. The dashed line indicates the spectrum by assuming the Sun to be a blackbody with a temperature of $T_S = 5800$ K. The simulations were performed with MODTRAN4 (Version 3 Revision 1, MOD4v3r1) with the Sun in the zenith.

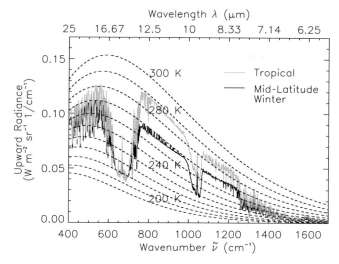

Figure 1.4 Simulated IR upward radiance spectrum at the TOA. "Tropical" and "Mid-Latitude Winter" profiles of atmospheric parameters were applied and taken from Anderson et al. (1986). The dashed lines represent blackbody radiances for various temperatures (200–300 K in 10-K increments). Absorption bands corresponding to CO_2 (15 µm), H_2O (6.3 µm), and O_3 (9.6 µm) are evident. The simulations were performed with MODTRAN4 (Version 3 Revision 1, MOD4v3r1).

Since CO_2 is well mixed and relatively stable in the atmosphere, this absorption band is very useful for the retrieval of the vertical temperature profile.

1.4
The Greenhouse Effect

Spectra of the atmospheric transmissivity along with the downward radiances at the surface are plotted in Figure 1.5. Atmospheric gases (mainly O_2, H_2O, O_3, and CO_2) absorb strongly in certain wavelength regions. However, the atmosphere is quite transparent for a large portion of the solar spectrum, and most of the solar radiation passes through the atmosphere. This is not true for the terrestrial spectral region where much less of the emitted radiation is transmitted through the atmosphere, see Figure 1.6, which shows the spectral signatures of both the transmissivity and the outgoing terrestrial radiance. The spectral signatures result from the variations of the atmospheric gas absorption.

Figure 1.6a,b depict the atmospheric transmissivity in the largest part of the terrestrial wavelength region, and Figure 1.6c,d show the simulated upward emitted radiances at the TOA. In the atmospheric window region between 8 and 13 µm, the atmosphere is essentially transparent to surface emitted radiation. This is more obvious in the comparatively dry (low water vapor amount) vertical profile of the "Subarctic Winter" case (Figure 1.6a) rather than the more humid situation of a "Trop-

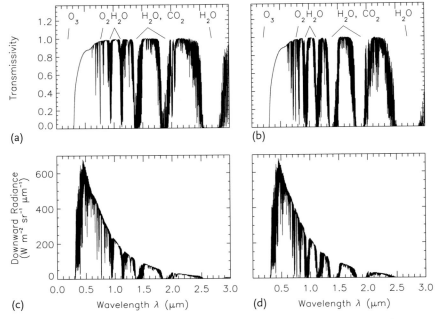

Figure 1.5 Spectral distribution of simulated solar atmospheric transmissivity (a, b) and solar downward radiances at the surface (c, d). "Subarctic Winter" (a, c) and "Tropical" (b, d) profiles of atmospheric parameters were applied and taken from Anderson et al. (1986). Absorption bands corresponding to: O_3: $< 0.31 \, \mu m$; O_2: $0.76 \, \mu m$; H_2O: 0.72, 0.82, 0.94, 1.1, 1.38, 1.87, $2.7 \, \mu m$; CO_2: 1.4, 2.0, $2.7 \, \mu m$, see Table 8.2. The simulations were performed with MODTRAN4 (Version 3 Revision 1, MOD4v3r1) with the Sun in the zenith.

ical" atmosphere (Figure 1.6b). Opposite to the transmissivity, the emitted upward radiances simulated at the TOA (Figure 1.6c,d) are larger for the "Tropical" profile than the "Subarctic Winter" case because the surface temperature is much higher in the "Tropical" case. Outside the atmospheric window region, nearly all terrestrial radiant energy is absorbed by atmospheric trace gases known as greenhouse gases, for example, H_2O, CO_2, and O_3. These gases keep the global temperature approximately 33 K higher than the equilibrium temperature of an atmosphere-free Earth exposed to solar radiation and can be treated as sources of blackbody radiation in certain wavelength regions. Figure 1.6c,d present an illustration of the atmosphere acting as a nonblackbody emitter.

During the last century, the global CO_2 concentration in the atmosphere has been drastically increased by human activities (Keeling et al., 1976), causing a manmade global warming (note, there is an ongoing debate on this subject), the anthropogenic greenhouse effect. In order to stabilize the terrestrial climate system at the preindustrial temperature level, the anthropogenic output of greenhouse gases needs to be significantly decreased.

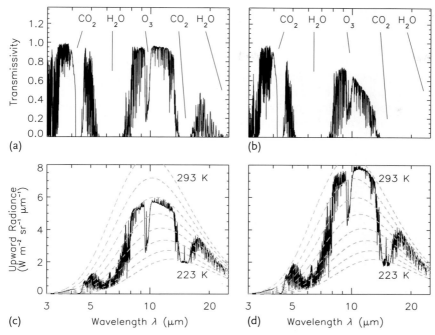

Figure 1.6 Spectral distribution of simulated terrestrial atmospheric transmissivity (a, b) and IR upward radiances emitted at the TOA (c, d). "Subarctic Winter" (a, c) and "Tropical" (b, d) profiles of atmospheric parameters were applied and taken from Anderson et al. (1986). The dashed lines indicate Planck's functions for 293 K (upper curve) and 223 K (lower curve) in 10-K increments. Major absorption/emission bands corresponding to: CO_2: 4.3, 15 µm; H_2O: 6.3 µm; O_3: 9.6 µm. The simulations were performed with MODTRAN4 (Version 3 Revision 1, MOD4v3r1).

1.5
Relevance to the Interpretation of Spaceborne Observations

Global scale observations of the Earth's atmosphere, oceans, and land surfaces are needed to study the planet as an unseparated system. Satellite observations provide unprecedented data sets to study numerous dynamic, physical, and chemical processes in the Earth-atmosphere system. The satellite instruments are designed to detect EM radiation in various spectral regions. The EM radiation signals received by satellite sensors need interpretation to decipher the processes that govern the transfer of EM radiation through the atmosphere and to separate the influences of ocean and land surfaces.

Two major types of spaceborne instruments, passive receivers and active sensors, are deployed on orbiting satellites. Passive receivers detect radiation reflected or emitted by atmospheric constituents, such as, gas molecules, aerosol particles, and cloud or precipitation elements. The signals recorded by the passive satellite receivers can be used to retrieve atmospheric parameters, such as, temperature profiles, wind and humidity fields, aerosol properties, and cloud characteristics. Additionally, they are able to measure the components of the solar and terrestrial radiation budget. The active instruments, for example, LIDAR and RADAR, emit EM radiation and measure the backscatter from the initial radiation beams, which contains atmospheric property information. Examples of advanced spaceborne sensors are those on the "A-train" satellite platforms (Stephens et al., 2002).

To properly interpret spaceborne observations made by either passive or active instruments, knowledge of atmospheric radiative transfer and of the interactions between radiation and atmospheric constituents of interest is required. The basic principles of atmospheric radiation will be covered in this text to help one understand the fundamental principles of spaceborne atmospheric remote sensing.

2
Notation and Math Refresher

The study of atmospheric radiation is a discipline that requires quantitative descriptions of various physical processes involved in radiative transfer. Subrahmanya Chandrasekhar said "Physical laws must have mathematical beauty" (Wali, 1997), and Leonardo da Vinci stated "No human investigation can be termed true knowledge if it does not proceed to mathematical demonstration" (Nuland, 2000). Thus, the usage of mathematical equations and expressions is both essential and inevitable in the study of atmospheric phenomena. To avoid potential confusion, we devote this chapter to a description of the notation conventions and physical dimensions to be used in the remainder of the book. We briefly review vector algebra, the Dirac δ-function, solid angle, Legendre functions, and the quadrature formula. Vector algebra has a pivotal role in the study of the EM fields and the Dirac δ-function specifies either a collimated radiation beam or the direct radiation incident on the TOA. The solid angle is important to the definition of a fundamental radiometric quantity, the radiance. Legendre functions are necessary for expanding the phase function and will be discussed in Chapter 7. A quadrature formula, or Gaussian quadrature, is normally used to discretize a radiation field in some numerical techniques for solving the radiative transfer equation.

2.1
Physical Dimensions and Prefixes

The dimension, or unit, of a physical quantity is necessary to give a clear meaning to the quantity. For an equation that involves a number of physical quantities, the consistency of the units determines whether the equation is appropriately posed. For a given physical equation, the dimension associated with the left side of the equation must equal its counterpart on the right side. In other words, the equation $x = y$ is true only if $[x] = [y]$, where $[\ldots]$ indicates the dimension of the associated physical quantity. The terms x and y may be combined differences, products, or the powers of several physical quantities.

Unless explicitly stated, the SI system (Le Système International d'Unités) is used throughout this book. The SI system was established in 1960 during the

Theory of Atmospheric Radiative Transfer, First Edition. Manfred Wendisch and Ping Yang
© 2012 WILEY-VCH Verlag GmbH & Co. KGaA. Published 2012 by WILEY-VCH Verlag GmbH & Co. KGaA.

eleventh General Assembly of Measures and Weights. Examples of the SI-basic units are given in Table 2.1.

Examples of SI-derived units, expressed in terms of the SI-basic units, are given in Table 2.2. Table 2.3 lists prefixes that indicate multiples and fractions of units.

Table 2.1 SI-basic units.

Quantity	SI-basic unit	Symbol
Length	Meter	m
Mass	Kilogram	kg
Time	Second	s
Electrical current	Ampere	A
Temperature	Kelvin	K
Amount of substance	Mol	mol
Light intensity	Candela	cd

Table 2.2 Some SI-derived units.

Quantity	SI-derived unit	Symbol
Area	Square meter	m^2
Volume	Cubic meter	m^3
Speed	Meter per second	$m\,s^{-1}$
Acceleration	Meter per square second	$m\,s^{-2}$
Frequency	Hertz	$Hz = s^{-1}$
Force	Newton	$N = kg\,m\,s^{-2}$
Pressure	Pascal	$Pa = N\,m^{-2} = kg\,m^{-1}\,s^{-2}$
Energy	Joule	$J = N\,m = kg\,m^2\,s^{-2}$
Power (flux)	Watt	$W = J\,s^{-1} = kg\,m^2\,s^{-3}$
Electrical voltage	Volt	$V = W\,A^{-1} = kg\,m^2\,s^{-3}\,A^{-1}$

Table 2.3 Examples of prefixes.

Multiple	Prefix	Symbol	Portion	Prefix	Symbol
10^{24}	Yotta	Y	10^{-24}	Yokto	y
10^{21}	Zetta	Z	10^{-21}	Zepto	z
10^{18}	Exa	E	10^{-18}	Atto	a
10^{15}	Peta	P	10^{-15}	Femto	f
10^{12}	Tera	T	10^{-12}	Pico	p
10^{9}	Giga	G	10^{-9}	Nano	n
10^{6}	Mega	M	10^{-6}	Micro	μ
10^{3}	Kilo	k	10^{-3}	Milli	m
10^{2}	Hecto	h	10^{-2}	Centi	c
10^{1}	Deca	da	10^{-1}	Deci	d

2.2
Some Rules and Conventions

Several conventions dictate the notations in this book. A space is always placed between different physical units; however, no space is put between a physical unit and its prefix, be it a multiple or a portion; otherwise, confusion may arise. For example, milliwatt is written in the form of mW without a space. The term "m" stands for "milli" which indicates the prefix 10^{-3}, see Table 2.3, and the "W" is the SI-derived unit of "watt," see Table 2.2. In contrast, "meter" multiplied by "watt" is written with a space in between, that is m W. In this case, the term "m" indicates the SI-basic unit "meter" (see Table 2.1) and not the prefix "milli." A similar example is millimeter (mm without a space), which is distinguished from square meter (m m with space is $m \cdot m = m^2$).

In this book, a physical quantity is denoted using the following conventions: a scalar quantity is written in *italic*, such as, radiance I (see Section 3.2); a vector is written in **bold** with an overhead arrow, for example, the Euklidic position vector is indicated by $\vec{r}$ (see Section 3.2); a unit vector is marked with a widehat, such as, $\hat{e}_1$, a unit vector in a Cartesian coordinate system (see Section 2.3.1); a matrix is identified with a unique font, such as, the Mueller matrix IS (see Section 4.7).

The unit of a physical quantity and a mathematical operator are not set in *italic*, while all scalar variables are written in *italic*. This convention eliminates potential confusion, as illustrated by the following example:

$$\frac{\mathrm{d}d}{\mathrm{d}r} \, .$$

The expression above is the differentiation of the scalar variable, d, with respect to the scalar variable, r. The differentiation operator "d" is not written in *italic*, but the scalar variables "d" and "r" are italicized. Physical constants are written in *italic*, mathematical constants are not set *italic*.

2.3
Vector Algebra Brief

2.3.1
Major Vector Operations

Vector calculation with respect to a Cartesian coordinate system is straightforward, although other coordinate systems, such as, spherical and cylindrical coordinates, are necessary for some physical problems. This subsection is a brief review of vector calculus in a Cartesian coordinate system.

The scattering of EM radiation by small particles, for example, aerosol particles, ice crystals, and water droplets, is fundamental to the study of radiative transfer. In the physical process of EM scattering, Maxwell's equations describe the spatial and temporal variations of electric and magnetic fields and the relationship between

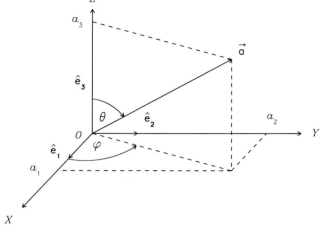

Figure 2.1 A vector $\vec{a}$ in a Cartesian coordinate system.

them. Vector calculus plays an important role in the study of EM scattering as EM waves are vector waves. Unlike a scalar quantity completely specified by its magnitude and its physical units, a vector is a quantity which, by definition, requires both direction and magnitude.

As shown in Figure 2.1, we use three unit vectors, $\hat{\mathbf{e}}_1$, $\hat{\mathbf{e}}_2$, and $\hat{\mathbf{e}}_3$, to denote the directions of the X, Y, and Z axes of a Cartesian coordinate system with coordinates (x, y, z). The angles θ and φ are the zenith and azimuthal angles, respectively. The zenith angle, θ, is measured from the local zenith, and $\theta = \pi$ or $180°$ indicates the local nadir direction. In radiative transfer calculation, $\mu = \cos\theta$, is frequently used. The azimuthal angle, φ, is normally measured counterclockwise with respect to the X axis on the OXY plane (O indicates the origin of the Cartesian coordinate system).

An arbitrary vector, $\vec{a}$, can be represented in terms of its projections onto the three axes of the Cartesian coordinate system X, Y, and Z as

$$\vec{a} = a_1 \cdot \hat{\mathbf{e}}_1 + a_2 \cdot \hat{\mathbf{e}}_2 + a_3 \cdot \hat{\mathbf{e}}_3 . \tag{2.1}$$

We write two arbitrary vectors, $\vec{a}$ and $\vec{b}$, using their components in the form

$$\vec{a} = \begin{pmatrix} a_1 \\ a_2 \\ a_3 \end{pmatrix}, \quad \text{and} \quad \vec{b} = \begin{pmatrix} b_1 \\ b_2 \\ b_3 \end{pmatrix} . \tag{2.2}$$

The scalar product, or dot product, of two vectors is defined as

$$\vec{a} \cdot \vec{b} = a_1 \cdot b_1 + a_2 \cdot b_2 + a_3 \cdot b_3 . \tag{2.3}$$

The vector product, or cross product, is given by

$$\vec{a} \times \vec{b} = \begin{vmatrix} \hat{e}_1 & \hat{e}_2 & \hat{e}_3 \\ a_1 & a_2 & a_3 \\ b_1 & b_2 & b_3 \end{vmatrix}$$

$$= \hat{e}_1(a_2 \cdot b_3 - a_3 \cdot b_2) + \hat{e}_2(a_3 \cdot b_1 - a_1 \cdot b_3) + \hat{e}_3(a_1 \cdot b_2 - a_2 \cdot b_1) .$$

$$(2.4)$$

Several vector operations involve the del operator, $\vec{\nabla}$, defined by

$$\vec{\nabla} = \hat{e}_1 \frac{\partial}{\partial x} + \hat{e}_2 \frac{\partial}{\partial y} + \hat{e}_3 \frac{\partial}{\partial z} . \tag{2.5}$$

Following are three common operations associated with the del operator.

a) The divergence of a vector, $\vec{a}$, is given by

$$\vec{\nabla} \cdot \vec{a} = \frac{\partial a_1}{\partial x} + \frac{\partial a_2}{\partial y} + \frac{\partial a_3}{\partial z} . \tag{2.6}$$

b) The curl of a vector, $\vec{a}$, is defined as

$$\vec{\nabla} \times \vec{a} = \begin{vmatrix} \hat{e}_1 & \hat{e}_2 & \hat{e}_3 \\ \frac{\partial}{\partial x} & \frac{\partial}{\partial y} & \frac{\partial}{\partial z} \\ a_1 & a_2 & a_3 \end{vmatrix}$$

$$= \hat{e}_1 \left(\frac{\partial a_3}{\partial y} - \frac{\partial a_2}{\partial z} \right) + \hat{e}_2 \left(\frac{\partial a_1}{\partial z} - \frac{\partial a_3}{\partial x} \right) + \hat{e}_3 \left(\frac{\partial a_2}{\partial x} - \frac{\partial a_1}{\partial y} \right) .$$

$$(2.7)$$

c) The gradient of an arbitrary scalar function, ϕ, is obtained by

$$\vec{\nabla}\phi = \hat{e}_1 \frac{\partial \phi}{\partial x} + \hat{e}_2 \frac{\partial \phi}{\partial y} + \hat{e}_3 \frac{\partial \phi}{\partial z} . \tag{2.8}$$

2.3.2
Use of Index Notation

Instead of the traditional vector notation, it is common to use index notation in vector calculation. We use a_i, where $i = 1, 2$, or 3, to indicate the ith component of a vector, $\vec{a}$, and we use ∂_i to indicate the differential operation associated with the ith component of the del operator (namely, $\partial_1 = \partial/\partial x$, $\partial_2 = \partial/\partial y$, and $\partial_3 = \partial/\partial z$). Thus, we can write the del operator as

$$\vec{\nabla} = \hat{e}_1 \partial_1 + \hat{e}_2 \partial_2 + \hat{e}_3 \partial_3 . \tag{2.9}$$

With the introduction of index notation, the Einstein summation rule can further simplify vector calculation. By this rule, a subscript index [not a subscript number, such as in Eq. (2.9)] repeated twice in an expression denotes summation. Thus, the dot product of two vectors can be written in the form of

$$\vec{a} \cdot \vec{b} = a_i \cdot b_i = a_k \cdot b_k \, , \tag{2.10}$$

where indices i and k in Eq. (2.10) are dummy indices that denote summation. Thus, if i or k in Eq. (2.10) is replaced with any other symbol, the dot product's value remains the same.

It is convenient to introduce a quantity ε_{ijk} for vector calculations involving the cross product of two vectors. This quantity is defined as

$$\varepsilon_{ijk} = (\hat{\mathbf{e}}_i \times \hat{\mathbf{e}}_j) \cdot \hat{\mathbf{e}}_k \, . \tag{2.11}$$

ε_{ijk} is a tensor; however, in this book, we will not cover tensor calculus and will identify ε_{ijk} as the "permutation symbol". The texts written by Coburn (1955) and Simmonds (2000) are recommended for tensor analysis involving contravariant and covariant tensors. The first two chapters of the latter text offer a concise introduction to vector and tensor analysis at an undergraduate level.

From Eq. (2.11), we have the following relation:

$$\varepsilon_{ijk} = \begin{cases} 1 \, , & \text{if } (i, j, k) \text{ is from even permutations of } (1, 2, 3) \, , \\ -1 \, , & \text{if } (i, j, k) \text{ is from odd permutations of } (1, 2, 3) \, , \\ 0 \, , & \text{if two or three of } (i, j, k) \text{ are the same} \, . \end{cases} \tag{2.12}$$

Using the permutation symbol, we can express the three components of the cross product of two vectors in a unified form. For example, let $\vec{c} = \vec{a} \times \vec{b}$, then the ith ($i = 1, 2,$ or 3) component of $\vec{c}$ can be expressed in the form

$$c_i = \varepsilon_{ijk} \cdot a_j \cdot b_k \, . \tag{2.13}$$

To apply the permutation symbol to vector calculus, we introduce the Kronecker symbol, or Kronecker delta, δ_{ij}, defined as

$$\delta_{ij} = \begin{cases} 1 \, , & \text{if } i = j \, , \\ 0 \, , & \text{if } i \neq j \, . \end{cases} \tag{2.14}$$

Without providing a mathematical proof, we include a useful relation between the permutation symbol and the Kronecker delta:

$$\varepsilon_{ijk} \cdot \varepsilon_{\alpha\beta k} = \delta_{i\alpha} \cdot \delta_{j\beta} - \delta_{j\alpha} \cdot \delta_{i\beta} \, . \tag{2.15}$$

The preceding equations provide a basis for a convenient and efficient way to perform vector calculations. To illustrate this, we give three examples.

Example 2.1

Let us consider the expansion of $\vec{a} \times (\vec{b} \times \vec{c})$. From Eq. (2.13), we know that

$$(\vec{b} \times \vec{c})_k = \varepsilon_{k\alpha\beta} \cdot b_\alpha \cdot c_\beta . \tag{2.16}$$

Consequently, we obtain the ith component of the following vector product:

$$\begin{aligned}
\left[\vec{a} \times (\vec{b} \times \vec{c})\right]_i &= \varepsilon_{ijk} \cdot a_j \cdot (\vec{b} \times \vec{c})_k \\
&= \varepsilon_{ijk} \cdot a_j \cdot (\varepsilon_{k\alpha\beta} \cdot b_\alpha \cdot c_\beta) .
\end{aligned} \tag{2.17}$$

Applying Eq. (2.15) to Eq. (2.17) in conjunction with $\varepsilon_{k\alpha\beta} = \varepsilon_{\alpha\beta k}$ yields

$$\begin{aligned}
\left[\vec{a} \times (\vec{b} \times \vec{c})\right]_i &= (\delta_{i\alpha} \cdot \delta_{j\beta} - \delta_{i\beta} \cdot \delta_{j\alpha}) \cdot a_j \cdot b_\alpha \cdot c_\beta \\
&= b_i \cdot a_j \cdot c_j - a_j \cdot b_j \cdot c_i \\
&= b_i \cdot (\vec{a} \cdot \vec{c}) - (\vec{a} \cdot \vec{b}) \cdot c_i .
\end{aligned} \tag{2.18}$$

Thus, we have

$$\vec{a} \times (\vec{b} \times \vec{c}) = \vec{b} \,(\vec{a} \cdot \vec{c}) - (\vec{a} \cdot \vec{b}) \,\vec{c} . \tag{2.19}$$

Example 2.2

The index notation can easily be applied to the following expression:

$$\begin{aligned}
\left[\vec{\nabla} \times (\vec{\nabla} \times \vec{c})\right]_i &= \varepsilon_{ijk} \, \partial_j \, (\vec{\nabla} \times \vec{c})_k \\
&= \varepsilon_{ijk} \, \partial_j \, \varepsilon_{k\alpha\beta} \, \partial_\alpha c_\beta \\
&= (\delta_{i\alpha} \cdot \delta_{j\beta} - \delta_{i\beta} \cdot \delta_{j\alpha}) \, \partial_j \partial_\alpha c_\beta \\
&= \partial_i \partial_j c_j - \partial_j \partial_j c_i \\
&= \partial_i (\vec{\nabla} \cdot \vec{c}) - \nabla^2 c_i .
\end{aligned} \tag{2.20}$$

Thus, we have

$$\begin{aligned}
\vec{\nabla} \times (\vec{\nabla} \times \vec{c}) &= \vec{\nabla}(\vec{\nabla} \cdot \vec{c}) - (\vec{\nabla} \cdot \vec{\nabla}) \,\vec{c} \\
&= \vec{\nabla}(\vec{\nabla} \cdot \vec{c}) - \nabla^2 \vec{c} ,
\end{aligned} \tag{2.21}$$

where ∇^2 indicates the scalar product of two del operators, that is,

$$\nabla^2 = \vec{\nabla} \cdot \vec{\nabla} . \tag{2.22}$$

Example 2.3

We practice the preceding vector-calculating technique by expanding the expression $\vec{\nabla} \cdot (\vec{a} \times \vec{b})$:

$$
\begin{aligned}
\vec{\nabla} \cdot (\vec{a} \times \vec{b}) &= \partial_i (\vec{a} \times \vec{b})_i \\
&= \partial_i \varepsilon_{ijk} (a_j b_k) \\
&= \varepsilon_{ijk} \partial_i a_j b_k + \varepsilon_{ijk} a_j \partial_i b_k \\
&= b_k \cdot \varepsilon_{kij} \partial_i a_j - a_j \cdot \varepsilon_{jik} \partial_i b_k \ .
\end{aligned} \tag{2.23}
$$

Thus, we have

$$
\vec{\nabla} \cdot (\vec{a} \times \vec{b}) = \vec{b} \cdot (\vec{\nabla} \times \vec{a}) - \vec{a} \cdot (\vec{\nabla} \times \vec{b}) \ . \tag{2.24}
$$

2.4
Dirac δ-Function

The Earth is approximately 1.5×10^8 km away from the Sun, and the solar radiation reaching the top of the terrestrial atmosphere is essentially collimated. Rigorously speaking, collimated radiation is a beam that propagates in one specific direction without divergence. This feature of the incident solar radiation at the TOA is described using the Dirac δ-function. To introduce the Dirac δ-function, let us consider the function

$$
u_\sigma(x) = \begin{cases} 0 \ , & \text{if } x < -\sigma \ , \\ +1/\sigma^2 \cdot x + 1/\sigma \ , & \text{if } -\sigma \le x < 0 \ , \\ -1/\sigma^2 \cdot x + 1/\sigma \ , & \text{if } 0 \le x \le \sigma \ , \\ 0 \ , & \text{if } x > \sigma \ . \end{cases} \tag{2.25}
$$

The function defined in Eq. (2.25) is graphically illustrated in Figure 2.2a. Integrating this function over the entire domain yields

$$
\int_{-\infty}^{\infty} u_\sigma(x) \, dx = \int_{-\sigma}^{0} \left(\frac{x}{\sigma^2} + \frac{1}{\sigma} \right) dx + \int_{0}^{\sigma} \left(-\frac{x}{\sigma^2} + \frac{1}{\sigma} \right) dx = 1 \ . \tag{2.26}
$$

Thus, the integration of the function over the entire domain $(-\infty, \infty)$, the area in Figure 2.2 enclosed by the X axis and the dashed lines (in Figure 2.2a) representing the function $u_\sigma(x)$ is always unity regardless of the value of σ. However, the maximum value of the function $u_\sigma(x)$ for x between $-\sigma$ and σ increases with the decrease of σ, which is evident from the illustration in Figure 2.2b. When σ approaches zero, the function defined by Eq. (2.25) leads to the Dirac δ-function, that is

$$
\delta(x) = \lim_{\sigma \to 0} u_\sigma(x) \ . \tag{2.27}
$$

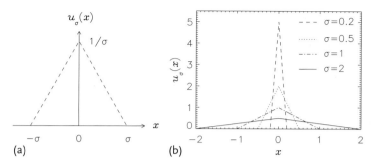

Figure 2.2 (a) Illustration of the function defined in Eq. (2.25). (b) Demonstration of the effect of decreasing σ on $u_\sigma(x)$.

The Dirac δ-function has the following properties:

$$\delta(x) = \begin{cases} \infty, & \text{if } x = 0, \\ 0, & \text{if } x \neq 0, \end{cases} \tag{2.28}$$

and

$$\int_{-\infty}^{\infty} \delta(x - x_0) \cdot \phi(x) \, dx = \phi(x_0), \tag{2.29}$$

where $\phi(x)$ is a well-behaved, arbitrary scalar function, and x_0 is a given point on the X axis. As a special case of Eq. (2.29), the following relation holds:

$$\int_{-\infty}^{\infty} \delta(x) \, dx = 1. \tag{2.30}$$

The Dirac δ-function can be expressed in terms of other well-behaved functions subject to certain conditions. For example,

$$\delta(x) = \lim_{\eta \to 0} \left(\frac{1}{\pi} \cdot \frac{\eta}{x^2 + \eta^2} \right), \tag{2.31}$$

$$\delta(x) = \lim_{\eta \to 0} \left\{ \frac{1}{\eta \cdot \sqrt{\pi}} \cdot \exp\left[-\left(\frac{x}{\eta} \right)^2 \right] \right\}, \tag{2.32}$$

$$\delta(x) = \lim_{\eta \to 0} \begin{cases} \dfrac{1}{\eta}, & \text{if } -\dfrac{\eta}{2} \leq x \leq \dfrac{\eta}{2}, \\ 0, & \text{if } x < -\dfrac{\eta}{2}, \text{ or } x > \dfrac{\eta}{2}. \end{cases} \tag{2.33}$$

The continuous functions on the right-hand sides of Eqs. (2.31) and (2.32) can be used to describe the broadening of absorption lines and will be discussed in Chapter 8. As an example, the angular distribution of the radiance of the incident solar radiation can be expressed in terms of the Dirac δ-function and will be described in detail in Chapter 6.

2.5
Geometry

2.5.1
Directions

To discuss the transfer of radiation, we must specify the propagation direction of a pencil of radiation. A unit vector, such as $\hat{s}$, is convenient to specify the radiation propagation direction. The three components of $\hat{s}$ can be written as

$$\hat{s} = \begin{pmatrix} s_1 \\ s_2 \\ s_3 \end{pmatrix} = \begin{pmatrix} \hat{e}_1 \cdot \hat{s} \\ \hat{e}_2 \cdot \hat{s} \\ \hat{e}_3 \cdot \hat{s} \end{pmatrix} = \begin{pmatrix} \sin\theta \cdot \cos\varphi \\ \sin\theta \cdot \sin\varphi \\ \cos\theta \end{pmatrix} , \tag{2.34}$$

where θ and φ are the zenith and azimuthal angles, see Figure 2.1.

2.5.2
Solid Angle

To define the solid angle, consider an observer positioned at the center (the origin O in Figure 2.3) of a large imaginary sphere with radius r. In addition, consider an arbitrarily shaped object located between the observer and the surface of the sphere. If the observer illuminates the object, the shadow (projected area) of the object on the spherical surface has a differential area element d^2A (schematically illustrated by the shaded area in Figure 2.3). The solid angle, $d^2\Omega$, subtended by the object, as viewed by the observer, is given by

$$d^2\Omega = \frac{d^2A}{r^2} . \tag{2.35}$$

The square of the differentiation operator denotes this to be a two-dimensional (2D) differential element. The unit of the solid angle is "steradian" and is usually abbreviated as "sr." Since the total surface area of the sphere is $4\pi \cdot r^2$, the solid angle subtended by the entire spherical surface, as viewed by the observer, is 4π sr. Similarly, the surface area of one half of the sphere is $2\pi \cdot r^2$, and the corresponding solid angle is 2π sr. The definition of the solid angle involves two elements: (i) an entity by which a solid angle is subtended; and, (ii) a location from which the entity is viewed. Thus, the solid angle subtended by an object differs for two observers located at different places.

Hereafter, we omit the unit steradian (sr) if we use specific values for the solid angle, for example, 4π sr or 2π sr. This is not precisely correct and may cause some confusion when units are checked in derivations and equations. However, common practice leaves out the unit steradian, and we follow the convention in this book. Also, for planar angles (e.g., θ, φ), the strict unit of radian (rad) will be neglected in the remaining text.

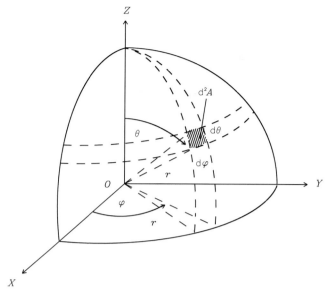

Figure 2.3 Illustration of the definition of the solid angle and polar coordinates.

For many practical applications, polar coordinates, θ and φ, are used to calculate the solid angle. As shown in Figure 2.3, the area of the shaded region on the sphere is

$$d^2 A = r^2 \cdot \sin \theta \, d\theta \, d\varphi \,. \tag{2.36}$$

The solid-angle element subtended by the shaded region, as viewed from the origin, is given by

$$d^2 \Omega = \frac{d^2 A}{r^2} = \sin \theta \, d\theta \, d\varphi \,. \tag{2.37}$$

To illustrate the application of Eq. (2.37) to the calculation of the solid angle, consider the solid angle subtended by a circular disk of radius a with respect to an observing point located at a distance of r along the symmetry axis of the disk (see Figure 2.4). A common mistake is to give the solid angle in terms of the ratio of the disk area to the square of r. Since the disk is not on any imaginary sphere centered at the observing point, the approach is incorrect. With Eq. (2.37), the correct way to calculate the solid angle is

$$\Omega(\theta) = \int_0^{2\pi} \int_0^{\theta} \sin \theta' \, d\theta' \, d\varphi = 2\pi \cdot (1 - \cos \theta) \,. \tag{2.38}$$

For an angle of $\theta = 45°$ ($a = r$ in Figure 2.4), we get

$$\Omega(\theta = 45°) = 2\pi \cdot \left(1 - \frac{\sqrt{2}}{2}\right) \approx 0.5858\pi \,. \tag{2.39}$$

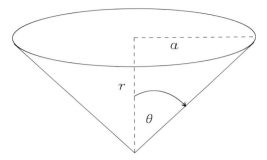

Figure 2.4 Configuration of a circular disk viewed from an observing point along the symmetry axis of the disk.

If the disk is infinitely large, then $\theta = 90°$ and we have

$$\Omega\,(\theta = 90°) = 2\pi \cdot (1 - 0) = 2\pi \; . \tag{2.40}$$

2.5.3
Angle between Two Directions

In radiative transfer calculations, the angle between two directions often needs to be specified, for example, the scattering angle defined as the angle between the incident and scattered radiation beams. As an example, consider two directions denoted by unit vectors $\hat{s}$ and $\hat{s}'$, having polar coordinates (θ, φ) and (θ', φ'), respectively. Let the angle between the two directions be the scattering angle, ϑ, calculated by

$$\cos \vartheta = \hat{s} \cdot \hat{s}' = (\sin \theta \cdot \cos \varphi, \sin \theta \cdot \sin \varphi, \cos \theta) \cdot \begin{pmatrix} \sin \theta' \cdot \cos \varphi' \\ \sin \theta' \cdot \sin \varphi' \\ \cos \theta' \end{pmatrix}$$

$$= \cos \theta \cdot \cos \theta' + \sqrt{1 - \cos^2 \theta} \cdot \sqrt{1 - \cos^2 \theta'} \cdot \cos(\varphi - \varphi')$$

$$= \mu \cdot \mu' + \sqrt{1 - \mu^2} \cdot \sqrt{1 - \mu'^2} \cdot \cos(\varphi - \varphi') \; , \tag{2.41}$$

where $\mu = \cos \theta$ and $\mu' = \cos \theta'$. From Eq. (2.41), it is evident that the scattering angle, ϑ, is a function of θ, φ, θ', and φ'. A useful equation involving both the scattering angle, ϑ, and the Dirac δ-function, see Joseph et al. (1976), is

$$\delta(1 - \cos \vartheta) = 2\pi \cdot \delta(\mu - \mu') \cdot \delta(\varphi - \varphi') \; . \tag{2.42}$$

2.6
Orthogonal Functions

Legendre polynomials and Legendre functions are very useful in the study of radiative transfer. For example, some numerical methods for solving the RTE expand the phase function in terms of Legendre polynomials to calculate irradiance, or in terms of Legendre functions to calculate radiance.

2.6.1
Legendre Polynomials

In spherical coordinates, Legendre polynomials are part of the solution of the Laplace equation. The polynomials are usually represented by P_n where the subscript n denotes the order. The Legendre polynomials are defined by

$$P_n(x) = \frac{1}{2^n \cdot n!} \frac{d^n}{dx^n} [(x^2 - 1)^n] ; \quad x \in [-1, 1] . \tag{2.43}$$

The orthogonality of the Legendre polynomials is expressed by

$$\int_{-1}^{1} P_j(x) \cdot P_l(x) \, dx = \delta_{jl} \cdot \left(\frac{2}{2j + 1} \right) ; \quad x \in [-1, 1] . \tag{2.44}$$

For the first three Legendre polynomials, it follows that

$$P_0 = 1 , \tag{2.45}$$

$$P_1(x) = x , \tag{2.46}$$

$$P_2(x) = \frac{1}{2} \cdot (3x^2 - 1) . \tag{2.47}$$

We can apply a recurrence formula to generate higher-order Legendre polynomials, that is,

$$\frac{d}{dx}[P_{n+1}(x)] = (2n + 1) \cdot P_n(x) + \frac{d}{dx}[P_{n-1}(x)] . \tag{2.48}$$

Another type of recurrence formula ($n \geq 2$) is given by

$$P_n(x) = \frac{(2n - 1) \cdot x \cdot P_{n-1}(x) - (n - 1) \cdot P_{n-2}(x)}{n} . \tag{2.49}$$

Legendre polynomials have the following properties:

$$P_n(1) = 1 , \quad P_n(-1) = (-1)^n , \quad \text{and} \quad P_n(-x) = (-1)^n \cdot P_n(x) . \tag{2.50}$$

The Dirac δ-function can be expressed in terms of Legendre polynomials as (Morse and Feshbach, 1953)

$$\delta(x - x_0) = \sum_{n=0}^{\infty} \left(\frac{2n + 1}{2} \right) \cdot P_n(x) \cdot P_n(x_0) ; \quad x, x_0 \in [-1, 1] . \tag{2.51}$$

2.6.2
Legendre Functions

The Legendre functions, also called associated Legendre polynomials, are part of the solution of the Helmholtz equation in spherical coordinates. The Legendre functions of *j*th order are defined by

$$P_n^{(j)}(x) = (1 - x^2)^{j/2} \cdot \frac{d^j}{dx^j}[P_n(x)] \; ; \quad x \in [-1, 1] \; ; \quad |j| \le n \; ; \tag{2.52}$$

satisfying the orthogonality relation

$$\int_{-1}^{1} P_n^{(j)}(x) \cdot P_{n'}^{(j)}(x) \, dx = \delta_{nn'} \cdot \left(\frac{2}{2n + 1} \right) \cdot \frac{(n + j)!}{(n - j)!} \; ; \quad x \in [-1, 1] \; . \tag{2.53}$$

When $j = 0$, the Legendre functions reduce to the Legendre polynomials

$$P_n^{(0)}(x) \equiv P_n(x) \; ; \quad x \in [-1, 1] \; . \tag{2.54}$$

The first three Legendre functions are given by

$$P_0^{(0)} = P_0 = 1 \; , \tag{2.55}$$

$$P_1^{(1)}(x) = (1 - x^2)^{1/2} \; , \tag{2.56}$$

$$P_2^{(1)}(x) = (1 - x^2)^{1/2} \cdot 3x \; . \tag{2.57}$$

For $x = \mu = \cos \theta$, it follows that

$$P_1^{(1)}(\mu) = \sin \theta \; , \tag{2.58}$$

$$P_2^{(1)}(\mu) = \frac{3}{2} \cdot \sin 2\theta \; . \tag{2.59}$$

Two useful recurrence formulas are

$$P_n^{(j+1)}(x) = 2j \cdot x \cdot (1 - x^2)^{-1/2} \cdot P_n^{(j)}(x) - [n \cdot (n + 1) - j \cdot (j - 1)] \cdot P_n^{(j-1)}(x) \tag{2.60}$$

and

$$P_{n+1}^{(j)}(x) = \frac{(2n + 1) \cdot x \cdot P_n^{(j)}(x) - (n + j) \cdot P_{n-1}^{(j)}(x)}{n - j + 1} \; . \tag{2.61}$$

In numerical computation, it is convenient to use the renormalized Legendre functions first introduced by Dave and Armstrong (1970) and defined as

$$\tilde{P}_n^{(j)}(x) = \sqrt{\frac{(n - j)!}{(n + j)!}} \cdot P_n^{(j)}(x) \; ; \quad x \in [-1, 1] \; . \tag{2.62}$$

The recurrence relation for the renormalized Legendre is

$$\tilde{P}_{n+1}^{(j)}(x) = \frac{2n+1}{\sqrt{(n+j+1)\cdot(n-j+1)}} \cdot x \cdot \tilde{P}_n^{(j)}(x)$$
$$- \sqrt{\frac{(n+j)\cdot(n-j)}{(n+j+1)\cdot(n-j+1)}} \cdot \tilde{P}_{n-1}^{(j)}(x) . \tag{2.63}$$

Two initial values for the preceding recurrence relation are

$$\tilde{P}_j^{(j)}(x) = (-1)^j \cdot \sqrt{\frac{(2j-1)!!}{(2j)!!}} \cdot (1-x^2)^{j/2} , \tag{2.64}$$

$$\tilde{P}_{j+1}^{(j)}(x) = x\sqrt{2j+1} \cdot \tilde{P}_j^{(j)}(x) . \tag{2.65}$$

It can be proven that

$$\int_{-1}^{1} \tilde{P}_j^{(m)}(x) \cdot \tilde{P}_l^{(m)}(x) \, dx = \delta_{jl} \cdot \left(\frac{2}{2j+1} \right); \quad x \in [-1,1] . \tag{2.66}$$

The motivation for using the renormalized Legendre functions is stability in numerical computation (Thomas and Stamnes, 1999). In radiative transfer calculations, the phase function, $\mathcal{P}$, see Section 4.8.1, is usually expanded as

$$\mathcal{P}(\cos\vartheta) = \mathcal{P}(\mu,\varphi,\mu',\varphi')$$
$$= \sum_{m=0}^{N-1} (2-\delta_{m0}) \cdot \left\{ \sum_{n=m}^{N-1} C_n \cdot \frac{(n-m)!}{(n+m)!} \cdot P_n^{(m)}(\mu) \cdot P_n^{(m)}(\mu') \right\}$$
$$\times \cos m(\varphi - \varphi') , \tag{2.67}$$

with the dimensionless Legendre coefficients C_n

$$C_n = \frac{2n+1}{2} \int_{-1}^{1} \mathcal{P}(\mu) \cdot P_n(\mu) \, d\mu , \tag{2.68}$$

with $C_0 = 1$. The factor $(n-m)!/(n+m)!$ in Eq. (2.67) rapidly reduces to zero while the values of the Legendre functions may be excessive for a large m with $n \geq m$. Thus, numerical computation that uses Eq. (2.67) to expand the phase function is not stable, particularly in cases where the phase function has a strong forward peak, for example, an ice-cloud phase function at a visible wavelength. To stabilize numerical computation, the expansion in Eq. (2.67) can be performed in terms of the renormalized Legendre functions as

$$\mathcal{P}(\cos\vartheta) = \sum_{m=0}^{N-1} (2-\delta_{m0}) \cdot \left\{ \sum_{n=m}^{N-1} C_n \cdot \tilde{P}_n^{(m)}(\mu) \cdot \tilde{P}_n^{(m)}(\mu') \right\} \cdot \cos m(\varphi - \varphi') . \tag{2.69}$$

2.7
Quadrature Formula

The RTE is an integro-differential equation. In some numerical methods for simulating the transfer of radiation, the continuous integral involved in the radiative transfer equation is approximated in terms of a discrete sum. The integral of an arbitrary scalar function $\phi(x)$ may be discretized by the use of an even-order Gauss–Lobatto quadrature in the form of

$$\int_{-1}^{1} \phi(x)\,dx = \sum_{i=-s}^{s} \phi(x_i) \cdot c_i \,, \tag{2.70}$$

where $i = 0$ is explicitly excluded. The Gaussian abscissas (roots) of the even-numbered Legendre polynomials are x_i and are not arbitrary. The positive Gaussian weights of the quadrature are c_i. These weights are defined in terms of the Legendre polynomials $P_s(x)$ (Chandrasekhar, 1950):

$$c_i = \left(\frac{dP_s}{dx} \right)^{-1} \int_{-1}^{1} \frac{P_s(x)}{x - x_i}\,dx \,, \tag{2.71}$$

$$\sum_{i} c_i = 1 \,. \tag{2.72}$$

The $(2s)$-point Gaussian quadrature is exact for an integrand function that is a polynomial with a degree no higher than $(4s - 1)$. The proof of this feature of the Gaussian quadrature can be found in Press et al. (1992), where a computational program is provided for calculating the Gaussian abscissas, x_i, and Gaussian weights, c_i. Table 2.4 lists the abscissas and weights for two cases, $s = 1$ and $s = 2$. In numerical simulations of the transfer of radiation, the radiation field is approximated as two streams when $s = 1$ and is referred to as the two-stream approximation. Similarly, $s = 2$ leads to the four-stream approximation.

As an example, let us consider the following integral:

$$\int_{-1}^{1} x^6\,dx = \frac{2}{7} \approx 0.285\,714\,285\,714\,29 \,. \tag{2.73}$$

Table 2.4 Gaussian abscissas x_i and Gaussian weights c_i.

	Gaussian abscissas x_i	Gaussian weights c_i
$s = 1$		
$i = \pm 1$	$x_{\pm 1} = \pm 0.577\,350\,269\,189\,63$	$c_{\pm 1} = 1$
$s = 2$		
$i = \pm 1$	$x_{\pm 1} = \pm 0.339\,981\,043\,584\,86$	$c_{\pm 1} = 0.652\,145\,154\,862\,55$
$i = \pm 2$	$x_{\pm 2} = \pm 0.861\,136\,311\,594\,05$	$c_{\pm 2} = 0.347\,854\,845\,137\,45$

Using the formula in Eq. (2.70) and the values in Table 2.4 for $s = 2$, we obtain

$$\int_{-1}^{1} x^6 \, dx = \sum_{i=-2}^{2} (x_i)^6 \cdot c_i = 0.285\,714\,285\,714\,28 \, . \tag{2.74}$$

A finite number of digits were used for the Gaussian abscissas and weights, causing a difference between the last digit in the results of Eqs. (2.73) and (2.74). This example illustrates the four-point Gaussian quadrature to be exact for an integral with an integrand specified by a six-degree polynomial. In the RTE, the Gaussian abscissas represent the zenith angles ($x = \mu = \cos\theta$, $x_i = \mu_i = \cos\theta_i$) and do not include the coordinates, $\mu_i = -1$ and $\mu_i = 1$, corresponding to zenith-upward and nadir-downward radiation streams. To obtain the radiances in these two directions using the Gaussian quadrature, an extrapolation of the radiance versus the zenith angle is required; however, the extrapolation may incur numerical errors. The Labatto quadrature (Hildebrand, 1974), also known as the Radau formula, can be used to avoid extrapolation errors. The Labatto abscissas and weights for orders of up to nine can be found in the classic treatise on radiative transfer by Chandrasekhar (1950).

Problems

Problem 2.1 Rule of Thumb

a) Verify on yourself the rule of thumb that your thumb appears to be 2 degrees wide at arm's length.
b) In German, the expression equivalent to "rule of thumb" literally translates to "rule of fist." Under which angle does your fist appear at arm's length?

Problem 2.2 Solid Angle

a) For a solar eclipse to occur, the Moon must be located between the Earth and the Sun. However, both the Earth's and the Moon's orbits are not circles but ellipses. The distance between the Earth and the Moon varies between 357 000 and 407 000 km. The distance between the Earth and the Sun varies between 147.1×10^6 and 152.1×10^6 km. Calculate the range of solid angles under which the Moon and the Sun can appear, and decide which orbit is more important to make a solar eclipse total or not anywhere on Earth. (Use these radii: Earth: 6371 km; Moon: 1737 km; and Sun: 696 000 km.)
b) Even if the combination of distances should make a solar eclipse total, it is not always total. Why could that be?
c) Due to tidal friction, the mean Earth–Moon distance increases by 3.8 cm/year. How many more years will total solar eclipses be possible on Earth? (Source of numerical values: NASA Planetary Fact Sheets)

3
Fundamentals

3.1
Electromagnetic (EM) Radiation

3.1.1
Maxwell's Equations and Plane-Wave Solutions

The Maxwell equations describe the propagation of an EM wave. For this text, we use the source-free Maxwell equations in the form

$$\vec{\nabla} \cdot \vec{D} = 0 \,, \tag{3.1}$$

$$\vec{\nabla} \times \vec{H} = \frac{\partial \vec{D}}{\partial t} \,, \tag{3.2}$$

$$\vec{\nabla} \cdot \vec{B} = 0 \,, \tag{3.3}$$

$$\vec{\nabla} \times \vec{E} = -\frac{\partial \vec{B}}{\partial t} \,, \tag{3.4}$$

where $\vec{D}$ is the dielectric displacement vector in units of $A\,s\,m^{-2}$, $\vec{H}$ is the magnetic field vector in units of $A\,m^{-1}$, $\vec{B}$ is the magnetic induction vector in units of tesla (T) ($1\,T = 1\,V\,s\,m^{-2}$), $\vec{E}$ represents the electric field vector in units of $V\,m^{-1}$, see Jackson (1999), and t is time. For a uniform, isotropic, and linear medium, we have the following relations:

$$\vec{D} = \epsilon \vec{E} \,, \tag{3.5}$$

$$\vec{B} = \kappa \vec{H} \,, \tag{3.6}$$

where ϵ is the electric permittivity in units of farad (F) per meter ($1\,F\,m^{-1} = 1\,A\,s\,V^{-1}\,m^{-1}$) and κ is the magnetic permeability in units of henry (H) per meter ($1\,H\,m^{-1} = 1\,V\,s\,A^{-1}\,m^{-1}$). Both ϵ and κ are complex quantities characteristic of the medium in which the fields are concerned. In general textbooks, the symbol μ is used to indicate the magnetic permeability. However, in order to avoid

Theory of Atmospheric Radiative Transfer, First Edition. Manfred Wendisch and Ping Yang.
© 2012 WILEY-VCH Verlag GmbH & Co. KGaA. Published 2012 by WILEY-VCH Verlag GmbH & Co. KGaA.

confusion with the cosine of the zenith angle for which we use the symbol μ in this book, we decided to use the symbol κ to represent the magnetic permeability instead of μ. Furthermore, we use $\epsilon_0 = 8.8542 \times 10^{-12}$ A s V^{-1} m^{-1} and $\kappa_0 = 1.257 \times 10^{-6}$ V s A^{-1} m^{-1} to indicate the electric permittivity and magnetic permeability of a vacuum. Thus, for a homogeneous medium, for which permittivity and permeability are independent of the spatial location, the source-free Maxwell equations reduce to

$$\vec{\nabla} \cdot \vec{E} = 0 \,, \tag{3.7}$$

$$\vec{\nabla} \times \vec{H} = \epsilon \frac{\partial \vec{E}}{\partial t} \,, \tag{3.8}$$

$$\vec{\nabla} \cdot \vec{H} = 0 \,, \tag{3.9}$$

$$\vec{\nabla} \times \vec{E} = -\kappa \frac{\partial \vec{H}}{\partial t} \,. \tag{3.10}$$

Applying a curl operation to Eq. (3.10) leads to

$$\vec{\nabla} \times \vec{\nabla} \times \vec{E} = -\kappa \frac{\partial}{\partial t} (\vec{\nabla} \times \vec{H}) \,. \tag{3.11}$$

Furthermore, using the result from Eq. (2.21) and substituting the curl of the magnetic field vector from Eq. (3.8) into the preceding equation, we obtain

$$\vec{\nabla}(\vec{\nabla} \cdot \vec{E}) - \nabla^2 \vec{E} = -\kappa \cdot \epsilon \frac{\partial^2 \vec{E}}{\partial t^2} \,. \tag{3.12}$$

Because the divergence of the electric field vector is zero from Eq. (3.7), we have

$$\nabla^2 \vec{E} - \kappa \cdot \epsilon \frac{\partial^2 \vec{E}}{\partial t^2} = 0 \,. \tag{3.13}$$

Equation (3.13) represents a typical wave equation. For a medium, the wave equation for the electric field vector can be written as

$$\nabla^2 \vec{E} - \frac{1}{c_m^2} \frac{\partial^2 \vec{E}}{\partial t^2} = 0 \,, \tag{3.14}$$

where c_m is given by

$$c_m^2 = \frac{1}{\epsilon \cdot \kappa} \,. \tag{3.15}$$

If the wave propagates in a vacuum, then c_m in Eq. (3.14) reduces to the speed of light in a vacuum, denoted by

$$c = \frac{1}{\sqrt{\epsilon_0 \cdot \kappa_0}} = 2.997\,925 \times 10^8 \text{ m s}^{-1} \,. \tag{3.16}$$

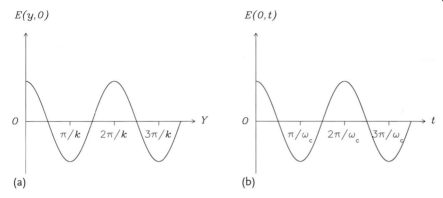

$E(y,0)$

$E(0,t)$

(a)

(b)

Figure 3.1 (a) Snapshot of a plane EM wave at time $t = 0$. (b) The variation of the wave, as observed at point $y = 0$.

To understand the properties of an EM wave, consider a plane EM wave in a vacuum, propagating along the Y axis, and the corresponding electric field vector points along the Z axis of a Cartesian coordinate system, see Figure 3.1a. This is a one-dimensional (1D) wave propagation problem and the electric field vector can be expressed as

$$\vec{E}(y, t) = E(y, t)\,\hat{\mathbf{e}}_3 \,, \tag{3.17}$$

where $\hat{\mathbf{e}}_3$ is a unit vector pointing along the Z axis. In this case, the EM wave equation in a vacuum can be written in a scalar form as

$$\frac{\partial^2 E}{\partial y^2} - \frac{1}{c^2}\frac{\partial^2 E}{\partial t^2} = 0 \,, \tag{3.18}$$

and a solution is given by

$$E(y, t) = E_0 \cdot \cos\left(k \cdot y - \omega_c \cdot t\right) \,, \tag{3.19}$$

where E_0 is a constant denoting the amplitude of the EM wave, k is the modified wavenumber, and ω_c is the circular frequency, see the following section for more explanation.

3.1.2
Wavelength, Frequency, Wavenumber, Dispersion Relation, and Phase Speed

A snapshot taken of the EM wave at a time of $t = 0$ reveals that the wave pattern repeats every spatial increment $2\pi/k$, as shown in Figure 3.1a. Thus, $2\pi/k$ is the wavelength and, hereafter, will be designated by λ. Observing the electric field vector at a fixed point ($y = 0$), we see that the wave pattern will repeat at every time interval $2\pi/\omega_c$, as shown in Figure 3.1b. Thus, $2\pi/\omega_c$ is the period of the wave. The frequency, ν, of the wave is the inverse of its period and is given by $\omega_c/(2\pi)$.

The quantities k and ω_c are the wave constant, or modified wavenumber in units of inverse meters (m^{-1}), and the circular, or angular frequency in units of inverse seconds (s^{-1}), respectively. The relationship between k and ω_c is usually referred to as the dispersion relation. From Eqs. (3.18) and (3.19), we obtain the dispersion relation

$$k^2 - \frac{\omega_c^2}{c^2} = 0 \,, \tag{3.20}$$

or

$$k = \pm \frac{\omega_c}{c} \,, \tag{3.21}$$

where the positive and negative signs indicate waves propagating towards the positive and negative Y axis directions, respectively. In the following discussion, we only consider a wave propagating towards the positive Y axis. If we follow a maximum or minimum of the electric field vector, we have the constant phase condition

$$k \cdot y - \omega_c \cdot t = \text{constant} \,. \tag{3.22}$$

Thus, we have the phase speed

$$\frac{\mathrm{d}y}{\mathrm{d}t} = \frac{\omega_c}{k} = c \,. \tag{3.23}$$

From the preceding discussion, the wavelength, λ, can be expressed in terms of c and ν as

$$\lambda = \frac{2\pi}{k} = \frac{2\pi}{\omega_c} \cdot c = \frac{c}{\nu} \,. \tag{3.24}$$

For IR radiation, it is convenient to use the wavenumber, $\tilde{\nu}$, defined as

$$\tilde{\nu} = \frac{1}{\lambda} \,. \tag{3.25}$$

Note, as a common convention, we use cm^{-1} for the units of the wavenumbers. m^{-1} or nm^{-1} should not be used for the units of wavenumbers, although they have the same dimension as cm^{-1}, that is, length^{-1}.

3.1.3
Coherence, Incoherence, and Polarization

Coherent radiation can be explained by an example. The output from a single oscillating electric dipole, or from an ensemble of electric dipoles with synchronized oscillations, is coherent EM radiation. If the phases of EM waves are coherent, phase interference must be considered in the contributions from wave sources. Incoherent radiation originates from a set of independent sources, and phase interference can be neglected. Atmospheric radiation can be considered incoherent for essentially all practical purposes.

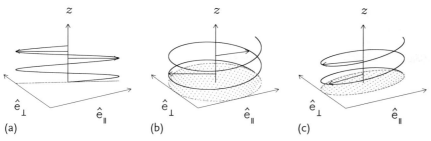

Figure 3.2 Illustration of linear (a), circular (b), and elliptic polarization (c).

The orientation of the oscillating electrical field vector associated with a transverse plane EM wave (see Figure 3.1) may be in an arbitrary direction as long as the electric vector is perpendicular to the direction of propagation. The vibration of the electric vector defines the polarization state of the corresponding radiation field and may distinguish two EM waves of the same frequency and amplitude (see Section 3.2.1 for the definition). Three types of polarization can be distinguished and are illustrated in Figure 3.2.

a) Linear Polarization: The electric field vector vibrates backward and forward within a fixed plane. The projection of the vibrating electric vector onto a plane, which is perpendicular to the direction of wave propagation, yields a straight line (see Figure 3.2a).
b) Circular Polarization: The electric field vector oscillates in a spiral fashion, either clockwise or counterclockwise, around the direction of propagation. If the spiral rotation of the electric field vector is clockwise, as viewed by an observer looking towards the light source, the polarization is right circularly polarized; whereas, its counterpart with counterclockwise rotation is defined as left circular polarization. This definition was adopted by Born and Wolf (2003) and Bohren and Huffman (1983). An opposite definition was used by van de Hulst (1957) and many others. The projection of the end point of the electric vector onto a plane perpendicular to the direction of wave propagation, yields a circle, see Figure 3.2b.
c) Elliptic Polarization: Both linear and circular polarization are special cases of elliptical polarization. The projection of the vibration onto a plane, which is perpendicular to the direction of wave propagation, yields an ellipse (see Figure 3.2c).

3.1.4
Wave–Particle Duality

Wave–particle duality is a unique characteristic of EM radiation. Specifically, in addition to its wave properties, a radiation beam may behave like a flow of discrete energy packages known as photons. The energy E_{phot} of each individual photon is

Table 3.1 Spectral photon number flux density $N_{phot,\lambda}$ for typical solar zenith angles and radiant energy flux densities.

Solar zenith angle (°)	Wavelength λ (μm)	$F_{phot,\lambda}$ (W m^{-2} μm^{-1})	$N_{phot,\lambda}$ (s^{-1} m^{-2} μm^{-1})
10	0.3	20	3.02×10^{19}
10	0.5	1620	4.08×10^{21}
10	0.85	840	3.59×10^{21}
75	0.3	0.019	2.94×10^{16}
75	0.5	270	6.80×10^{20}
75	0.85	170	7.27×10^{20}

determined by the frequency or wavelength of radiation, that is,

$$E_{phot} = h \cdot \nu = h \cdot \frac{c}{\lambda} = \hbar \cdot \omega_c , \tag{3.26}$$

where Planck's constant, h, is 6.6262×10^{-34} J s and $\hbar = h/2\pi = 1.05459 \times 10^{-34}$ J s. The lower the frequency, ν, or the longer the wavelength, λ, the less energy the photon carries.

The photon model helps to explain the photoelectric effect. Each photon has a discrete energy, E_{phot}, allowing it to release electrons from the surface of certain metal materials. The wave model cannot explain why the photoelectric effect starts only if the radiation exceeds a certain threshold frequency independent of its intensity. If we have a number of $N_{phot,\lambda}$ photons per unit time hitting a unit area with a distinct wavelength, λ, that is, a distinct energy E_{phot}, we have a spectral photon flux density of

$$F_{phot,\lambda} = N_{phot,\lambda} \cdot E_{phot} = N_{phot,\lambda} \cdot h \cdot \nu = N_{phot,\lambda} \cdot \frac{h \cdot c}{\lambda} . \tag{3.27}$$

The calculated spectral photon number flux density in units of s^{-1} m^{-2} μm^{-1}, $N_{phot,\lambda}$, for radiant energy flux densities normally encountered in the atmosphere are large values (see Table 3.1). Here, $N_{phot,\lambda}$ is the number flux density of photons corresponding to a specific frequency or wavelength. The total number flux density of photons resulting from the spectral (energetic) integration is even larger.

3.1.5
Atmospheric EM Radiation Spectrum

EM waves with a wide region of frequencies are omnipresent in the atmosphere. This frequency region ranges from a few cycles per second to more than 10^{26} Hz; correspondingly, λ may vary from lengths of hundreds of kilometers to the diameter of an atomic nucleus. Atmospheric radiation is the superposition of all EM waves in the atmosphere.

A monochromatic spectrum of radiation refers to an infinitesimally narrow spectral regime centered at one specific wavelength; a discrete spectrum of radiation in-

Table 3.2 Some numbers relating the wavelength, frequency, wavenumber, with photon energy for certain types of EM radiation.

Type	Wavelength λ	Frequency $\nu = c/\lambda$ (Hz)	Photon energy $E_{phot} = h \cdot \nu$ (eV)	Wavenumber $\tilde{\nu} = 1/\lambda$ (cm^{-1})
Radio waves	100–0.001 km	$3.00 \times 10^{3}-$ 3.00×10^{8}	$1.24 \times 10^{-11}-$ 1.24×10^{-6}	$1.00 \times 10^{-7}-$ 1.00×10^{-2}
Microwave (MW)	1–0.001 m	$3.00 \times 10^{8}-3.00 \times 10^{11}$	$1.24 \times 10^{-6}-1.24 \times 10^{-3}$	$1.00 \times 10^{-2}-10$
Infrared (IR)	1–0.002 mm	$3.00 \times 10^{11}-3.00 \times 10^{14}$	$1.24 \times 10^{-3}-0.62$	10–5000
Near IR (NIR)	2000–750 nm	$(1.50-4.00) \times 10^{14}$	0.62–1.65	$5000-1.33 \times 10^{4}$
Visible (VIS)	750–370 nm	$(4.00-8.10) \times 10^{14}$	1.65–3.35	$(1.33-2.70) \times 10^{4}$
Dark red	750–680 nm	$(4.00-4.41) \times 10^{14}$	1.65–1.82	$(1.33-1.47) \times 10^{4}$
Red	680–622 nm	$(4.41-4.82) \times 10^{14}$	1.82–1.99	$(1.47-1.61) \times 10^{4}$
Orange	622–597 nm	$(4.82-5.02) \times 10^{14}$	1.99–2.08	$(1.61-1.68) \times 10^{4}$
Yellow	597–576 nm	$(5.02-5.20) \times 10^{14}$	2.08–2.15	$(1.68-1.74) \times 10^{4}$
Green	576–492 nm	$(5.20-6.09) \times 10^{14}$	2.15–2.52	$(1.74-2.03) \times 10^{4}$
Blue	492–455 nm	$(6.09-6.59) \times 10^{14}$	2.52–2.73	$(2.03-2.20) \times 10^{4}$
Purple	455–370 nm	$(6.59-8.10) \times 10^{14}$	2.73–3.35	$(2.20-2.70) \times 10^{4}$
Ultraviolett (UV)	370–10 nm	$8.10 \times 10^{14}-3.00 \times 10^{16}$	3.35–124	$2.70 \times 10^{4}-10^{6}$
UV A	370–315 nm	$(8.10-9.52) \times 10^{14}$	3.35–3.94	$(2.70-3.17) \times 10^{4}$
UV B	315–280 nm	$(0.95-1.07) \times 10^{15}$	3.94–4.43	$(3.17-3.57) \times 10^{4}$
UV C	280–10 nm	$(0.11-3.00) \times 10^{16}$	4.43–124	$3.57 \times 10^{4}-10^{6}$
X-rays	3–0.03 nm	$10^{17}-10^{19}$	$414-4.14 \times 10^{4}$	$(0.03-3.33) \times 10^{8}$
Gamma rays	30–3 pm	$10^{19}-10^{20}$	$(0.41-4.14) \times 10^{5}$	$(0.33-3.33) \times 10^{9}$
Cosmic rays	< 50 fm	$> 6 \times 10^{21}$	$> 2.48 \times 10^{7}$	$> 2 \times 10^{11}$

volves narrow spectral intervals at several wavelengths; and a continuous spectrum of radiation consists of an infinite number of wavelengths. Atmospheric radiation can generally be considered continuous.

Solid and liquid substances normally emit a continuous spectrum of radiation. Gases, however, are independent, oscillating, and moving molecules that emit radiation in narrow spectral intervals or spectral lines. A conglomerate of spectral lines form a spectral band (line or band spectra).

The longest wavelengths encountered in the atmosphere are radio waves (see Table 3.2). The shortest wavelengths, photons with the highest energy, are carried by cosmic radiation reaching the highest atmospheric layers but usually not the ground. Most of the ultraviolet (UV) radiation is absorbed in the stratospheric ozone layer. Radiation at the surface is generally observed at wavelengths longer than 0.3 μm. Two important spectral regions are used for energy budget and remote sensing applications. The solar spectrum spans from 0.2 to 5 μm wavelengths. The terrestrial thermal emission spectral region refers to wavelengths between 5 and 100 μm; whereas, the thermal infrared (TIR) spectral region usually refers to wavelengths of $\sim$ 3–50 μm. For microwave (MW) applications, the TIR wavelength range is short; whereas, the TIR seems to have rather long wavelengths in comparison with visible (VIS) radiation. Because short and long are always relative terms, we prefer to apply the terms solar and terrestrial, instead of shortwave and longwave.

3.2
Basic Radiometric Quantities

3.2.1
Radiant Energy Flux, Flux Density, and Radiance

To introduce basic radiometric quantities, consider a small area element, $d^2 A$, whose location (defined in terms of the location of the center of the area element) is indicated by the Euklidic position vector, $\vec{r}$, and a pencil of radiation passing through the area element along a direction indicated by a unit vector $\hat{s}$. The spectrum associated with the radiation beam ranges from λ to $(\lambda + d\lambda)$. The radiant energy passing through the area element within a time interval from t to $(t + dt)$ is E_{rad} in units of joule. We define the spectral radiant energy flux, Φ_λ, at a specific location, $\vec{r}$, and time, t, as

$$\Phi_\lambda(\vec{r}, t) = \frac{d^2 E_{rad}}{dt\, d\lambda} \; . \tag{3.28}$$

The units for the spectral radiant energy flux, Φ_λ, are $J\, s^{-1}\, \mu m^{-1} = W\, \mu m^{-1}$. The corresponding spectral radiant energy flux density, F_λ, is given by

$$F_\lambda(\vec{r}, t) = \frac{d^2 \Phi_\lambda}{d^2 A} = \frac{d^4 E_{rad}}{dt\, d\lambda\, d^2 A} \; . \tag{3.29}$$

The units for the spectral radiant energy flux density, F_λ, are $W\, m^{-2}\, \mu m^{-1}$. If the radiant energy is confined within a solid angle element, $d^2 \Omega$, pointing along a specific direction $\hat{s}$, the corresponding spectral radiance, I_λ, is defined as

$$I_\lambda(\vec{r}, \hat{s}, t) = \frac{d^4 \Phi_\lambda}{d^2 A_\perp\, d^2 \Omega} = \frac{d^6 E_{rad}}{\cos\theta\, dt\, d\lambda\, d^2 A\, d^2 \Omega} \; , \tag{3.30}$$

where θ is the angle between the propagation direction of the radiation $\hat{s}$ and the normal unit vector $\hat{n}$ of the area element, see Figure 3.3. In the preceding equation, the quantity $(\cos\theta\, d^2 A)$ is the projection of the area element onto a plane that is perpendicular to the propagation direction of the beam. The units of the spectral radiance I_λ are $W\, m^{-2}\, sr^{-1}\, \mu m^{-1}$.

According to this definition, radiance is invariant along the propagation of radiant energy within a nonabsorptive medium. To understand the characteristics of radiance, let us consider a narrow radiation beam shown in Figure 3.4. At location P, the area of the beam's cross section normal to the propagation direction, is assumed to be $d^2 A_\perp$. The radiant energy passing through $d^2 A_\perp$, within time interval dt, is $d^6 E_{rad}$. Similarly, the corresponding quantities at location P' are assumed to be $d^2 A'_\perp$ and $d^6 E'_{rad}$. As imposed by the principle of energy conservation, $d^6 E_{rad}$ and $d^6 E'_{rad}$ are equal because the medium is nonabsorptive.

From the geometry shown in Figure 3.4, the solid angle element subtended by $d^2 A'_\perp$, as viewed from P, is

$$d^2 \Omega = \frac{d^2 A'_\perp}{s^2} \; , \tag{3.31}$$

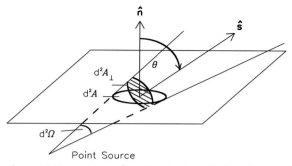

Figure 3.3 Geometric configuration for the definition of spectral radiance.

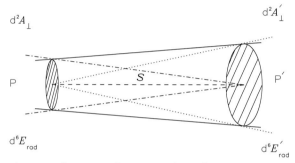

Figure 3.4 Illustration of a narrow radiation beam.

where s is the distance between P and P′. Similarly, the solid angle element subtended by $d^2 A_\perp$, as viewed from P′, is

$$d^2 \Omega' = \frac{d^2 A_\perp}{s^2} .$$ (3.32)

The spectral radiance observed at location P is

$$I_{P,\lambda} = \frac{d^6 E_{\mathrm{rad}}}{dt \, d^2 A_\perp \, d^2 \Omega \, d\lambda}$$

$$= \frac{d^6 E_{\mathrm{rad}}}{dt \, d^2 A_\perp \, d^2 A'_\perp \, s^{-2} \, d\lambda} ,$$ (3.33)

and the spectral radiance observed at location P′ is

$$I'_{P,\lambda} = \frac{d^6 E'_{\mathrm{rad}}}{dt \, d^2 A'_\perp \, d^2 \Omega' \, d\lambda}$$

$$= \frac{d^6 E'_{\mathrm{rad}}}{dt \, d^2 A'_\perp \, d^2 A_\perp \, s^{-2} \, d\lambda} ,$$ (3.34)

as evident from the preceding equations in conjunction with the energy conservation relation, that is,

$$d^6 E'_{rad} = d^6 E_{rad} \ .$$ (3.35)

Hence, the spectral radiances $I_{P,\lambda}$ and $I'_{P,\lambda}$ are equal.

From the preceding discussion, the radiance of the radiation field originating from a point source is invariant; whereas, the corresponding irradiance is inversely proportional to the square of the distance between the source and the observational point. For instance, the solar radiance observed at the Earth's TOA is the same as that observed at the top of the Martian atmosphere, although the solar flux density at the latter location is smaller.

Once the spectral radiometric quantities are defined, the corresponding spectrally integrated counterparts are quite straightforward to derive. For example, the spectrally integrated (broadband) radiance for a spectral interval (λ_1, λ_2) is written as

$$I(\bar{\mathbf{r}}, \hat{\mathbf{s}}, t) = \int_{\lambda_1}^{\lambda_2} I_\lambda(\bar{\mathbf{r}}, \hat{\mathbf{s}}, t) \, d\lambda \ .$$ (3.36)

Another example comprises the spectrally integrated (broadband) radiative flux density for a spectral interval (λ_1, λ_2), that is,

$$F(\bar{\mathbf{r}}, t) = \int_{\lambda_1}^{\lambda_2} F_\lambda(\bar{\mathbf{r}}, t) \, d\lambda \ .$$ (3.37)

In the following discussions, we may not always refer to spectral radiometric quantities (indicated by the subscript λ), even though most can be treated as either spectral or broadband. Thus, in some equations, we will omit the subscript λ. If the units of the referred quantity contain μm^{-1}, the respective quantity is spectral; otherwise, the quantity is broadband and obtained by integration over a certain spectral region. Alternatively, instead of the subscript λ, frequency, ν, or wavenumber, $\tilde{\nu}$, can be used to indicate spectral radiometric quantities as well. A quantity is called monochromatic if the wavelength appears as an explicit independent argument; it does not necessarily contain the unit μm^{-1}.

3.2.2
Radiant Energy Density and Radiance

With the radiance, we can also define the radiant energy density. Let $u_\lambda(\bar{\mathbf{r}}, t) \, d\lambda$ indicate the radiant energy per unit volume and per steradian within a spectral interval $d\lambda$ measured at spatial location, $\bar{\mathbf{r}}$, and at time, t. To obtain an explicit expression of $u_\lambda(\bar{\mathbf{r}}, t) \, d\lambda$ in terms of spectral radiance, I_λ, let us consider a narrow beam of radiation passing location, $\bar{\mathbf{r}}$, whose cross section, normal to the propagation direction, is $d^2 A_\perp$ at location $\vec{r}$. A volume element of the beam, which is measured from $\vec{r}$ within the time interval dt, is given by

$$d^3 V = c\, d^2 A_\perp\, dt \,, \tag{3.38}$$

where c is the speed of light in a vacuum. According to the definition of spectral radiance, see Eq. (3.30), the radiant energy propagating within the solid angle element $d^2 \Omega$ and within $d^3 V$ is

$$\begin{aligned} d^6 E_{\mathrm{rad}} &= I_\lambda(\vec{r}, \hat{s}, t) \cdot \cos\theta\; dt\, d\lambda\, d^2 A\, d^2 \Omega \\ &= I_\lambda(\vec{r}, \hat{s}, t)\, dt\, d\lambda\, d^2 A_\perp\, d^2 \Omega \,. \end{aligned} \tag{3.39}$$

Thus, the portion of the spectral radiant energy density, corresponding to the radiant energy confined within $d^2 \Omega$, is given by

$$\begin{aligned} u_\lambda(\vec{r}, \hat{s}, t)\, d\lambda\, d^2 \Omega &= \frac{d^6 E_{\mathrm{rad}}}{d^3 V} \\ &= I_\lambda \frac{dt\, d\lambda\, d^2 A_\perp\, d^2 \Omega}{c\, d^2 A_\perp\, dt} \\ &= \frac{I_\lambda}{c}\, d\lambda\, d^2 \Omega \,. \end{aligned} \tag{3.40}$$

The unit of u_λ is $\mathrm{J\, m^{-3}\, sr^{-1}\, \mu m^{-1}}$. If we consider radiant energy propagating in all directions, by integrating the spectral energy density over a 4π solid angle, we obtain the average spectral energy density

$$\begin{aligned} \overline{u}_\lambda(\vec{r}, t) &= \iint\limits_{4\pi} u_\lambda(\vec{r}, \hat{s}, t)\, d^2 \Omega \\ &= \frac{1}{c} \iint I_\lambda(\vec{r}, \hat{s}, t)\, d^2 \Omega \\ &= \frac{4\pi}{c} \cdot \overline{I}_\lambda(\vec{r}, t) \,, \end{aligned} \tag{3.41}$$

where $\overline{I}_\lambda(\vec{r}, t)$ is the averaged radiance given by

$$\overline{I}_\lambda(\vec{r}, t) = \frac{1}{4\pi} \iint\limits_{4\pi} I_\lambda(\vec{r}, \hat{s}, t)\, d^2 \Omega \,. \tag{3.42}$$

The broadband (between λ_1 and λ_2) energy density is given by

$$\overline{u}(\vec{r}, t) = \int\limits_{\lambda_1}^{\lambda_2} \overline{u}_\lambda(\vec{r}, t)\, d\lambda = \frac{4\pi}{c} \int\limits_{\lambda_1}^{\lambda_2} \overline{I}_\lambda(\vec{r}, t)\, d\lambda \,.$$

Using the energy of a photon $E_{\mathrm{phot}} = h \cdot c/\lambda$, see Eq. (3.26), we obtain the spectral and broadband number densities of photons by

$$n_{\mathrm{phot},\lambda}(\vec{r}, t) = \frac{\overline{u}_\lambda(\vec{r}, t)}{E_{\mathrm{phot}}} = \frac{4\pi \cdot \lambda}{c^2 \cdot h} \cdot \overline{I}_\lambda(\vec{r}, t) \,, \tag{3.43}$$

and

$$n_{\text{phot}}(\bar{\mathbf{r}}, t) = \frac{\overline{u}(\bar{\mathbf{r}}, t)}{E_{\text{phot}}} = \frac{4\pi}{c^2 \cdot h} \int_{\lambda_1}^{\lambda_2} \lambda \cdot \overline{I}_\lambda(\bar{\mathbf{r}}, t)\, d\lambda \; . \tag{3.44}$$

$n_{\text{phot},\lambda}$ is measured in units of $\text{m}^{-3}\,\text{sr}^{-1}\,\mu\text{m}^{-1}$; n_{phot} in units of $\text{m}^{-3}\,\text{sr}^{-1}$.

3.2.3
Irradiance, Emittance, Exitance, and Actinic Radiation

The radiant energy flux density, F, refers to the flux of radiant energy, which impinges, passes through (crosses), or exits a flat unit area. The radiant energy flux density, F, associated with a beam in the direction $\hat{\mathbf{s}}$ and through a surface element, $d^2 A$, is proportional to $\hat{\mathbf{n}} \cdot \hat{\mathbf{s}}$, where $\hat{\mathbf{n}}$ is the normal unit vector of the surface. In the discussion of the transfer of radiation in the atmosphere, a horizontal surface is usually selected as a reference plane, such that $\hat{\mathbf{n}} = \hat{\mathbf{e}}_3$ and

$$\hat{\mathbf{n}} \cdot \hat{\mathbf{s}} = \hat{\mathbf{e}}_3 \cdot \hat{\mathbf{s}} = \cos\theta = \mu \; . \tag{3.45}$$

If $\hat{\mathbf{n}} \cdot \hat{\mathbf{s}} > 0$, we talk of upward radiation which is indicated by an upward arrow. In case of $\hat{\mathbf{n}} \cdot \hat{\mathbf{s}} < 0$, downward radiation is considered and a downward arrow is included. In general, radiant energy flux density is referred to as irradiance, a quantity frequently used in the study of radiant energy budget, if a real horizontal unit area is illuminated by radiation from all possible directions of a hemisphere. If the irradiance originates from an emitting surface, the quantity is usually referred to as emittance, and the corresponding radiance is commonly called brightness. If the radiative flux density is associated with a beam emerging from an area element either by reflection, transmission from underneath, or emission, the radiative flux density is an exitance. Only if the flux density is associated with a beam that crosses or penetrates through an imaginary area element, is it called a flux density.

A spherical, rather than a planar receiver area, requires the term actinic flux density, F_{act}, often wrongly quoted as actinic flux, to be applied. This quantity is not proportional to $\cos\theta$ (see Figure 3.5) and is an important quantity in atmospheric chemistry. The actinic flux density represents the energy flux onto a unit sphere, normalized to the cross section of the sphere, and may sometimes be referred to as spheradiance. The weights associated with the angular distribution of a radiation field for computing irradiance and actinic flux density are illustrated in Figure 3.5.

The surface to which the radiant energy is related may be real, for example, the ground, the top of a cloud layer, or the surface of a cloud droplet. On the other hand, it may also be imaginary, such as, an arbitrary level in the atmosphere. For atmospheric applications, the surfaces for irradiance, exitance, or flux density are considered to be horizontal.

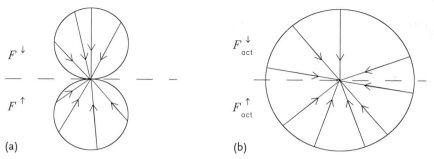

Figure 3.5 (a) Irradiance is proportional to integrating over $\cos\theta$; (b) actinic flux density is not.

3.2.4
Relation between Upward, Downward, and Net Actinic Flux Densities and Radiance

Consider a unit area on a horizontal reference plane, the spectral irradiance is obtained by integration over the lower or the upper hemisphere to obtain the upward irradiance, $F_\lambda^\uparrow$, or the downward irradiance, $F_\lambda^\downarrow$, respectively. For example, the upward spectral irradiance, $F_\lambda^\uparrow$, is found from

$$F_\lambda^\uparrow(\mathbf{r}, t) = \iint\limits_{2\pi} I_\lambda(\mathbf{r}, \hat{\mathbf{s}}, t)\, \hat{\mathbf{e}}_3 \cdot \hat{\mathbf{s}}\, \mathrm{d}^2\Omega \ , \tag{3.46}$$

where the unit vector normal to the assumed horizontal surface is $\hat{\mathbf{e}}_3$. In polar coordinates, we obtain

$$\hat{\mathbf{e}}_3 \cdot \hat{\mathbf{s}} = \cos\theta \ . \tag{3.47}$$

Thus, by additionally applying Eq. (2.37), we get

$$F_\lambda^\uparrow(\mathbf{r}, t) = \int\limits_0^{2\pi} \int\limits_0^{\pi/2} I_\lambda(\mathbf{r}, \theta, \varphi, t) \cdot \cos\theta \cdot \sin\theta\, \mathrm{d}\theta\, \mathrm{d}\varphi \ . \tag{3.48}$$

A similar expression holds for the downward irradiance $F_\lambda^\downarrow$

$$F_\lambda^\downarrow(\mathbf{r}, t) = -\int\limits_0^{2\pi} \int\limits_{\pi/2}^{\pi} I_\lambda(\mathbf{r}, \theta, \varphi, t) \cdot \cos\theta \cdot \sin\theta\, \mathrm{d}\theta\, \mathrm{d}\varphi \ . \tag{3.49}$$

The spectral net irradiance $F_{\mathrm{net},\lambda}$ is defined as

$$F_{\mathrm{net},\lambda} = F_\lambda^\downarrow - F_\lambda^\uparrow \ , \tag{3.50}$$

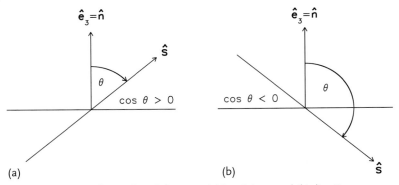

Figure 3.6 Change of sign of $\cos\theta$ for upward (a) and downward (b) directions.

which is equivalent to

$$F_{\text{net},\lambda}(\vec{r}, t) = -\int_0^{2\pi}\int_{\pi/2}^{\pi} I_\lambda(\vec{r}, \theta, \varphi, t) \cdot \cos\theta \cdot \sin\theta \, d\theta \, d\varphi$$

$$- \int_0^{2\pi}\int_0^{\pi/2} I_\lambda(\vec{r}, \theta, \varphi, t) \cdot \cos\theta \cdot \sin\theta \, d\theta \, d\varphi$$

$$= -\int_0^{2\pi}\int_0^{\pi} I_\lambda(\vec{r}, \theta, \varphi, t) \cdot \cos\theta \cdot \sin\theta \, d\theta \, d\varphi \ . \tag{3.51}$$

Often, instead of θ, the dimensionless variable $\mu = |\cos\theta|$ is used. In this case, the previous equations transform to

$$F_\lambda^\uparrow(\vec{r}, t) = \int_0^{2\pi}\int_0^1 I_\lambda(\vec{r}, \mu, \varphi, t) \cdot \mu \, d\mu \, d\varphi \ , \tag{3.52}$$

and

$$F_\lambda^\downarrow(\vec{r}, t) = \int_0^{2\pi}\int_0^1 I_\lambda(\vec{r}, -\mu, \varphi, t) \cdot \mu \, d\mu \, d\varphi \ . \tag{3.53}$$

The minus sign in front of μ in Eq. (3.53) indicates the downward direction and $\cos\theta$ would become negative, see Figure 3.6.

The spectral actinic flux density $F_{\text{act},\lambda}$ is obtained in a similar way without the cosine weighting, that is,

$$F_{\text{act},\lambda}(\vec{r}, t) = F_{\text{act},\lambda}^\downarrow(\vec{r}, t) + F_{\text{act},\lambda}^\uparrow(\vec{r}, t)$$

$$= \int_0^{2\pi}\int_0^1 I_\lambda(\vec{r}, -\mu, \varphi, t) \, d\mu \, d\varphi + \int_0^{2\pi}\int_0^1 I_\lambda(\vec{r}, \mu, \varphi, t) \, d\mu \, d\varphi \ . \tag{3.54}$$

3.2.5
Isotropic Radiation Field

A given radiation field is homogeneous if the corresponding radiometric quantities are independent of location $\vec{r}$. If the radiance is independent of direction $\hat{s}$, the radiation field is isotropic. Thus, by applying Eqs. (3.52) and (3.53) to an isotropic radiation field, we get

$$F_{\text{iso},\lambda}^{\downarrow}(\vec{r}, t) = 2\pi \cdot I_{\text{iso},\lambda}(\vec{r}, t) \int_0^1 \mu \, d\mu = \pi \cdot I_{\text{iso},\lambda}(\vec{r}, t)$$

$$= F_{\text{iso},\lambda}^{\uparrow}(\vec{r}, t) \, . \tag{3.55}$$

The spectral actinic flux density under isotropic conditions, $F_{\text{act,iso},\lambda}$, is obtained in a similar way using Eq. (3.54), that is,

$$F_{\text{act,iso},\lambda}(\vec{r}, t) = 2\pi \cdot I_{\text{iso},\lambda}(\vec{r}, t) \int_0^1 d\mu + 2\pi \cdot I_{\text{iso},\lambda}(\vec{r}, t) \int_0^1 d\mu$$

$$= 4\pi \cdot I_{\text{iso},\lambda}(\vec{r}, t) \, . \tag{3.56}$$

3.2.6
Reflectivity, Absorptivity, and Transmissivity

The three dimensionless optical quantities reflectivity, $\mathcal{R}$, absorptivity, $\mathcal{A}$, and transmissivity, $\mathcal{T}$, are defined to specify relative optical measures of surfaces or atmospheric layers. The three measures quantify the amount of radiant energy fluxes reflected, absorbed, or transmitted by a surface (or layer) relative to the incident radiant energy flux. These quantities are not independent of each other and satisfy the following relation:

$$\mathcal{R} + \mathcal{A} + \mathcal{T} = 1 \, . \tag{3.57}$$

3.3
Blackbody and Graybody Radiation: Basic Laws

3.3.1
Planck's Law

Absorption and emission are the processes of energy exchange between radiant and other forms of energy. Emitted radiation may be either absorbed at another atmospheric location or may escape into space. An object with a temperature, T, emits radiation at all wavelengths, frequencies, or wavenumbers; however, the total amount of emitted radiation has an upper boundary. The upper boundary is defined by Planck's law as the spectral radiance emitted by a blackbody, with a unit of $W\,m^{-2}\,sr^{-1}\,\mu m^{-1}$. A blackbody is defined as a perfect emitter, emitting the radiance predicted by Planck's law, and a perfect absorber, but does not exist in the real world.

Planck's law as a function of wavelength, λ, is given by

$$B_\lambda(T) = \frac{2h \cdot c^2}{\lambda^5} \frac{1}{\exp[h \cdot c/(k_B \cdot \lambda \cdot T)] - 1} \, , \tag{3.58}$$

with Planck's constant $h = 6.6262 \times 10^{-34}$ J s; Boltzmann's constant $k_B = 1.3805 \times 10^{-23}$ J K^{-1}; the speed of light in a vacuum c is defined in Eq. (3.16); the absolute temperature, T, in kelvin; and, the wavelength, λ, in meters. The speed of light in air can be considered with sufficient accuracy to be equal to c. Examples of Planck's function for different temperatures are given in Figure 3.7. The radiation emitted by a blackbody only depends on its temperature; therefore, the emitted radiance can be inverted to get the temperature of a blackbody that emits an equal amount of radiation at the wavelength of interest. The blackbody temperature, thus calculated, is referred to as the brightness temperature.

Let us change the domains of Planck's function. We assume

$$B_\lambda |d\lambda| = B_\nu |d\nu| = B_{\tilde{\nu}} |d\tilde{\nu}| \, . \tag{3.59}$$

Since $\lambda = c/\nu = 1/\tilde{\nu}$, we have

$$|d\lambda| = \frac{c}{\nu^2} |d\nu| \, , \tag{3.60}$$

$$= \frac{1}{\tilde{\nu}^2} |d\tilde{\nu}| \, . \tag{3.61}$$

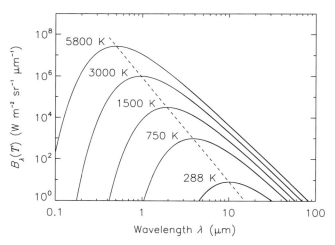

Figure 3.7 Spectral distribution of terrestrial blackbody radiation emitted by the surface for different values of temperature (solid lines) given by Eq. (3.58). The maxima of the Planck curves are indicated by a dashed line, see Wien's law.

Using these equalities, we obtain

$$B_\nu = \frac{c}{\nu^2} \cdot B_\lambda \,, \tag{3.62}$$

$$B_{\tilde{\nu}} = \frac{1}{\tilde{\nu}^2} \cdot B_\lambda \,. \tag{3.63}$$

With these transformations, we obtain Planck's law as a function of frequency ($\nu = c/\lambda$)

$$B_\nu(T) = \frac{2h \cdot \nu^3}{c^2} \cdot \frac{1}{\exp[h \cdot \nu/(k_B \cdot T)] - 1} \,, \tag{3.64}$$

and as a function of wavenumber ($\tilde{\nu} = 1/\lambda$),

$$B_{\tilde{\nu}}(T) = \frac{2h \cdot \tilde{\nu}^3 \cdot c^2}{\exp[h \cdot \tilde{\nu} \cdot c/(k_B \cdot T)] - 1} \,. \tag{3.65}$$

3.3.2
Wien's Displacement Law

Planck's function peaks at a wavelength inversely proportional to the absolute temperature T (see Figure 3.7). The maximum value of Planck's function can be quantitatively described by calculating the first-order derivative of Planck's function and zeroing the resultant expression. The manipulation gives rise to Wien's displacement law, given by

$$\lambda_{\max} = \frac{k_W}{T} \,, \tag{3.66}$$

where $k_W = 2897 \; \mu m \, K$. Thus, the emission from the Sun (with $T_S \approx 5800 \, K$) peaks at a wavelength of $\lambda_{\max} = 0.5 \, \mu m$. For the typical Earth surface and atmospheric temperatures of $\sim 200{-}300 \, K$, Planck's function peaks between 9.6 and 14.4 μm wavelengths.

To derive Wien's displacement law, we begin with Planck's law in the form of Eq. (3.58). Taking the derivative of Planck's function with respect to the wavelength and applying the product rule yields

$$\begin{aligned}
\frac{dB_\lambda}{d\lambda} = &-\frac{10h \cdot c^2}{\lambda^6} \cdot \frac{1}{\exp[h \cdot c/(k_B \cdot \lambda \cdot T)] - 1} \\
&- \frac{2h \cdot c^2}{\lambda^5} \frac{1}{\{\exp[h \cdot c/(k_B \cdot \lambda \cdot T)] - 1\}^2} \\
&\times \exp\left(\frac{h \cdot c}{k_B \cdot \lambda \cdot T}\right) \cdot \left(-\frac{h \cdot c}{k_B \cdot \lambda^2 \cdot T}\right).
\end{aligned} \tag{3.67}$$

Let the derivative be zero, thus yielding the equation for λ_{max}, that is,

$$\frac{5}{\lambda_{max}} = \frac{1}{\exp[h \cdot c/(k_B \cdot \lambda_{max} \cdot T)] - 1} \cdot \exp\left(\frac{h \cdot c}{k_B \cdot \lambda_{max} \cdot T}\right) \cdot \left(\frac{h \cdot c}{k_B \cdot \lambda_{max}^2 \cdot T}\right) . \tag{3.68}$$

Substitute

$$x_{max} = \frac{h \cdot c}{k_B \cdot \lambda_{max} \cdot T} , \tag{3.69}$$

into Eq. (3.68) in order to yield

$$5 = \frac{1}{\exp(x_{max}) - 1} \cdot \exp(x_{max}) \cdot x_{max} . \tag{3.70}$$

The solution is $x_{max} = 4.9651$. Thus, we get

$$\lambda_{max} \cdot T = \frac{h \cdot c}{k_B \cdot x_{max}} = \text{constant} = k_W = 2897 \, \mu m \, K . \tag{3.71}$$

In the frequency domain, a similar procedure yields

$$\frac{d B_\nu}{d\nu} = 0 \Rightarrow \nu_{max} \cdot \frac{1}{T} = \text{constant} = k_{W,\nu} = 5.8789 \times 10^{10} \, Hz \, K^{-1} , \tag{3.72}$$

where ν_{max} is given in units of Hz. If the exponential term in Planck's law, Eq. (3.58), is neglected, we obtain the maximum of Planck's function

$$B_{\lambda_{max}}(T) \approx \frac{a_1}{\lambda_{max}^5} , \tag{3.73}$$

with $a_1 = \text{constant}$. Hence, it follows that

$$\log B_{\lambda_{max}}(T) \approx \log a_1 - 5 \cdot \log \lambda_{max} . \tag{3.74}$$

The above equation indicates that the maximum of Planck's function decreases by five orders in magnitude if the wavelength increases by one order in magnitude. This becomes obvious when Planck's function is plotted on a logarithmic scale in Figure 3.7. For example, Planck's curve corresponding to $T = 3000$ K has a maximum at about 1 µm wavelength, where the value of Planck's curve is approximately $10^6 \, W \, m^{-2} \, sr^{-1} \, \mu m^{-1}$. Planck's curve for $T = 288$ K has a value of approximately $10^1 \, W \, m^{-2} \, sr^{-1} \, \mu m^{-1}$ at roughly 10 µm, which is five orders of magnitude less than at $T = 3000$ K.

Note, the maximum of Planck's curve in the frequency domain, λ'_{max}, is different from that in the wavelength domain, that is,

$$\lambda'_{max} = \lambda(\nu_{max}) \neq \lambda_{max} . \tag{3.75}$$

This inequality occurs because of the nonlinear relation between the wavelength and the frequency. To be more specific,

$$\lambda'_{max} = \frac{c}{\nu_{max}} = \frac{c}{k_{W,\nu} \cdot T} = \frac{5099\ \mu m\ K}{T} \neq \lambda_{max} = \frac{k_W}{T} = \frac{2897\ \mu m\ K}{T}\ .$$

(3.76)

For example, at the same temperature $T_S \approx 6000\ K$ (temperature of the Sun), we calculate different wavelengths: $\lambda'_{max} = 0.8\ \mu m$ (near infrared, NIR) and $\lambda_{max} = 0.5\ \mu m$ (green).

3.3.3
Stefan–Boltzmann Law

The Stefan–Boltzmann law gives the broadband irradiance, an important quantity for energy budget consideration, emitted by a blackbody. It is obtained by integrating Planck's function over wavelength and over 2π of a hemispheric solid angle. Blackbody radiation is isotropic, and solid angular integration yields the value of π, allowing the desired broadband blackbody irradiance F_{BB} to be obtained by

$$F_{BB}(T) = \pi \int_0^\infty B_\lambda(T)\, d\lambda = \sigma \cdot T^4\ ,$$

(3.77)

where σ is the Stefan–Boltzmann constant given by

$$\sigma = \frac{2\pi^5 \cdot k_B^4}{15 c^2 \cdot h^3} = 5.671 \times 10^{-8}\ W\,m^{-2}\,K^{-4}\ .$$

(3.78)

To get Eq. (3.77), we calculate the following integral:

$$F_{BB}(T) = \pi \int_0^\infty \frac{2h \cdot c^2}{\lambda^5} \frac{1}{\exp[h \cdot c/(k_B \cdot \lambda \cdot T)] - 1}\, d\lambda\ .$$

(3.79)

We substitute

$$x = \frac{h \cdot c}{k_B \cdot \lambda \cdot T}\ ,$$

(3.80)

and

$$\frac{dx}{d\lambda} = -\frac{h \cdot c}{k_B \cdot \lambda^2 \cdot T}$$

(3.81)

into Eq. (3.79) to yield

$$F_{BB}(T) = \pi \cdot \frac{2k_B^4 \cdot T^4}{h^3 \cdot c^2} \int_0^\infty \frac{x^3}{\exp(x) - 1}\, dx\ .$$

(3.82)

The value of the integral is exactly $\pi^4/15$ (Bronstein and Semendjajev, 1985), and we get the broadband blackbody irradiance by

$$
\begin{aligned}
F_{\mathrm{BB}}(T) &= \frac{2k_{\mathrm{B}}^4 \cdot \pi^5}{15 h^3 \cdot c^2} \cdot T^4 \\
&= \sigma \cdot T^4 .
\end{aligned}
\tag{3.83}
$$

3.3.4
Rayleigh–Jeans and Wien's Approximations

The Rayleigh–Jeans approximation is important in MW remote sensing ($\lambda > 1\,\mathrm{mm}$). It refers to the approximation of Planck's function in the case of large wavelength ($\lambda \longrightarrow \infty$) or low frequency ($\nu \longrightarrow 0$). Deriving the asymptotic behavior of the Planck function in both limits is straightforward. For $\lambda \longrightarrow \infty$ ($\nu \longrightarrow 0$), Planck's law can be approximated by

$$
B_\lambda(T) \approx \frac{2c \cdot k_{\mathrm{B}}}{\lambda^4} \cdot T ,
\tag{3.84}
$$

or in the frequency domain

$$
B_\nu(T) \approx \frac{2k_{\mathrm{B}} \cdot \nu^2}{c^2} \cdot T .
\tag{3.85}
$$

It is evident from Eqs. (3.84) and (3.85) that Planck's function is linearly proportional to the temperature in the Rayleigh–Jeans approximation. This feature is very useful in many applications.

The other extreme of Planck's function is given by Wien's approximation, which assumes that $\lambda \longrightarrow 0$ ($\nu \longrightarrow \infty$). This approximation may hold for short wavelengths and high temperatures. With this limit, the Planck function reduces to

$$
B_\lambda(T) \approx \frac{2h \cdot c^2}{\lambda^5} \cdot \exp[-h \cdot c/(\lambda \cdot k_{\mathrm{B}} \cdot T)] ,
\tag{3.86}
$$

or as a function of frequency ν

$$
B_\nu(T) \approx \frac{2h \cdot \nu^3}{c^2} \cdot \exp[-h \cdot \nu/(k_{\mathrm{B}} \cdot T)] .
\tag{3.87}
$$

3.3.5
Emissivity and Kirchhoff's Law

Emissivity of Surfaces and the Atmosphere
The Planck law hypothetically describes the maximum value of radiation theoretically emitted by a blackbody because real bodies do not exhibit the idealized behavior. Monochromatic emissivity, $\varepsilon(\lambda)$, describes the degree of deviation from blackbody emission. An emissivity of unity is characteristic for a blackbody. A spectrally constant emissivity, less than unity, defines a graybody. The broadband emissivity

is the monochromatic emissivity integrated over a particular wavelength region. In this sense, emissivity is the ratio of actual emission by a given body to the emission if the respective body were a blackbody.

The emissivity of land surfaces, for example, snow, bare soil, water, and grass must be distinguished from the emissivity of an atmospheric volume element consisting of atmospheric gases, aerosol particles, and/or clouds. In the case of surface emissivity, the emitted radiation depends on direction, wavelength, surface temperature, and on the surface material physical properties, such as, the complex refractive index, surface roughness parameters, etc. In the TIR spectral region ($\lambda > 3\ \mu\text{m}$), many natural surfaces emit radiation with $\varepsilon > 0.8$, but their emissivities do not depend on the direction. Thus, the radiance emitted by a unit surface area element at wavelength λ is

$$I_\lambda = \varepsilon(\lambda) \cdot B_\lambda(T)\,. \tag{3.88}$$

In general, the monochromatic emissivity is a function of the wavelength. If a body emits less radiation than predicted by Planck's law, monochromatic emissivity can be easily derived by measuring the emitted radiance I_λ

$$\varepsilon(\lambda) = \frac{I_\lambda}{B_\lambda(T)}\,. \tag{3.89}$$

The spectral emissivity of typical surfaces in the wavelength region $3-15\ \mu\text{m}$ is shown in Figure 3.8.

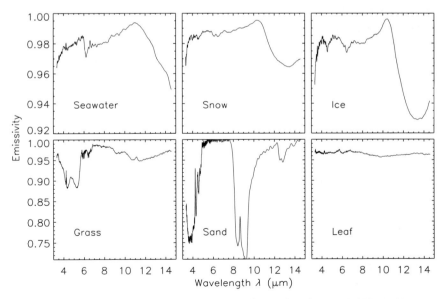

Figure 3.8 Spectral emissivities of typical surfaces in the wavelength region $3-15\ \mu\text{m}$. Data from http://g.icess.ucsb.edu/modis/EMIS/html/em.html (accessed November 2011).

Table 3.3 Examples of the broadband emissivity $\varepsilon(\lambda_1, \lambda_2)$ with $\lambda_1 = 3\,\mu m, \lambda_2 = 50\,\mu m$. Adapted from Petty (2006).

Surface type	$\varepsilon(\lambda_1, \lambda_2)$ (%)
Water	92–96
Fresh, dry snow	82–99.5
Ice	96
Green grass	97.5–98.6
Sand, dry	84–90
Soil, moist	95–98
Skin, human	95
Soil, dry plowed	90
Desert	90–91
Forest and shrubs	90
Concrete	71–88
Granite	89.8
Aluminum, polished	1–5

For certain purposes, it is appropriate to use the broadband emissivity ε defined by using irradiances or radiances:

$$\varepsilon = \frac{F}{F_{BB}} = \frac{\int_0^\infty F_\lambda \, d\lambda}{\pi \int_0^\infty B_\lambda(T) \, d\lambda} = \frac{F}{\sigma \cdot T^4} \, . \tag{3.90}$$

In the real world, no object is uniformly gray. Each body has a certain emissivity wavelength dependence over the full EM spectrum. If we restrict the wavelength region to (λ_1, λ_2), the broadband emissivity is obtained by:

$$\varepsilon(\lambda_1, \lambda_2) = \frac{F(\lambda_1, \lambda_2)}{F_{BB}(\lambda_1, \lambda_2)} \, . \tag{3.91}$$

$F(\lambda_1, \lambda_2)$ is the irradiance actually emitted by the object (surface). The corresponding irradiance emitted by a blackbody is given by:

$$F_{BB}(\lambda_1, \lambda_2) = \pi \int_{\lambda_1}^{\lambda_2} B_\lambda(T) \, d\lambda \, . \tag{3.92}$$

For example, the broadband emissivities ($\lambda_1 = 3\,\mu m$, $\lambda_2 = 50\,\mu m$; TIR) for various surfaces are given in percentages for various surfaces in Table 3.3.

Kirchhoff's Law
Kirchhoff's law states that the monochromatic emissivity $\varepsilon(\lambda)$ of a certain medium (gas, liquid, or solid) in thermodynamic equilibrium is equal to its monochromatic absorptivity $\mathcal{A}(\lambda)$ introduced in Section 3.2.6:

$$\varepsilon(\lambda, \theta, \varphi) = \mathcal{A}(\lambda, \theta, \varphi) \, . \tag{3.93}$$

This is strictly valid for monochromatic radiation and only for specified viewing directions. The common practice is to apply Kirchhoff's law as an approximation to graybody emissivity and absorptivity specified for a certain wavelength region.

For a blackbody, a nonblackbody, and a graybody, we have the following relations, namely,

$$\varepsilon(\lambda) = A(\lambda) = 1 \qquad \text{blackbody} ,$$
$$\varepsilon(\lambda) = A(\lambda) < 1 \qquad \text{nonblackbody} ,$$
$$\varepsilon = A < 1 \qquad \text{graybody} .$$

For most applications, Kirchhoff's law can be strictly used; however, the law only applies for the "local thermodynamic equilibrium". This occurs if, for example, the molecules in the atmosphere exchange energies with each other by means of collisions much more rapidly than the variation of the radiation field. Local thermodynamic equilibrium breaks down at high altitudes where collisions between the molecules are very rare, prohibiting the application of Kirchhoff's law.

Problems

Problem 3.1 Starlight and Moonlight

a) Calculate the irradiance received from a full moon at the TOA, and compare it to the extraterrestrial irradiance from the Sun. The lunar albedo is 0.11.

b) In astronomy, the brightness of a star is measured on the logarithmic apparent magnitude (mag) scale, defined as follows: If the mag-value of star A is 5 mag larger than that of star B, its irradiance for an observer on Earth is 100 times less than that of star B (star A is dimmer than star B). The irradiance received on Earth from the Sun corresponds to an apparent magnitude of −26.73 mag. The brightest star of the night sky, Sirius, has an apparent magnitude of −1.46 mag. Estimate its irradiance incident on the TOA of the Earth (ignore any spectral issues). The lowest irradiance visible to the human eye is approximately $10^{-10} \, \text{W} \, \text{m}^{-2}$. What is the apparent magnitude of the faintest stars that humans can see with the naked eye? Place the full moon from (a) on the mag scale.

c) The star Alpha Centauri A is the fourth brightest star in the night sky with an apparent magnitude of 0.00 mag. It is 1.22 times larger in diameter than the Sun but has the same surface temperature. How far away is Alpha Centauri A?

Problem 3.2 Habitable Zone

a) For the star Betelgeuse, calculate the radius (in AU) of the orbit of a planet that can sustain life. Such a planet must be located in the Habitable Zone where it is neither too hot nor too cold. For the solar system, the Habitable Zone lies between 0.95 and 1.5 AU distance from the Sun. Betelgeuse is a variable star that changes its brightness between 0.2 and 1.2 mag over the years. The distance from Earth is assumed to be 6.1×10^{15} km.

Hints: One Astronomical Unit (AU) corresponds to the average Earth–Sun distance. The mag scale is based on the visible part of the spectrum; it only accounts for 13% of Betelgeuse's radiant energy. Calculate the Habitable Zone for the extreme cases of the brightness cycle and look for an overlap. On the way, compare the total radiative energy emitted by Betelgeuse to that of the Sun. Ignore astrophysical and astrobiological subtleties (e.g., Betelgeuse will die young, so there won't be enough time for life to develop as it did on Earth).

b) There are several theories about the mechanism behind the variability; the two most basic parameters are the surface temperature and radius of the star. Adhere to one of the theories which says that the radius is constant and only the temperature varies, and estimate the wavelength shift of the spectral maximum of Betelgeuse's emissions (you'll need an estimate of the radius, 950 times that of the Sun). What would the result mean for hypothetical plants and people on the planet?

Problem 3.3 Geometry and Irradiance

Consider an observer on the stage of a semicircular amphitheater. What is the maximum elevation angle of the amphitheater walls if the irradiance measured on stage may be only 1% lower than that in the case of an unobstructed sky? Assume isotropic radiation.

Problem 3.4 Planck's Function

The Planck function (a spectral radiance) is given by Eq. (3.58).

a) Plot B_λ for the wavelength range of 0–20 μm for two temperatures T of 5770 and 270 K. Be careful with the units!
b) Convert the Planck function into the frequency domain $B_\nu(\nu, T)$.
c) For $T = 5770$ K, plot both B_λ (0.3–1 μm), and B_ν (0–1500 THz). Determine at which wavelength the maxima in the curves of B_λ and B_ν occur for this temperature. Please interpret the difference.
d) Find the maximum of the Planck curves in the wavelength and frequency domains for the general case.

Problem 3.5 Approximations of Planck's Law

There are two approximations of the Planck distribution: Rayleigh–Jeans and Wien.

a) Take your results from Problem 3.4 and add the two approximations (only for $T = 5770$ K and only as a function of wavelength) into the graphs.

b) Calculate and plot the relative deviations between the Planck distribution and the two approximations and give the wavelength ranges for which the relative differences exceed 5%.

c) For Wien's approximation, derive an analytic equation for the relative deviation from the Planck distribution and verify your result from (b).

Problem 3.6 A Simple Climate Model

Design a simple climate model by applying the following three crude assumptions: There are no atmospheric gases nor aerosol particles. The Earth's surface albedo is zero and the surface is a blackbody. Assume a cloud (transmissivity of $\mathcal{T}_C = 0.5$, emissivity of $\varepsilon = 0.75$) situated in the atmosphere at a certain temperature T_C.

a) Quantify the radiative energy budgets (solar and terrestrial) at the surface and at the cloud level. Calculate the equilibrium temperatures for the surface T_E and the cloud layer T_C.

b) What happens qualitatively if the three assumptions are dropped? For each case, estimate the tendency of the corresponding surface-temperature change.

c) Satellites measure emitted IR radiance to derive cloud properties by remote sensing. From this measurement, a cloud top temperature T_{top} (not to be mixed with T_C) can be calculated assuming blackbody radiation. With a given atmospheric temperature profile, T_{top} can be translated into a cloud-top altitude z_{top}. Calculate T_{top} for case (a) and estimate z_{top} using the US standard atmosphere.

d) Use the cloud layer temperature T_C to estimate z_{top} and compare this to your results from (c). What is the reason for the difference?

Problem 3.7 A More Realistic Climate Model

Establish the radiative energy budget for an atmosphere with an Earth's surface albedo of γ_E, a cloud transmissivity for solar radiation of $\mathcal{T}_C$ (no solar cloud absorption), a cloud emissivity (for thermal wavelengths) of ε_C, and an emissivity of the Earth surface of ε_E. Assume the atmosphere to be one homogeneous layer, and starting from the flux densities that fall onto and depart from both the surface and the "atmosphere," please determine, as functions of the incoming solar flux density, the surface albedo and emissivity, and the cloud transmissivity and emissivity, using the data in Table 3.4.

Specifically, please derive a formula for the

a) Cloud temperature T_C,
b) Earth's surface temperature T_E, and
c) Ratio of equilibrium cloud to surface temperature.

Table 3.4 Some data for Venus, Earth, and Mars.

Planet	Distance to Sun	Albedo
Venus	0.72 AU	0.9
Earth	1.00 AU	0.3
Mars	1.52 AU	0.25

Problem 3.8 Greenhouse Effect

Consider a simple thermal equilibrium of a celestial body that is illuminated by the Sun.

a) Without taking into account atmospheric processes, calculate the expected equilibrium temperatures of Venus, Earth, and Mars.

Hint: You will need the values given in Table 3.4.
The solar radiation at 1 AU distance from the Sun yields an incoming flux density of 1367 W m^{-2}. Compare your results to the observed mean surface temperatures of these planets: 737 K on Venus, 288 K on Earth, 210 K on Mars. Source: http://nssdc.gsfc.nasa.gov/planetary/planetfact.html (accessed November 2011)

b) The atmospheres of Venus and Mars both consist of about 96% carbon dioxide. How does this relate to the greenhouse effect calculated for these two planets in (a)?

Problem 3.9 Cooling a Satellite

A satellite orbiting around the Earth is heated by solar radiation and by the operation of the instruments it carries. The only way to control a satellite's temperature in outer space is radiative cooling. Let the instruments release heat of a power of 200 W. Assume the satellite to be a 1 m^3 cube with one side of the cube oriented perpendicular to the Sun.

a) Calculate the emissivity for thermal radiation of the satellite housing required to establish a thermal equilibrium at a temperature of 10 °C if the satellite surface has a solar albedo of 0.2 or 0.8.

b) How does the satellite temperature change if you take into account the thermal emissions by the Earth? Assume the Earth's and its atmosphere's effective emission temperature to be −31 °C and the satellite to orbit at 300 km above the Earth's surface such that 1 m^2 of the satellite receives the thermal radiation from the Earth.

Problem 3.10 Photon Energy

a) Find an equation for the radiant pressure p_{rad} to which a spacecraft with a highly reflecting solar sail (solar panel) of an area A_{sail} is exposed when orbiting around the Sun. The solar panel is aligned perpendicular to the Sun. The spacecraft orbits the Sun at the average Sun–Earth distance. Keep in mind that each photon has a certain momentum which changes when the photon hits the solar panel. How large is p_{rad} compared to the atmospheric sea-level pressure?

b) Calculate the time a spacecraft would need to accelerate to a speed of $10\,\mathrm{m\,s^{-1}}$ by the photons that hit the solar panel. The spacecraft has a solar sail of $A_{sail} = 100 \times 100\,\mathrm{m^2}$ area and a mass of $1000\,\mathrm{kg}$.

c) How does this time change for a solar sail made of absorbing material?

Problem 3.11 Radiation without Atmosphere

Remember the following relations for the upward and downward irradiances, see Eqs. (3.48) and (3.49):

$$F^\uparrow = \int_0^{2\pi} \int_0^{\pi/2} I(\theta,\varphi) \cdot \cos\theta \cdot \sin\theta \, d\theta \, d\varphi , \tag{3.94}$$

$$F^\downarrow = -\int_0^{2\pi} \int_{\pi/2}^{\pi} I(\theta,\varphi) \cdot \cos\theta \cdot \sin\theta \, d\theta \, d\varphi . \tag{3.95}$$

Also remember the expressions given for the net irradiance $F = F^\downarrow - F^\uparrow$, and the upward and downward actinic flux densities given in Eq. (3.54):

$$F_{act}^\uparrow = \int_0^{2\pi} \int_0^{\pi/2} I(\theta,\varphi) \cdot \sin\theta \, d\theta \, d\varphi , \tag{3.96}$$

$$F_{act}^\downarrow = \int_0^{2\pi} \int_{\pi/2}^{\pi} I(\theta,\varphi) \cdot \sin\theta \, d\theta \, d\varphi , \tag{3.97}$$

where $F_{act} = F_{act}^\downarrow + F_{act}^\uparrow$ (actinic flux density). Assume there is no scattering and no absorption in the atmosphere. The radiance from the Sun is given by

$$I(-\mu,\varphi) = I_0 \cdot \delta(\mu - \mu_0) \cdot \delta(\varphi - \varphi_0) , \tag{3.98}$$

with the solar zenith angle θ_0, the solar zenith distance $\mu_0 = \cos\theta_0$, and the solar azimuth angle φ_0. The Earth's surface has an albedo $\gamma_E \neq 0$. Assume that the reflected upward radiance field is isotropic and calculate:

a) The downward irradiance and the downward component of the actinic flux density.

b) The upward irradiance and the upward component of the actinic flux density.

c) Use the definition of the albedo to derive a relation between I_0 and the isotropic reflected radiance $I_{iso}^\uparrow$. For which solar zenith angle is an illuminated surface the brightest?

d) Use the relation from (c) to calculate the net irradiance and the total actinic flux density.

e) Give two examples of what could be different in reality.

Problem 3.12 Net, Actinic, and Isotropic Radiation

Review the meaning of solid angle, radiance, irradiance, and the solar constant ($F_k = 1367\,\mathrm{W\,m^{-2}}$). Show that for an isotropic radiation source, the net irradiance $F_{net,\,iso}$ and the actinic flux density $F_{act,\,iso}$ are given by

$$F_{net,\,iso} = 0\,, \tag{3.99}$$

$$F_{act,\,iso} = 4\pi \cdot I_{iso}\,. \tag{3.100}$$

Problem 3.13 Heat Budget of a Human Body

Calculate the heating power of a human body with the following recipe:

a) Construct a "human body" as follows: (i) head = sphere of 20 cm diameter; (ii) remainder of the body = block (cuboid) of $1.5\,\mathrm{m} \times 0.6\,\mathrm{m} \times 0.3\,\mathrm{m}$. Consider the arms glued to the body and legs glued together. Calculate the surface area of that body.

b) Assuming that this is an ideal blackbody and that its temperature is exactly $36\,^\circ\mathrm{C}$, calculate the radiance and irradiance that is emitted by the body. Assume the radiation is emitted isotropically and the body does not reabsorb any of the radiation.

c) What is the power (in Watts) emanating from that body? How many $\mathrm{kW\,h}$ would a human body produce in terms of heat throughout a day and how does that relate to the food energy of 2500 kcal/day recommended to be consumed by an average human being? What is wrong here? Give two reasons.

d) Why do slim people always start feeling cold earlier than people of average weight, that is, from the point of view of radiative transfer?

e) By how much (in percent) does the emitted power increase if the person has a fever ($40\,^\circ\mathrm{C}$)?

Problem 3.14 City Lights

Assume a common winter scene: Fresh snow with a reflectivity of $\mathcal{R}_E$ has fallen during the day. At night, low clouds cover the sky with a reflectivity of $\mathcal{R}_C$. The street lights provide nighttime illumination in the form of a mean downward irradiance F_{mean}. Derive a formula for the resulting total downward and upward irradiance in the city. Then calculate the ratios $F^\downarrow/F_{mean}$ and $F^\uparrow/F_{mean}$ for $\mathcal{R}_E = 0.9$ and $\mathcal{R}_C = 0.8$.

4
Interactions of EM Radiation and Individual Particles

4.1
Overview

Particulate matter, for example, ice crystals within cirrus clouds, liquid water droplets within warm clouds, and airborne dust particles, are abundant in the atmosphere. To understand the radiative characteristics of an atmospheric layer consisting of small particles, we must begin with the fundamental optical properties of the particles. When a beam of radiation impinges on a particle, a part of the incident EM energy is converted into other forms of energy by the absorption process; the remaining amount is redistributed in all directions by the scattering process. Absorption is closely linked to emission, particularly in the near NIR and IR parts of the EM spectrum. The absorption and emission processes associated with atmospheric gases will be discussed in Chapter 8. In this book, we will confine our discussion to elastic scattering, that is, the frequency of the scattered EM wave is the same as that of the incident wave. The attenuation of the incident radiation by both absorption and scattering is referred to as extinction.

The single-scattering properties of a particle encompass the entire absorption and scattering characteristics of the scattering particle and are essential to the quantification of the angular distribution and polarization state of the scattered radiation. Radiative quantities depend on the wavelength of the incident EM wave, the inherent optical properties (e.g., dielectric constants), the shape, the orientation (unless the particle is a sphere), and the individual particle size.

The study of the single-scattering properties of particles is a very active discipline with several excellent monographs dedicated to the subject: Bohren and Huffman (1983); Mishchenko et al. (2002); van de Hulst (1981), and Borghese et al. (2007). The early studies of particulate single-scattering properties were focused on spheres. The fundamental work in this aspect was reported by eminent scientists, such as, Lorenz (1890), Love (1899), Mie (1908), and Debye (1909), and the results of their work have become known as either the Lorenz–Mie theory or Mie theory. Numerical codes to simulate the scattering properties of spherical particles are available from Wiscombe (1980) and Bohren and Huffman (1983). Significant research has been directed towards understanding the single-scattering properties of nonspherical and inhomogeneous particles existing in nature. The reader is referred to reviews by Mishchenko et al. (2002), Kahnert (2003), Yang and Liou (2006), Kokhanovsky (2006), Wriedt (2009), and Mishchenko (2009) to find techniques for computing the single-scattering properties of nonspherical particles.

As EM scattering by particles is very complicated, it is impossible to cover a comprehensive and indepth description of EM scattering phenomena in one chapter. However, we intend to introduce the basic physical characteristics necessary to quantitatively describe the optical properties of a dielectric particle.

4.2
Complex Index of Refraction

Individual particulate single-scattering properties depend on the inherent optical constants of the particle's molecular material. One such constant, the complex index of refraction or the refractive index of the particle $\tilde{n}$, is wavelength-dependent, dimensionless, defined by its real and imaginary parts in the form

$$\tilde{n} = \tilde{n}_{\mathrm{re}} + \tilde{n}_{\mathrm{im}} \cdot \mathrm{i} \,, \tag{4.1}$$

where $\mathrm{i} = \sqrt{-1}$. In this book, we chose $\mathrm{e}^{-\mathrm{i}\cdot\omega_c\cdot t}$ for the temporal variation of a harmonic plane wave to give rise to a positive imaginary part of the refractive index. If $\mathrm{e}^{\mathrm{i}\cdot\omega_c\cdot t}$ is selected, the imaginary part of the refractive index is negative; however, both $\mathrm{e}^{-\mathrm{i}\cdot\omega_c\cdot t}$ and $\mathrm{e}^{\mathrm{i}\cdot\omega_c\cdot t}$ have been used in the literature. Throughout the analysis, consistency is critical once either $\mathrm{e}^{-\mathrm{i}\cdot\omega_c\cdot t}$ or $\mathrm{e}^{\mathrm{i}\cdot\omega_c\cdot t}$ is selected. The real and imaginary parts of the refractive index, $\tilde{n}_{\mathrm{re}}$ and $\tilde{n}_{\mathrm{im}}$, are related to each other by the Kramers–Kronig relation, Eqs. (4.26) and (4.27) in Lucarini et al. (2005),

$$\tilde{n}_{\mathrm{re}}(\nu) - 1 = \frac{2}{\pi} \mathbb{P} \int_0^\infty \frac{\nu' \cdot \tilde{n}_{\mathrm{im}}(\nu')}{\nu'^2 - \nu^2} \, \mathrm{d}\nu' \,, \tag{4.2}$$

$$\tilde{n}_{\mathrm{im}}(\nu) = -\frac{2\nu}{\pi} \mathbb{P} \int_0^\infty \frac{\tilde{n}_{\mathrm{re}}(\nu') - 1}{\nu'^2 - \nu^2} \, \mathrm{d}\nu' \,, \tag{4.3}$$

where $\mathbb{P}$ indicates the Cauchy principal value of the integral.

If the particle is embedded in a vacuum, the refractive index of the particle, relative to a vacuum, is

$$\tilde{n} = \sqrt{\frac{\epsilon \cdot \kappa}{\epsilon_0 \cdot \kappa_0}} = c \cdot \sqrt{\epsilon \cdot \kappa} \,, \tag{4.4}$$

where ϵ and κ are the electric permittivity and the magnetic permeability of the particle material. Note, ϵ_0 and κ_0 have been defined in Section 3.1.1. The phase speed of EM radiation in a vacuum, c, is given by Eq. (3.16). In this book, we restrict ourselves to the case of a particle embedded in air and assume that the phase speed is approximately the same as in a vacuum. In general, both ϵ and κ are complex quantities. The phase speed of light within the particle is given by the real part of c_{m} defined in Eq. (3.15).

The real part, $\tilde{n}_{\mathrm{re}}$, of the refractive index of a particle can be interpreted as the ratio of the phase speed in the surrounding medium, in this case either vacuum or air, to its counterpart within the particle. If no absorption is involved (i.e., $\tilde{n}_{\mathrm{im}} = 0$), we have

$$\tilde{n}_{\mathrm{re}} = \frac{c}{c_{\mathrm{m}}} \, . \tag{4.5}$$

The imaginary part, $\tilde{n}_{\mathrm{im}}$, of the refractive index describes the rate of absorption of the EM wave within the particle. If no EM radiation is absorbed within the particle, then $\tilde{n}_{\mathrm{im}} = 0$.

Figure 4.1 gives some examples of the monochromatic values of the refractive index. The data for the refractive indices of liquid water and ice are taken from Segelstein (1981) and from Warren and Brandt (2008). The data for the refractive indices of dust and ammonium sulfate are from Levoni et al. (1997). In the VIS, NIR region of the EM wave spectrum, the real part of the refractive index shows moderate spectral variation, but the imaginary part varies over several orders of magnitude.

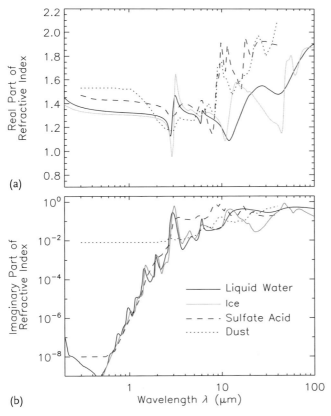

Figure 4.1 Spectra of refractive indices for different materials. (a) Real part $\tilde{n}_{\mathrm{re}}$; (b) imaginary part $\tilde{n}_{\mathrm{im}}$.

4.3

Decomposition of Electric Field Vector

Let us only consider the incident and the scattered electric waves. The magnetic field can be obtained from Maxwell's equations and is based on the electric field. In general, the oscillating electric field is quantified in terms of the complex electric field vector, $\vec{E}$,

$$\vec{E} = \vec{E}_0 \cdot \exp(\mathrm{i} \cdot k \cdot z - \mathrm{i} \cdot \omega_c \cdot t) , \tag{4.6}$$

where t is time in units of seconds, $\vec{E}$ is the complex electric field vector in units of $\mathrm{V\,m^{-1}}$, $\vec{E}_0$ represents the complex electric amplitude vector in units of $\mathrm{V\,m^{-1}}$, $\mathrm{i} = \sqrt{-1}$, and, in units of $\mathrm{m^{-1}}$, k is the modified wavenumber of the EM wave in either air or vacuum. In general, the modified wavenumber in a medium k_m is defined by

$$k_\mathrm{m} = \frac{2\pi \cdot \tilde{n}}{\lambda} = \frac{\omega_c \cdot \tilde{n}}{c} , \tag{4.7}$$

where $\tilde{n}$ is the complex refractive index of either the medium or the particle, see Eq. (4.1), in which the electric wave propagates, z is the Cartesian coordinate in the direction of the propagation of the incident EM wave, and ω_c is the circular frequency, see Eq. (3.24). The oscillating electric field vector associated with a plane EM wave can be decomposed into two components parallel and perpendicular to a reference plane in the form

$$\vec{E} = E_\parallel \cdot \hat{e}_\parallel + E_\perp \cdot \hat{e}_\perp , \tag{4.8}$$

where the subscripts "$\parallel$" and "$\perp$" indicate parallel and perpendicular components with respect to a reference plane, respectively. $E_\parallel$ and $E_\perp$ are scalar but generally complex quantities. $\hat{e}_\parallel$ and $\hat{e}_\perp$ are the base vectors parallel and perpendicular to the reference plane.

The geometric configuration for the scattering of a plane EM wave by a particle is shown in Figure 4.2. The reference plane containing the incident and scattering directions is referred to as the scattering plane. The angle between the incident and scattering directions is called the scattering angle ϑ (not θ, which is used to indicate the zenith angle). The scattered electric field vector is transverse in the radiation zone, which refers to a region of observation at a substantial distance, R, from the particle. $\hat{e}_{\perp\mathrm{inc}}$ and $\hat{e}_{\parallel\mathrm{inc}}$ represent the base vectors for the incident EM wave, and $\hat{e}_{\perp\mathrm{sca}}$ and $\hat{e}_{\parallel\mathrm{sca}}$ are the base vectors for the scattered EM wave. $(\hat{e}_{\perp\mathrm{inc}} \times \hat{e}_{\parallel\mathrm{inc}})$ points along the incident direction; $(\hat{e}_{\perp\mathrm{sca}} \times \hat{e}_{\parallel\mathrm{sca}})$ points along the scattering direction. By applying Eq. (4.8) separately to the incident and the scattered EM waves, and by considering Figure 4.2, we can decompose both the incident and the scattered electric field vectors into two components parallel and perpendicular to the scattering plane in the forms of

$$\vec{E}_\mathrm{inc} = E_{\parallel\mathrm{inc}} \cdot \hat{e}_{\parallel\mathrm{inc}} + E_{\perp\mathrm{inc}} \cdot \hat{e}_{\perp\mathrm{inc}} \tag{4.9}$$

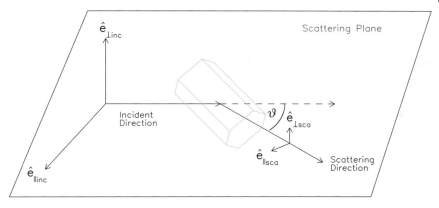

Figure 4.2 The geometric configuration associated with the scattering of a plane EM wave by a particle.

and

$$\vec{E}_{sca} = E_{\|sca} \cdot \hat{e}_{\|sca} + E_{\perp sca} \cdot \hat{e}_{\perp sca} ,$$ (4.10)

where the subscripts "inc" and "sca" indicate the incident and scattered EM waves. $E_{\|inc}$ and $E_{\perp inc}$ are scalar, complex electric amplitudes of the parallel and perpendicular components of the incident electric field vectors, and $E_{\|sca}$ and $E_{\perp sca}$ are the counterparts of the scattered electric field vectors.

4.4
Complex Amplitude Scattering Matrix

The complex amplitude scattering matrix describes the scattering and absorption processes in detail. If we treat the incident and scattered complex electric field vectors separately, the scattering process is described by the matrix equation, given by

$$\begin{pmatrix} E_\| \\ E_\perp \end{pmatrix}_{sca} = \frac{\exp[i \cdot k \cdot (R - z)]}{-i \cdot k \cdot R} \cdot \begin{pmatrix} A_{11} & A_{12} \\ A_{21} & A_{22} \end{pmatrix} \cdot \begin{pmatrix} E_\| \\ E_\perp \end{pmatrix}_{inc} ,$$ (4.11)

where R is the radial distance from the scattering particle. The matrix

$$\mathsf{IA} = \begin{pmatrix} A_{11} & A_{12} \\ A_{21} & A_{22} \end{pmatrix}$$ (4.12)

is called either the complex scattering matrix or the complex amplitude scattering matrix and essentially contains all the single-scattering property information about the particle.

Note that in Eq. (4.11), the far-field condition is assumed to be fulfilled for the scattered electric field. The complex scattering matrix depends on the particle's inherent optical properties, that is, the complex refractive index, the size and shape, and the wavelengths of the incident electric field. If the scattering particle is non-spherical, the complex amplitude scattering matrix also depends on the orientation of the particle with respect to the incident direction. For spheres, the complex scattering matrix is diagonal and the elements A_{12} and A_{21} are zero. The elements of the complex amplitude scattering matrix $\mathbb{A}$, $A_{ij}(i, j = 1, 2)$, are the complex scattering amplitudes and depend on the scattering angle ϑ. In this book, we assume rotational symmetric scattering and just one angular coordinate, the scattering angle, ϑ, is sufficient for specifying the single-scattering properties, that is, we assume scattering has no azimuthal angular dependence. The amplitude scattering matrix describes the transformation of the incident electric field into the scattered electric field, and its elements are dimensionless, scalar, and generally complex.

4.5
Stokes Vector

Complex electric and magnetic fields are difficult to measure; therefore, the concept of the real Stokes vector and real scattering (or Mueller) matrix (see Section 4.7) will be introduced. These concepts involve only real variables and relate the elements of the complex amplitude scattering matrix to real quantities that are much easier to measure.

A plane EM wave is a transverse vector wave. Four real quantities known as the Stokes parameters, originally introduced by G. Stokes in 1852, are used to completely specify the radiation field including its polarization state. To define the Stokes parameters, consider the configuration shown in Figure 4.3, where a beam propagates in the direction pointing out of the paper. The electric field vector, $\vec{\mathbf{E}}$, associated with the beam can be decomposed with respect to the unit vectors $\hat{\mathbf{e}}_\perp$ and $\hat{\mathbf{e}}_\parallel$ using Eq. (4.8).

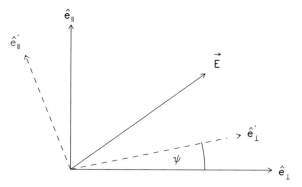

Figure 4.3 Electric field vector associated with a beam pointing out of the paper. Two reference systems specified in terms of $(\hat{\mathbf{e}}_\perp, \hat{\mathbf{e}}_\parallel)$ and $(\hat{\mathbf{e}}'_\perp, \hat{\mathbf{e}}'_\parallel)$ are used to decompose the field vector into two components.

The Stokes vector $\vec{\mathbf{S}}$ is defined by four components, each of them carrying the units of irradiance (W m^{-2}), that is,

$$\vec{\mathbf{S}} = \begin{pmatrix} F \\ Q \\ U \\ V \end{pmatrix}, \tag{4.13}$$

where F is the total irradiance. In some textbooks, the term F is denoted by I and called either intensity or specific intensity, although the quantity actually is irradiance. For an EM wave within a non-absorbing medium, F is defined in the form

$$F = \frac{1}{2}\sqrt{\frac{\epsilon}{\kappa}} \cdot (E_\parallel \cdot E_\parallel^* + E_\perp \cdot E_\perp^*) = F_\parallel + F_\perp$$

$$= \frac{1}{2}\sqrt{\frac{\epsilon}{\kappa}} \left(|E_{0,\parallel}|^2 + |E_{0,\perp}|^2 \right) . \tag{4.14}$$

The asterisk indicates the complex conjugate. Note that the constant factor $(1/2)\sqrt{\epsilon/\kappa}$ is usually omitted because the Stokes parameters are often discussed in a relative sense (Bohren and Huffman, 1983; Hovenier et al., 2004; Mishchenko et al., 2002).

In the case of a linearly polarized EM wave, Q quantifies the irradiance of parallel minus perpendicular components, both with respect to a certain reference plane, see Figure 4.3. Linear polarization limits the electric field vector in its polarized plane to oscillation in a single polarization plane. The parallel configuration of linearly polarized radiation indicates the electric vector in its polarization plane oscillates parallel to the reference plane. If the electric field vector oscillates in a direction perpendicular to the reference plane, then the corresponding radiation field is referred to as perpendicular for the linearly polarized radiation. Thus, Q describes the difference between the linear polarization parallel and perpendicular components with respect to the reference plane. A distinction between Q and the degree of polarization, a dimensionless and relative measure of polarization, will be introduced in Section 4.6. Quantitatively, Q is defined as

$$Q = \frac{1}{2}\sqrt{\frac{\epsilon}{\kappa}} \cdot (E_\parallel \cdot E_\parallel^* - E_\perp \cdot E_\perp^*) = F_\parallel - F_\perp . \tag{4.15}$$

The Stokes vector component U describes the linearly polarized irradiance in reference to a polarization plane that is tilted by $\pm 45°$ with respect to the reference plane

$$U = \frac{1}{2}\sqrt{\frac{\epsilon}{\kappa}} \cdot (E_\parallel \cdot E_\perp^* + E_\perp \cdot E_\parallel^*). \tag{4.16}$$

When $U = 0$, there is no linear polarization along the $\pm 45°$ directions, but there may be a $\pm 45°$ components of the electric field vector. For example, if $E_\perp = 0$, then $U = 0$. However, $E_\parallel$ can be decomposed into two components parallel to $\pm 45°$ directions in the reference plane.

V represents the circularly polarized irradiance. The electric field vector does not oscillate in a plane, but either on a circular or elliptical rotating spiral, see Figure 3.2. In the case of an electric field vector spiralling parallel to a plane normal to the propagation direction of the wave, the circular or elliptical polarized irradiance V is defined to quantify the state of circular polarization in the form,

$$V = i \cdot \frac{1}{2} \sqrt{\frac{\epsilon}{\kappa}} \cdot \left(E_{\parallel} \cdot E_{\perp}^* - E_{\perp} \cdot E_{\parallel}^* \right). \tag{4.17}$$

Q and U depend on the choice of the reference plane. If the base vectors $\hat{e}_{\perp}$ and $\hat{e}_{\parallel}$ are replaced by another set of base vectors $\hat{e}_{\perp}'$ and $\hat{e}_{\parallel}'$ that are rotated in a counterclockwise direction by an angle, ψ, from the original base, see Figure 4.3, the components of the Stokes vector are transformed to their counterparts with respect to the new reference plane as

$$\begin{pmatrix} F' \\ Q' \\ U' \\ V' \end{pmatrix} = \begin{pmatrix} 1 & 0 & 0 & 0 \\ 0 & \cos 2\psi & -\sin 2\psi & 0 \\ 0 & \sin 2\psi & \cos 2\psi & 0 \\ 0 & 0 & 0 & 1 \end{pmatrix} \cdot \begin{pmatrix} F \\ Q \\ U \\ V \end{pmatrix}. \tag{4.18}$$

4.6
Degree of Polarization

In general, the dimensionless degree of polarization P is given by

$$P = \frac{1}{F} \cdot \sqrt{Q^2 + U^2 + V^2} \leq 1. \tag{4.19}$$

For an unpolarized EM wave, $P = 0$, and for a completely polarized EM wave, $P = 1$. We define the degree of linear polarization, P_{lin}, as

$$P_{\text{lin}} = \frac{1}{F} \cdot \sqrt{Q^2 + U^2} \leq 1. \tag{4.20}$$

If an electric field vector linearly oscillates parallel or perpendicular to the reference plane, the tilt of the polarization direction with respect to the reference plane is 0 or 90°, causing U to be equal to zero. In this case, the degree of linear polarization with respect to the reference plane is also defined in the literature as

$$P_{\text{lin},0°} = -\frac{Q}{F}, \tag{4.21}$$

where the negative sign is a matter of convention. If $Q = 0$, the degree of linear polarization for a 45° tilt between the polarization and the reference plane is

$$P_{\text{lin},45°} = \frac{U}{F}. \tag{4.22}$$

Furthermore, the degree of circular polarization P_{cir} is introduced by

$$P_{\text{cir}} = \frac{V}{F}. \tag{4.23}$$

From the definitions, we obtain

$$P = \sqrt{P_{\text{lin}}^2 + P_{\text{cir}}^2} = \sqrt{P_{\text{lin},0°}^2 + P_{\text{lin},45°}^2 + P_{\text{cir}}^2} \,, \tag{4.24}$$

and we can rewrite the Stokes vector as

$$\vec{S} = \begin{pmatrix} F_{\text{unp}} \\ 0 \\ 0 \\ 0 \end{pmatrix} + \begin{pmatrix} F_{\text{lin}} \\ Q \\ U \\ 0 \end{pmatrix} + \begin{pmatrix} F_{\text{cir}} \\ 0 \\ 0 \\ V \end{pmatrix} \,, \tag{4.25}$$

where the three components on the right side of Eq. (4.25) represent unpolarized, linearly, and circularly polarized irradiance. The three components are defined as

$$F = F_{\text{unp}} + F_{\text{lin}} + F_{\text{cir}} \,, \tag{4.26}$$

$$F_{\text{lin}} = \sqrt{Q^2 + U^2 + V^2} - V \,, \tag{4.27}$$

$$F_{\text{cir}} = V \,, \tag{4.28}$$

where F_{unp} is referred to as unpolarized irradiance given by

$$F_{\text{unp}} = F - \sqrt{Q^2 + U^2 + V^2} \,. \tag{4.29}$$

4.7
Mueller Matrix

Instead of using the complex amplitude scattering matrix IA, scattering and absorption processes can also be described alternatively by the real Mueller matrix, also called scattering matrix IS. The scattering problem is explained in Figure 4.4. With the Stokes vectors of the incident radiation, $\vec{S}_{\text{inc}}$, and scattered radiation, $\vec{S}_{\text{sca}}$, we define the linear transformation

$$\vec{S}_{\text{sca}} = \left(\frac{1}{k \cdot R} \right)^2 \cdot \text{IS} \cdot \vec{S}_{\text{inc}} \tag{4.30}$$

with $\text{IS} = (S_{ij})$ and $i, j = 1, 2, 3, 4$. The Mueller matrix elements S_{ij} are dimensionless, k represents the modified wavenumber defined by Eq. (3.21), and R is the radial distance from the scattering particle. The validity of Eq. (4.30) is in the radiation zone or the far-field region with $k \cdot R \gg 1$, $R \gg d_c$ and $R \gg k \cdot d_c^2/2$, where $2 d_c$ is the characteristic dimension of the scattering particle (Mishchenko, 2006).

The Stokes vector of the incident radiation is related to the counterpart of the scattered radiation via the Mueller matrix in the form

$$\begin{pmatrix} F \\ Q \\ U \\ V \end{pmatrix}_{\text{sca}} = \left(\frac{1}{k \cdot R} \right)^2 \cdot \text{IS} \cdot \begin{pmatrix} F \\ Q \\ U \\ V \end{pmatrix}_{\text{inc}} \,. \tag{4.31}$$

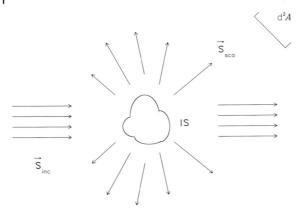

Figure 4.4 Illustration of the scattering of radiation by a particle. Redrawn from Jackson (1975) with modification.

The dimensionless elements, (S_{jl}); $j, l = 1, 2, 3, 4$, of the real Mueller matrix IS, depend on the complex index of refraction, $\tilde{n}$, the habit or shape, the size of the scattering particle, the wavelength of the incident EM wave, and the scattering angle ϑ. In Eq. (4.31), IS is given as

$$\text{IS} = \begin{pmatrix} S_{11} & S_{12} & S_{13} & S_{14} \\ S_{21} & S_{22} & S_{23} & S_{24} \\ S_{31} & S_{32} & S_{33} & S_{34} \\ S_{41} & S_{42} & S_{43} & S_{44} \end{pmatrix} . \tag{4.32}$$

The Mueller matrix IS in Eq. (4.32) is usually presented in a normalized form in the literature and is referred to as the phase matrix given by

$$\text{IP} = \begin{pmatrix} P_{11} & P_{12} & P_{13} & P_{14} \\ P_{21} & P_{22} & P_{23} & P_{24} \\ P_{31} & P_{32} & P_{33} & P_{34} \\ P_{41} & P_{42} & P_{43} & P_{44} \end{pmatrix} , \tag{4.33}$$

where

$$P_{11} = \frac{S_{11}}{1/(4\pi) \iint_{4\pi} S_{11} \, d^2 \Omega} , \tag{4.34}$$

and

$$P_{ij} = P_{11} \cdot \frac{S_{ij}}{S_{11}} \quad \text{for } i, j = 1, 2, 3, 4 . \tag{4.35}$$

To be explained in Section 4.8.1, see Eq. (4.63), the scattering cross section for unpolarized incident radiation is given by

$$C_{\text{sca,unp}} = \frac{1}{k^2} \iint_{4\pi} S_{11} \, d^2 \Omega . \tag{4.36}$$

Thus,

$$P_{11} = \frac{4\pi \cdot S_{11}}{k^2 \cdot C_{\text{sca,unp}}} . \tag{4.37}$$

For unpolarized incident radiation, P_{11} is the same as the phase function discussed in detail in Section 4.8.1.

Consider an ensemble of particles and assume that the particles have their mirror positions in equal number and in random orientations. In this case, the Mueller matrix only has six independent elements (van de Hulst, 1980), that is,

$$\mathsf{IS}_{\text{LMS}} = \begin{pmatrix} S_{\text{LMS},11} & S_{\text{LMS},12} & 0 & 0 \\ S_{\text{LMS},12} & S_{\text{LMS},22} & 0 & 0 \\ 0 & 0 & S_{\text{LMS},33} & S_{\text{LMS},34} \\ 0 & 0 & -S_{\text{LMS},34} & S_{\text{LMS},44} \end{pmatrix} . \tag{4.38}$$

This is a special form of the Mueller matrix called the Lorenz–Mie structure. Note that two relations are implied in Eq. (4.38), namely,

$$S_{\text{LMS},21} = S_{\text{LMS},12} , \tag{4.39}$$

$$S_{\text{LMS},43} = -S_{\text{LMS},34} . \tag{4.40}$$

The elements of the Mueller matrix (S_{ij}) are connected to those of the complex amplitude scattering matrix (A_{ij}) introduced in Section 4.4, see Eq. (4.12), by

$$S_{11} = \frac{1}{2} \cdot \left(|A_{11}|^2 + |A_{22}|^2 + |A_{21}|^2 + |A_{12}|^2 \right), \tag{4.41}$$

$$S_{12} = \frac{1}{2} \cdot \left(|A_{11}|^2 - |A_{22}|^2 + |A_{21}|^2 - |A_{12}|^2 \right), \tag{4.42}$$

$$S_{22} = \frac{1}{2} \cdot \left(|A_{11}|^2 + |A_{22}|^2 - |A_{21}|^2 - |A_{12}|^2 \right), \tag{4.43}$$

$$S_{33} = \Re\mathrm{e} \left\{ A_{11}^* \cdot A_{22} + A_{12} \cdot A_{21}^* \right\} , \tag{4.44}$$

$$S_{34} = \Im\mathrm{m} \left\{ A_{11} \cdot A_{22}^* + A_{21} \cdot A_{12}^* \right\} , \tag{4.45}$$

$$S_{44} = \Re\mathrm{e} \left\{ A_{11}^* \cdot A_{22} - A_{12} \cdot A_{21}^* \right\} , \tag{4.46}$$

where $\Re\mathrm{e}\{\dots\}$ and $\Im\mathrm{m}\{\dots\}$ indicate the real and imaginary parts of the argument, respectively. For nonspherical particles, S_{22} usually differs from S_{11}; however, for spherical and nonspherical particles with certain orientations, S_{22} and S_{11} are the same. For this reason, the ratio S_{22}/S_{11} is widely regarded as an indicator of particle nonsphericity (Bohren and Huffman, 1983; Mishchenko et al., 2002).

4.8
Optical Properties of Individual Particles

4.8.1
Optical Parameters

Optical cross sections: C_{ext}, C_{abs}, C_{sca}   The optical cross sections are measures of how effective an individual particle interacts with incident EM radiation in extinction, absorption, and scattering processes. They are relative measures related to the incident flux density F_{inc} and have the form

$$C_{ext} = \frac{\Phi_{ext}}{F_{inc}} , \quad C_{abs} = \frac{\Phi_{abs}}{F_{inc}} , \quad C_{sca} = \frac{\Phi_{sca}}{F_{inc}} , \tag{4.47}$$

where Φ is a radiant energy flux (in watts) subject to extinction, absorption, or scattering. C_{ext} is the extinction cross section, C_{abs} is the absorption cross section, and C_{sca} is the scattering cross section of an individual particle. Because F_{inc} carries the units of $W\,m^{-2}$, the optical cross sections C_{ext}, C_{abs}, and C_{sca} have the units of m^2.

The sum of the scattering and absorption cross sections is equal to the extinction cross section

$$C_{ext} = C_{abs} + C_{sca} . \tag{4.48}$$

Efficiency factors: Q_{ext}, Q_{sca}, Q_{abs}   The scattering, absorption, and extinction cross sections are normalized with the geometric cross section of the particle to yield the dimensionless efficiency factors for scattering Q_{sca}, absorption Q_{abs}, and extinction Q_{ext}:

$$Q_{sca} = \frac{C_{sca}}{A_{proj}} , \quad Q_{abs} = \frac{C_{abs}}{A_{proj}} , \quad Q_{ext} = \frac{C_{ext}}{A_{proj}} , \tag{4.49}$$

where A_{proj} is the geometric cross section of the particle projected onto a plane perpendicular to the incident direction. For spherical particles, the projected geometric cross section, A_{proj}, corresponds to $\pi \cdot r^2$, where r is the radius of the spherical particle. The term Q in Eq. (4.49) should not be confused with the second component of the Stokes vector, in particular not for scattered radiation.

Single-scattering albedo: $\tilde{\omega}$ The single-scattering albedo $\tilde{\omega}$ is an optical particle property defined as

$$\tilde{\omega} = \frac{C_{sca}}{C_{ext}} = \frac{Q_{sca}}{Q_{ext}} . \tag{4.50}$$

When $\tilde{\omega} = 1$, there is no absorption, and the case is called conservative scattering; however, when $\tilde{\omega} = 0$, absorption occurs with no scattering.

Scattering function: f To describe the angular distribution of scattered EM radiation, we define a dimensionless, scalar scattering function $f(\vartheta)$ such that

$$F_{\text{sca}}(\vartheta) = \left(\frac{1}{k \cdot R} \right)^2 \cdot f(\vartheta) \cdot F_{\text{inc}} \,, \tag{4.51}$$

where ϑ represents the scattering angle. Note that since we are assuming axially rotational symmetric scattering, we have omitted any azimuthal angular dependence. F_{sca} is the scattered irradiance and F_{inc} is the incident irradiance. Using Eqs. (4.31) and (4.32), we obtain

$$F_{\text{sca}} = \left(\frac{1}{k \cdot R} \right)^2 \cdot (S_{11} \cdot F_{\text{inc}} + S_{12} \cdot Q_{\text{inc}} + S_{13} \cdot U_{\text{inc}} + S_{14} \cdot V_{\text{inc}}) \,. \tag{4.52}$$

A comparison of coefficients with Eq. (4.51) yields the scattering function $f(\vartheta)$, that is,

$$f(\vartheta) = S_{11} + S_{12} \cdot \frac{Q_{\text{inc}}}{F_{\text{inc}}} + S_{13} \cdot \frac{U_{\text{inc}}}{F_{\text{inc}}} + S_{14} \cdot \frac{V_{\text{inc}}}{F_{\text{inc}}} \,. \tag{4.53}$$

Using the definitions of linear and circular polarization from Eqs. (4.21)–(4.23) with Eq. (4.53) produces

$$f(\vartheta) = S_{11} - S_{12} \cdot P_{\text{lin},0°,\text{inc}} + S_{13} \cdot P_{\text{lin},45°,\text{inc}} + S_{14} \cdot P_{\text{cir,inc}} \,. \tag{4.54}$$

For the scattering by a particle, the source of the scattered radiation is a point source as viewed from the far-field perspective $(k \cdot R) \gg 1$. Consider a unit area on a large sphere centered at the scattering particle. The radius, R, of the sphere equals the distance between the scattering particle and a far-field observational point of interest. The solid angle, $\Delta \Omega$, subtended by the unit area, as viewed from the scattering particle, see Eq. (2.35), for the radial distance, R, is then given by

$$\Delta \Omega = \frac{1}{R^2} \, (\text{sr}) \,. \tag{4.55}$$

Thus, the radiance corresponding to the irradiance is given by (Hovenier et al., 2004)

$$I_{\text{sca}} = \frac{F_{\text{sca}}}{\Delta \Omega} = F_{\text{sca}} \cdot R^2 \,. \tag{4.56}$$

Note that sr^{-1} is part of the I_{sca} units. From Eq. (4.51), we have

$$I_{\text{sca}} = \frac{1}{k^2} \cdot f(\vartheta) \cdot F_{\text{inc}} \,. \tag{4.57}$$

Phase function: p or $\mathcal{P}$ The scalar phase function, $p(\vartheta)$, in units of sr^{-1}, is defined by

$$p(\vartheta) = \frac{f(\vartheta)}{k^2 \cdot C_{\text{sca}}} \, (\text{sr}^{-1}) \,, \tag{4.58}$$

where C_{sca}, in units of m², is the scattering cross section of the corresponding scattering particle defined by Eq. (4.47).

The phase function, p, represents the relative angular distribution of the scattered radiation. For a given phase function $p(\vartheta)$, the integration of the phase function over the entire solid angle domain yields a value of one, that is,

$$\iint\limits_{4\pi} p(\vartheta)\,d^2\Omega = 1 . \tag{4.59}$$

We can also define the dimensionless phase function, $\mathcal{P}(\vartheta)$, normalized to 4π as

$$\iint\limits_{4\pi} \mathcal{P}(\vartheta)\,d^2\Omega = 4\pi . \tag{4.60}$$

The relationship between the phase function, $p(\vartheta)$, in units of sr^{-1} and the dimensionless phase function, $\mathcal{P}(\vartheta)$, is given by

$$\mathcal{P}(\vartheta) = 4\pi \cdot p(\vartheta) . \tag{4.61}$$

Inserting Eq. (4.58) into Eq. (4.59) yields

$$\iint\limits_{4\pi} \frac{f(\vartheta)}{k^2 \cdot C_{sca}}\,d^2\Omega = 1 , \tag{4.62}$$

where C_{sca} is defined as

$$C_{sca} = \frac{1}{k^2} \iint\limits_{4\pi} f(\vartheta)\,d^2\Omega . \tag{4.63}$$

Asymmetry factor: g   The asymmetry factor, g, of the phase function associated with an individual particle is a coarse measure of the angular distribution of the radiation scattered by the particle. It is defined using the phase function by

$$g = \iint\limits_{4\pi} p(\vartheta) \cdot \cos\vartheta\,d^2\Omega = \frac{1}{2} \int\limits_{-1}^{1} \cos\vartheta \cdot \mathcal{P}(\cos\vartheta)\,d\cos\vartheta , \tag{4.64}$$

where we assume $\mathcal{P}(\cos\vartheta)$ to be independent of the scattering azimuthal angle (axially rotational symmetric scattering). For pure forward scattering ($\vartheta = 0°$), $\cos\vartheta = 1$ and we obtain, using the normalization condition of the phase function given by Eq. (4.59), $g = 1$. Similarly, we can show that a value of the asymmetry factor of $g = -1$ means specular reflection in the backward direction ($\vartheta = 180°$, $\cos\vartheta = -1$). A value of the asymmetry factor of $g = 0$ implies that the scattered energy flux density in the forward hemisphere is the same as that in the backward hemisphere (e.g., in the case of Rayleigh scattering to be discussed in Section 4.10).

4.8.2
Optical Theorem

The optical theorem is a fundamental physical principle related to the scattering of light by a particle and quantitatively specifies extinction in terms of the amplitude scattering matrix in the forward direction ($\vartheta = 0°$). The most general expression of the optical theorem in a matrix form was reported by Mishchenko et al. (2002).

To explain the optical theorem, let us consider an incident plane EM wave propagating along the z axis of a coordinate system. In the far-field region, the condition $(x^2 + y^2)/z^2 \ll 1$ holds in all directions around the forward scattering direction. Thus, we have

$$R = \sqrt{x^2 + y^2 + z^2}$$

$$= z \cdot \sqrt{1 + \frac{x^2 + y^2}{z^2}} \approx z + \frac{x^2 + y^2}{2z} + O(1/z^3) \,, \tag{4.65}$$

and

$$R - z = \frac{x^2 + y^2}{2z} + O(1/z^3) \,.$$

With the preceding approximation, the scattered electric field, see Eq. (4.11), in the forward direction ($\vartheta = 0°$) becomes

$$\begin{pmatrix} E_\parallel \\ E_\perp \end{pmatrix}_{\text{sca}} = \frac{\exp[i \cdot k \cdot (x^2 + y^2)/(2z)]}{-i \cdot k \cdot R}$$

$$\times \begin{pmatrix} A_{11}(\vartheta = 0°) \cdot E_{\parallel\text{inc}} + A_{12}(\vartheta = 0°) \cdot E_{\perp\text{inc}} \\ A_{21}(\vartheta = 0°) \cdot E_{\parallel\text{inc}} + A_{22}(\vartheta = 0°) \cdot E_{\perp\text{inc}} \end{pmatrix} \,, \tag{4.66}$$

where R is the distance from the scattering particle.

According to the definition of the Stokes parameters, see Eq. (4.14), the irradiance of the total (incident plus scattered) wave, if the particle is embedded in a vacuum, in the forward direction is

$$F = \frac{1}{2} \sqrt{\frac{\epsilon_0}{\kappa_0}} \cdot (E_\parallel \cdot E_\parallel^* + E_\perp \cdot E_\perp^*)$$

$$= \frac{1}{2} \sqrt{\frac{\epsilon_0}{\kappa_0}} \cdot \left\{ \left(E_{\parallel\text{inc}} + \frac{\exp[i \cdot k \cdot (x^2 + y^2)/(2z)]}{-i \cdot k \cdot R} \right.\right.$$

$$\times \left[A_{11}(0°) \cdot E_{\parallel\text{inc}} + A_{12}(0°) \cdot E_{\perp\text{inc}} \right] \Big)$$

$$\times \left(E_{\parallel\text{inc}}^* + \frac{\exp[-i \cdot k \cdot (x^2 + y^2)/(2z)]}{i \cdot k \cdot R} \right.$$

$$\times \left[A_{11}^*(0°) \cdot E_{\parallel\text{inc}}^* + A_{12}^*(0°) \cdot E_{\perp\text{inc}}^* \right] \Big)$$

$$+ \left(E_{\perp\text{inc}} + \frac{\exp[i \cdot k \cdot (x^2 + y^2)/(2z)]}{-i \cdot k \cdot R} \right.$$

$$\times \left[A_{21}(0°) \cdot E_{\|\text{inc}} + A_{22}(0°) \cdot E_{\perp\text{inc}} \right] \Big)$$

$$\times \left(E_{\perp\text{inc}}^* + \frac{\exp[-i \cdot k \cdot (x^2 + y^2)/(2z)]}{i \cdot k \cdot R} \right.$$

$$\left. \times \left[A_{21}^*(0°) \cdot E_{\|\text{inc}}^* + A_{22}^*(0°) \cdot E_{\perp\text{inc}}^* \right] \right) \Big\}$$

$$\approx \frac{1}{2} \sqrt{\frac{\epsilon_0}{\kappa_0}} \cdot \left\{ E_{\|,\text{inc}} \cdot E_{\|,\text{inc}}^* + E_{\perp,\text{inc}} \cdot E_{\perp,\text{inc}}^* \right\}$$

$$- \sqrt{\frac{\epsilon_0}{\kappa_0}} \cdot \Re\mathrm{e} \left\{ \frac{\exp[i \cdot k \cdot (x^2 + y^2)/(2z)]}{i \cdot k \cdot R} \right.$$

$$\times \left[A_{11}(0°) \cdot E_{\|\text{inc}} \cdot E_{\|\text{inc}}^* + A_{12}(0°) \cdot E_{\perp\text{inc}} \cdot E_{\|\text{inc}}^* \right.$$

$$\left. + A_{21}(0°) \cdot E_{\|\text{inc}} \cdot E_{\perp\text{inc}}^* + A_{22}(0°) \cdot E_{\perp\text{inc}} \cdot E_{\perp\text{inc}}^* \right] \Big\}$$

$$= F_{\text{inc}} - F_{\text{ext}} , \tag{4.67}$$

where

$$F_{\text{ext}} = \sqrt{\frac{\epsilon_0}{\kappa_0}} \cdot \Re\mathrm{e} \left\{ \frac{\exp[i \cdot k \cdot (x^2 + y^2)/(2z)]}{i \cdot k \cdot R} \right.$$

$$\times \left[A_{11}(0°) \cdot E_{\|\text{inc}} \cdot E_{\|\text{inc}}^* + A_{12}(0°) \cdot E_{\perp\text{inc}} \cdot E_{\|\text{inc}}^* \right.$$

$$\left. + A_{21}(0°) \cdot E_{\|\text{inc}} \cdot E_{\perp\text{inc}}^* + A_{22}(0°) \cdot E_{\perp\text{inc}} \cdot E_{\perp\text{inc}}^* \right] \Big\}$$

$$= \Re\mathrm{e} \left\{ \frac{\exp[i \cdot k \cdot (x^2 + y^2)/(2z)]}{i \cdot k \cdot R} \right.$$

$$\times \left(F_{\text{inc}} \cdot \left[A_{11}(0°) + A_{22}(0°) \right] + Q_{\text{inc}} \cdot \left[A_{11}(0°) - A_{22}(0°) \right] \right.$$

$$\left. + U_{\text{inc}} \cdot \left[A_{12}(0°) + A_{21}(0°) \right] + i \cdot V_{\text{inc}} \cdot \left[A_{12}(0°) - A_{21}(0°) \right] \right) \Big\} .$$

$$\tag{4.68}$$

The quantity F_{ext} is the irradiance associated with extinction. Thus, the extinction cross section is

$$C_{\text{ext}} = \int_{-\infty}^{\infty} \int_{-\infty}^{\infty} F_{\text{ext}} \, dx \, dy . \tag{4.69}$$

Using Eq. (4.68) and the Fresnel integral given by

$$\int_{-\infty}^{\infty} \exp(i \cdot x^2) \, dx = \left(\frac{\sqrt{2}}{2} + i \cdot \frac{\sqrt{2}}{2} \right) \cdot \sqrt{\pi} , \tag{4.70}$$

we obtain the extinction cross section, see Yang et al. (2011),

$$
C_{ext} = \frac{2\pi}{k^2} \cdot \Re \left\{ \left[A_{11}(0°) + A_{22}(0°) \right] + \left[A_{11}(0°) - A_{22}(0°) \right] \cdot \frac{Q_{inc}}{F_{inc}} \right.
$$
$$
\left. + \left[A_{12}(0°) + A_{21}(0°) \right] \cdot \frac{U_{inc}}{F_{inc}} + i \cdot \left[A_{12}(0°) - A_{21}(0°) \right] \cdot \frac{V_{inc}}{F_{inc}} \right\} .
$$

$$
(4.71)
$$

The dependence of the extinction on the polarization state of incident radiation is referred to as dichroism (Mishchenko et al., 2002). In general, the dichroism effect is small (Yang et al., 2011). For natural, that is, unpolarized incident EM radiation, Eq. (4.71) reduces to

$$
C_{ext,unp} = \frac{2\pi}{k^2} \cdot \Re \left\{ A_{11}(\vartheta = 0°) + A_{22}(\vartheta = 0°) \right\} . \tag{4.72}
$$

If $A_{11}(\vartheta = 0°)$ and $A_{22}(\vartheta = 0°)$ are the same, hereafter indicated by $A_{Mie}(\vartheta = 0°)$, as in the case of spheres, Eq. (4.72) reduces to the form of the optical theorem given in most texts, that is,

$$
C_{ext,unp,Mie} = \frac{4\pi}{k^2} \cdot \Re \left\{ A_{Mie}(\vartheta = 0°) \right\} . \tag{4.73}
$$

From Eq. (4.73), it is evident that the extinction cross is determined solely by the scattering amplitudes in the forward direction, although incident radiation is scattered in all directions and the extinction cross section is the sum of the scattering and absorption cross sections.

4.9
Spherical Particles (Lorenz–Mie Theory)

4.9.1
Assumptions and Goals

To begin, let polarized radiation impinge on a homogeneous, spherical particle with radius r and complex refractive index $\tilde{n}$, which is assumed to be embedded in a vacuum. If the surrounding medium is nonabsorbing with refractive index $\tilde{n}_m$, the relative refractive index

$$
\tilde{m} = \frac{\tilde{n}}{\tilde{n}_m} \tag{4.74}
$$

can be used in the Lorenz–Mie theory. If the surrounding medium is absorptive, calculations can be quite involved (Chylek, 1977; Fu and Sun, 2001; Yang et al., 2002) and will not be addressed in this book. Furthermore, we assume that the incident radiation has a wavelength λ and only elastic scattering will be considered, which requires the scattered EM wave to have the same frequency as the incident EM wave. The final assumption is that the scattered EM wave is observed at a sufficiently large distance from the spherical particle in order for the particle initiated spherical wave to be locally regarded as a plane wave with no curvature.

The goal is to simulate the dimensionless efficiency factors for scattering ($Q_{sca,Mie}$), absorption ($Q_{abs,Mie}$), and extinction ($Q_{ext,Mie}$), see Eq. (4.49). Furthermore, the elements of the complex amplitude scattering matrix, $IA_{Mie} = A_{Mie,ij}$, the elements of the Mueller matrix, $IS_{Mie} = S_{Mie,ij}$, the degree of polarization, and the scalar phase function are to be calculated. These quantities may be functions of the wavelength of the incident EM radiation, λ, the radius of the scattering spherical particle, r, the scattering angle, ϑ, the refractive index, $\tilde{n}$, or the size parameter of the particle

$$\alpha = \frac{2\pi \cdot r}{\lambda} .$$
(4.75)

Without detailing the derivation of the Lorenz–Mie theory, we will summarize the major results associated with the scattering of EM radiation by a homogeneous sphere.

4.9.2
Efficiency Factors: $Q_{ext,Mie}$, $Q_{sca,Mie}$, $Q_{abs,Mie}$

The dimensionless efficiency factors of extinction ($Q_{ext,Mie}$), scattering ($Q_{sca,Mie}$), and absorption ($Q_{abs,Mie}$) in the Lorenz–Mie theory are

$$Q_{ext,Mie} = \frac{2}{\alpha^2} \cdot \sum_{n=1}^{\infty} (2n + 1) \cdot \Re\{a_n + b_n\} ,$$
(4.76)

$$Q_{sca,Mie} = \frac{2}{\alpha^2} \cdot \sum_{n=1}^{\infty} (2n + 1) \cdot \left(|a_n|^2 + |b_n|^2 \right) ,$$
(4.77)

$$Q_{abs,Mie} = Q_{ext,Mie} - Q_{sca,Mie} .$$
(4.78)

The a_n and b_n in Eqs. (4.76) and (4.77) are the Lorenz–Mie coefficients of the nth order and can be computed by using the Riccati–Bessel functions and their recurrence relations (Bohren and Huffman, 1983).

An example of the extinction efficiency factor, $Q_{ext,Mie}$, as calculated from Eq. (4.76), is illustrated in Figure 4.5.

The extinction efficiency factor, $Q_{ext,Mie}$, approaches zero for small values of α and rises monotonically with increasing size parameter up to a value of $\alpha \approx 5$ ($r \approx 0.4\,\mu m$ at 550 nm wavelength) where it reaches a maximum of approximately 4.4. At this value, the spherical particle attenuates more than four times the radiation blocked by its projected cross section on a plane perpendicular to the incident direction. For increasing size parameters, oscillations of $Q_{ext,Mie}$ become obvious and approach an asymptotic value of two. Two types of oscillations occur in Figure 4.5, one coarse featured and one very fine featured.

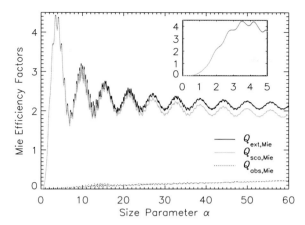

Figure 4.5 Extinction, scattering, and absorption efficiency factors ($Q_{ext,Mie}$, $Q_{sca,Mie}$, $Q_{abs,Mie}$) for a relative refractive index of $\tilde{n} = 1.55 + 0.001 \cdot i$ as a function of size parameter $\alpha = 2\pi \cdot r/\lambda$.

For the coarse interference oscillations, the spacing of the maxima $\triangle \alpha_{max}$ can be approximated by (Bohren and Huffman, 1983)

$$\triangle \alpha_{max} \approx \frac{\pi}{\tilde{n}_{re} - 1} . \tag{4.79}$$

For example, for $\tilde{n}_{re} = 1.55$, $\triangle \alpha_{max} \approx 5.7$, although this is only true for nonabsorbing spheres. The fine ripple oscillations are due to resonance features and are smoothed even by slight absorption, but for nonabsorbing spheres, the fine ripple structures are more pronounced. However, smoothing by the size distribution effects renders the rippling unimportant in the atmosphere for both absorbing and nonabsorbing particles.

Figure 4.5 also shows the efficiency factors for scattering $Q_{sca,Mie}$ and absorption $Q_{abs,Mie}$ calculated from the Lorenz–Mie theory using Eqs. (4.77)–(4.78). The difference between extinction and scattering increases as α increases. The absorption efficiency factor, $Q_{abs,Mie}$, does not show the coarse oscillations and continuously increases with an increasing size parameter α.

The extinction efficiency factor, $Q_{ext,Mie}$, as a function of the size parameter, α, for different values of refractive index, $\tilde{n}$, is shown in Figure 4.6. The coarse oscillations depend on the real part of the refractive index, see Eq. (4.79), and become smaller as the real part of the relative refractive index increases. For nonabsorbing spheres, the distinct fine ripple structure, obvious in the $Q_{ext,Mie}$ curves, is smoothed by absorption. Increasing absorption, corresponding to increasing the imaginary part of $\tilde{n}$, suppresses the small fluctuations in the $Q_{ext,Mie}$ curve; nevertheless, the asymptotic value of the extinction efficiency approaches 2 for very large values of α.

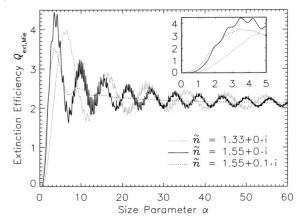

Figure 4.6 Extinction efficiency factor $Q_{\text{ext,Mie}}$ for different values of the refractive index $\tilde{n}$ as a function of the size parameter $\alpha = 2\pi \cdot r/\lambda$.

4.9.3
Single-Scattering Albedo: $\tilde{\omega}_{\text{Mie}}$

Using the scattering and extinction efficiency factors and Eq. (4.50), we derive the single-scattering albedo as

$$\tilde{\omega}_{\text{Mie}} = \frac{Q_{\text{sca,Mie}}}{Q_{\text{ext,Mie}}} \, . \tag{4.80}$$

An example of the calculated results is shown in Figure 4.7. When $\alpha \rightarrow 0$, $\tilde{\omega}_{\text{Mie}} \rightarrow 0$. The larger the imaginary relative refractive index or absorption strength, the stronger $\tilde{\omega}_{\text{Mie}}$ deviates from unity. An imaginary relative refractive index of larger than 0.1 is normally unrealistic for atmospheric particles.

4.9.4
Elements of the Complex Amplitude Scattering Matrix

The complex amplitude scattering matrix, see Section 4.4 Eq. (4.12), for spherical particles reduces to

$$\mathsf{IA}_{\text{Mie}} = \begin{pmatrix} A_{\text{Mie,11}} & 0 \\ 0 & A_{\text{Mie,22}} \end{pmatrix} , \tag{4.81}$$

where the nonzero complex scattering amplitudes for spherical particles are obtained by

$$A_{\text{Mie,11}} = \sum_{n=1}^{\infty} \frac{2n+1}{n \cdot (n+1)} \cdot [a_n \cdot \tau_n(\cos \vartheta) + b_n \cdot \pi_n(\cos \vartheta)] , \tag{4.82}$$

$$A_{\text{Mie,22}} = \sum_{n=1}^{\infty} \frac{2n+1}{n \cdot (n+1)} \cdot [a_n \cdot \pi_n(\cos \vartheta) + b_n \cdot \tau_n(\cos \vartheta)] , \tag{4.83}$$

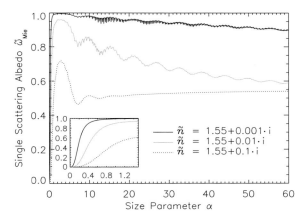

Figure 4.7 Single-scattering albedo $\tilde{\omega}_{\text{Mie}}$ for different values of the refractive index $\tilde{n}$ as a function of the size parameter $\alpha = 2\pi \cdot r/\lambda$.

where ϑ is the scattering angle. The Lorenz–Mie coefficients, a_n and b_n, correspond to Eqs. (4.76) and (4.77). The τ_n and π_n are defined using the associated Legendre polynomials P_n in the form of

$$\pi_n(\cos\vartheta) = \frac{P_n^{(1)}(\cos\vartheta)}{\sin\vartheta} \, , \tag{4.84}$$

$$\tau_n(\cos\vartheta) = \frac{\mathrm{d}}{\mathrm{d}\vartheta}[P_n^{(1)}(\cos\vartheta)]$$

$$= \cos\vartheta \cdot \pi_n(\cos\vartheta) + \sin\vartheta \cdot \frac{\mathrm{d}}{\mathrm{d}\vartheta}[\pi_n(\cos\vartheta)] \, . \tag{4.85}$$

4.9.5
Elements of the Mueller Matrix

The real Mueller matrix $\mathsf{IS}_{\text{Mie}} = S_{\text{Mie},jl}$, in its Lorenz–Mie structure, see Section 4.7 and Eq. (4.38), of spherical particles reduces to

$$\mathsf{IS}_{\text{Mie}} = \begin{pmatrix} S_{\text{Mie},11} & S_{\text{Mie},12} & 0 & 0 \\ S_{\text{Mie},12} & S_{\text{Mie},11} & 0 & 0 \\ 0 & 0 & S_{\text{Mie},33} & S_{\text{Mie},34} \\ 0 & 0 & -S_{\text{Mie},34} & S_{\text{Mie},33} \end{pmatrix} \, . \tag{4.86}$$

Note that the following relations are implied in Eq. (4.86), that is,

$$S_{\text{Mie},11} = S_{\text{Mie},22} \, , \tag{4.87}$$

$$S_{\text{Mie},33} = S_{\text{Mie},44} \, . \tag{4.88}$$

The Mueller matrix elements in Eq. (4.86) can be calculated from Eqs. (4.41)–(4.46) in conjunction with the relation $A_{\text{Mie},12} = A_{\text{Mie},21} = 0$:

$$S_{\text{Mie},11} = \frac{1}{2} \cdot \left(|A_{\text{Mie},11}|^2 + |A_{\text{Mie},22}|^2 \right), \tag{4.89}$$

$$S_{\text{Mie},12} = \frac{1}{2} \cdot \left(|A_{\text{Mie},11}|^2 - |A_{\text{Mie},22}|^2 \right), \tag{4.90}$$

$$S_{\text{Mie},33} = \frac{1}{2} \cdot \left(A^*_{\text{Mie},11} \cdot A_{\text{Mie},22} + A_{\text{Mie},11} \cdot A^*_{\text{Mie},22} \right), \tag{4.91}$$

$$S_{\text{Mie},34} = \frac{i}{2} \cdot \left(A^*_{\text{Mie},11} \cdot A_{\text{Mie},22} - A_{\text{Mie},11} \cdot A^*_{\text{Mie},22} \right). \tag{4.92}$$

The complex scattering amplitudes for spheres, $A_{\text{Mie},11}$ and $A_{\text{Mie},22}$, are simulated with Eqs. (4.82) and (4.83). The Mueller matrix elements are not independent of each other; particularly, we have

$$S_{\text{Mie},11} = \sqrt{S^2_{\text{Mie},12} + S^2_{\text{Mie},33} + S^2_{\text{Mie},34}} \,. \tag{4.93}$$

4.9.6
Polarization

If we assume that the incident radiation is linearly polarized parallel with respect to the scattering plane, then

$$\frac{Q_{\text{inc}}}{F_{\text{inc}}} = 1 \,, \quad U_{\text{inc}} = V_{\text{inc}} = 0 \,, \tag{4.94}$$

and it follows that

$$\begin{pmatrix} F_{\|\text{Mie}} \\ Q_{\|\text{Mie}} \\ U_{\|\text{Mie}} \\ V_{\|\text{Mie}} \end{pmatrix}_{\text{sca}} = \left(\frac{1}{k \cdot R} \right)^2$$

$$\times \begin{pmatrix} S_{\text{Mie},11} & S_{\text{Mie},12} & 0 & 0 \\ S_{\text{Mie},12} & S_{\text{Mie},11} & 0 & 0 \\ 0 & 0 & S_{\text{Mie},33} & S_{\text{Mie},34} \\ 0 & 0 & -S_{\text{Mie},34} & S_{\text{Mie},33} \end{pmatrix} \cdot F_{\|\text{inc}} \cdot \begin{pmatrix} 1 \\ 1 \\ 0 \\ 0 \end{pmatrix}$$

$$= \left(\frac{1}{k \cdot R} \right)^2 \cdot \begin{pmatrix} S_{\text{Mie},11} + S_{\text{Mie},12} \\ S_{\text{Mie},12} + S_{\text{Mie},11} \\ 0 \\ 0 \end{pmatrix} \cdot F_{\|\text{inc}} \,.$$

$$\tag{4.95}$$

From Eq. (4.95), the scattered irradiance is

$$F_{\|\text{Mie,sca}} = \left(\frac{1}{k \cdot R} \right)^2 \cdot \left(S_{\text{Mie},11} + S_{\text{Mie},12} \right) \cdot F_{\|\text{inc}} \,. \tag{4.96}$$

The relative angular distribution of the irradiance of the scattered radiation, under the assumption that the incident radiation is linearly polarized parallel with respect to the scattering plane, is obtained by applying Eqs. (4.89)–(4.90), that is,

$$f_{\|\text{Mie}} = S_{\text{Mie},11} + S_{\text{Mie},12} = |A_{\text{Mie},11}|^2 \,. \tag{4.97}$$

$f_{\|\text{Mie}}$ is the scalar scattering function for linearly parallel polarized incident radiation. From one result of the Lorenz–Mie theory, see Eq. (4.82), it follows that

$$\begin{aligned} f_{\|\text{Mie}} &= |A_{\text{Mie},11}|^2 \\ &= \left| \sum_{n=1}^{\infty} \frac{2n+1}{n \cdot (n+1)} \cdot [a_n \cdot \tau_n(\cos\vartheta) + b_n \cdot \pi_n(\cos\vartheta)] \right|^2 \,. \end{aligned} \tag{4.98}$$

In a similar way, perpendicularly polarized radiation impinging on a sphere with $Q_{\text{inc}}/F_{\text{inc}} = -1$, and $U_{\text{inc}} = V_{\text{inc}} = 0$ leads to

$$F_{\perp\text{Mie,sca}} = \left(\frac{1}{k \cdot R} \right)^2 \cdot (S_{\text{Mie},11} - S_{\text{Mie},12}) \cdot F_{\perp\text{inc}} \,, \tag{4.99}$$

with

$$f_{\perp\text{Mie}} = S_{\text{Mie},11} - S_{\text{Mie},12} = |A_{\text{Mie},22}|^2 \,. \tag{4.100}$$

$f_{\perp\text{Mie}}$ is the scalar scattering function for perpendicular polarized incident radiation, see Eq. (4.83),

$$\begin{aligned} f_{\perp\text{Mie}} &= |A_{\text{Mie},22}|^2 \\ &= \left| \sum_{n=1}^{\infty} \frac{2n+1}{n \cdot (n+1)} \cdot [a_n \cdot \pi_n(\cos\vartheta) + b_n \cdot \tau_n(\cos\vartheta)] \right|^2 \,. \end{aligned} \tag{4.101}$$

If we assume that unpolarized incident radiation strikes a sphere ($Q_{\text{inc}} = U_{\text{inc}} = V_{\text{inc}} = 0$), we obtain for the scalar, dimensionless scattering function

$$\begin{aligned} f_{\text{unp,Mie}} &= S_{\text{Mie},11} = \frac{1}{2} \cdot (|A_{\text{Mie},11}|^2 + |A_{\text{Mie},22}|^2) \\ &= \frac{1}{2} \cdot (f_{\|\text{Mie}} + f_{\perp\text{Mie}}) \,. \end{aligned} \tag{4.102}$$

If the incident radiation is unpolarized ($Q_{\text{inc}} = U_{\text{inc}} = V_{\text{inc}} = 0$), we obtain the degree of linear polarization $P_{\text{lin},0°,\text{sca,unp,Mie}}$ of the scattered radiation, see Eq. (4.21), by

$$P_{\text{lin},0°,\text{sca,unp,Mie}} = -\frac{Q_{\text{sca,unp,Mie}}}{F_{\text{sca,unp,Mie}}} = -\frac{S_{\text{Mie},12}}{S_{\text{Mie},11}} \,. \tag{4.103}$$

4.9.7
Phase Function for Unpolarized Incident Radiation: $\mathcal{P}_{\text{unp,Mie}}$

If the unpolarized incident radiation impinges on a sphere, Eq. (4.58) gives

$$p_{\text{unp,Mie}}(\vartheta) = \frac{f_{\text{unp,Mie}}(\vartheta)}{k^2 \cdot C_{\text{sca,Mie}}} \; (\text{sr}^{-1}) = \frac{S_{\text{Mie},11}}{k^2 \cdot C_{\text{sca,Mie}}} \; (\text{sr}^{-1}), \tag{4.104}$$

or by virtue of Eq. (4.61),

$$\mathcal{P}_{\text{unp,Mie}}(\vartheta) = 4\pi \cdot \frac{f_{\text{unp,Mie}}(\vartheta)}{k^2 \cdot C_{\text{sca,Mie}}} = 4\pi \cdot \frac{S_{\text{Mie},11}}{k^2 \cdot C_{\text{sca,Mie}}}, \tag{4.105}$$

where $C_{\text{sca,Mie}} = \pi \cdot r^2 \, Q_{\text{sca,Mie}}$ for spherical particles.

As an example, the phase function, $\mathcal{P}_{\text{unp,Mie}}(\vartheta)$, is plotted in Figure 4.8. The larger the particle size parameter, α, the more forward scattering is observed. Furthermore, we see an increase in asymmetry and complexity of the phase function with an increasing size parameter, α. For $\alpha = 1$, we note an almost symmetric Rayleigh phase function (see Section 4.10.3). For $\alpha = 4$, the forward scattering peak, or the forward diffraction peak, becomes more pronounced, more intense, and more narrow as the size parameter increases. From a practical perspective, the forward scattering peak for a very large size parameter essentially approaches the Dirac δ-function. With increasing size parameters, the forward scattering and the remainder of the phase function become more and more complex and ripples begin to form. For $\alpha = 100$, an enhanced scattering feature starts to form around $138°$,

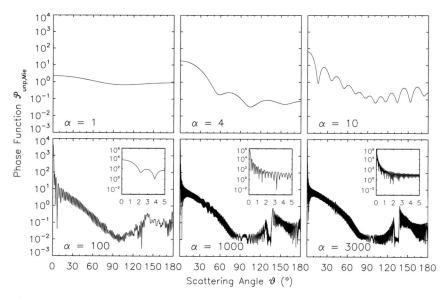

Figure 4.8 Phase function for different values of the size parameter $\alpha = 2\pi \cdot r/\lambda$ as a function of the scattering angle ϑ. In the simulations, the refractive index of liquid water was assumed.

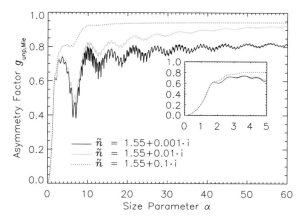

Figure 4.9 Asymmetry factor $g_{unp,Mie}$ for different values of the refractive index $\tilde{n}$ as a function of the size parameter $\alpha = 2\pi \cdot r/\lambda$.

which corresponds to the primary rainbow. A slightly weaker peak forms around 129°, corresponding to the secondary rainbow.

4.9.8
Asymmetry Factor: $g_{unp,Mie}$

The Lorenz–Mie asymmetry factor, $g_{unp,Mie}$, for unpolarized incident radiation is defined by Eq. (4.64), using the Lorenz–Mie phase function, and can be obtained analytically (van de Hulst, 1957) as

$$
g_{unp,Mie} = \frac{1}{2} \int_{-1}^{1} \cos\vartheta \cdot \mathcal{P}_{unp,Mie}(\cos\vartheta)\, d\cos\vartheta
$$

$$
= \frac{\left(2\sum_{n=2}^{\infty}\left[\Re\left\{ \frac{(n-1)\cdot(n+1)}{n} \cdot \left(a_{n-1}\cdot a_n^* + b_{n-1}\cdot b_n^*\right)\right\} + \frac{(2n-1)}{n\cdot(n-1)}\cdot \Re\left\{a_{n-1}\cdot b_{n-1}^*\right\}\right] \right)}{\sum_{n=1}^{\infty}(2n+1)\cdot\left(|a_n|^2 + |b_n|^2\right)} .
$$

$$(4.106)$$

The asterisk indicates the complex conjugate of the respective function. An example of the calculation results is given in Figure 4.9. As shown in previous figures, small fluctuations are smoothed by absorption.

4.10
Rayleigh Scattering and Oscillating Electric Dipole

4.10.1
Amplitudes Scattering Matrix and Mueller Matrix

When the scattering particle size is much smaller than the incident wavelength, the particle can be considered in a homogeneous electric field known as the "applied" field (van de Hulst, 1957), and the internal particle field is assumed to be homogeneous. Radiation is emitted if the applied field varies in time, inducing an oscillating dipole in conjunction with the presence of the particle. For simplicity, let us assume that the particle is placed at the origin of a coordinate system and the applied electric field is specified with respect to the coordinate system as

$$\vec{E}_{\text{inc}}(\vec{R}) \cdot \exp(-i \cdot \omega_c \cdot t) = \vec{E}_{0,\text{inc}} \cdot \exp(i \cdot k \cdot R - i \cdot \omega_c \cdot t) , \tag{4.107}$$

where $\vec{E}_{0,\text{inc}}$ is the complex amplitude vector of the applied (incident) electric field. The induced electric dipole moment vector is given by

$$\vec{P}(t) = \vec{P}_0 \cdot \exp(-i \cdot \omega_c \cdot t) , \tag{4.108}$$

where the amplitude vector of the dipole moment is given by

$$\vec{P}_0 = \epsilon_0 \cdot a \cdot \vec{E}_{0,\text{inc}} . \tag{4.109}$$

The electric dipole moment, $\vec{P}$, is in units of A s m, and the polarizability of the particle a is in units of m^3. The term a in Eq. (4.109), under the electrostatic approximation (Jackson, 1975), can be expressed in the form of

$$a = 4\pi \cdot \left(\frac{\tilde{n}^2 - 1}{\tilde{n}^2 + 2} \right) \cdot r^3 , \tag{4.110}$$

where $\tilde{n}$ is the complex refractive index and r is the radius of the scattering spherical particle size.

According to the theory of electrodynamics, the electric field associated with the radiation emitted from an oscillating dipole is given by (Jackson, 1975)

$$\vec{E}_{\text{sca}}(\vec{R}) = \frac{1}{4\pi \cdot \epsilon_0} \cdot \left\{ k^2 \, (\hat{s} \times \vec{P}_0) \times \hat{s} \frac{1}{R} \exp(i \cdot k \cdot R) \right.$$
$$\left. + [3\hat{s}(\hat{s} \cdot \vec{P}_0) - \vec{P}_0] \cdot \left(\frac{1}{R^3} - \frac{i \cdot k}{R^2} \right) \cdot \exp(i \cdot k \cdot R) \right\} , \tag{4.111}$$

where $\hat{s} = \vec{R}/R$ indicates the direction unit vector. In the radiation zone, or the far-field zone with $k \cdot R \gg 1$, the scattered field reduces to

$$\vec{E}_{\text{sca}}(\vec{R}) = \frac{1}{4\pi \cdot \epsilon_0} \cdot \left[k^2 \, (\hat{s} \times \vec{P}_0) \times \hat{s} \right] \cdot \frac{1}{R} \cdot \exp(i \cdot k \cdot R)$$
$$= \frac{k^2 \cdot a}{4\pi} \cdot \left[(\hat{s} \times \vec{E}_{0,\text{inc}}) \times \hat{s} \right] \cdot \frac{1}{R} \cdot \exp(i \cdot k \cdot R) . \tag{4.112}$$

The scattered electric field vector, $\vec{E}_{sca}$, can be decomposed into parallel and perpendicular components with respect to the scattering plane, see Figure 4.2, using Eq. (4.10). In a similar way, the incident electric amplitude vector is decomposed, that is,

$$\vec{E}_{0,inc} = E_{0,\|inc} \cdot \hat{e}_{\|inc} + E_{0,\perp inc} \cdot \hat{e}_{\perp inc} . \tag{4.113}$$

The unit vectors in Eqs. (4.10) and (4.113) satisfy the following relations:

$$\hat{e}_{\perp sca} \times \hat{e}_{\|sca} = \hat{s} , \tag{4.114}$$

$$\hat{e}_{\perp sca} \cdot \hat{e}_{\|sca} = 0 , \tag{4.115}$$

$$\hat{e}_{\|inc} \cdot \hat{e}_{\|sca} = \cos \vartheta , \tag{4.116}$$

where ϑ is the scattering angle. Using the technique learned in Chapter 2 for vector algebra, the proof is straightforward, that is,

$$\begin{aligned} E_{\perp sca} &= \vec{E}_{sca} \cdot \hat{e}_{\perp sca} \\ &= \frac{k^2 \cdot a}{4\pi} \cdot E_{0,\perp inc} \cdot \frac{1}{R} \cdot \exp(i \cdot k \cdot R) , \end{aligned} \tag{4.117}$$

$$\begin{aligned} E_{\|sca} &= \vec{E}_{sca} \cdot \hat{e}_{\|sca} \\ &= \frac{k^2 \cdot a}{4\pi} \cdot E_{0,\|inc} \cdot \cos \vartheta \cdot \frac{1}{R} \cdot \exp(i \cdot k \cdot R) . \end{aligned} \tag{4.118}$$

Thus, the scattered and incident fields are related in matrix form as, compare with Eq. (4.12),

$$\begin{pmatrix} E_\| \\ E_\perp \end{pmatrix}_{sca} = \frac{\exp(i \cdot k \cdot R)}{-i \cdot k \cdot R} \cdot \begin{pmatrix} A_{Rayl,11} & 0 \\ 0 & A_{Rayl,22} \end{pmatrix} \cdot \begin{pmatrix} E_\| \\ E_\perp \end{pmatrix}_{inc} , \tag{4.119}$$

where

$$A_{Rayl,11} = -\frac{i \cdot k^3}{4\pi} \cdot a \cdot \cos \vartheta = -i \cdot \alpha^3 \cdot \left(\frac{\tilde{n}^2 - 1}{\tilde{n}^2 + 2} \right) \cdot \cos \vartheta , \tag{4.120}$$

$$A_{Rayl,22} = -\frac{i \cdot k^3}{4\pi} \cdot a = -i \cdot \alpha^3 \cdot \left(\frac{\tilde{n}^2 - 1}{\tilde{n}^2 + 2} \right) . \tag{4.121}$$

In Eqs. (4.120) and (4.121), α indicates the size parameter. The Mueller matrix associated with the amplitude scattering matrix in Eq. (4.119) is given by

$$IS_{Rayl} = \begin{pmatrix} S_{Rayl,11} & S_{Rayl,12} & 0 & 0 \\ S_{Rayl,12} & S_{Rayl,11} & 0 & 0 \\ 0 & 0 & S_{Rayl,33} & S_{Rayl,34} \\ 0 & 0 & -S_{Rayl,34} & S_{Rayl,33} \end{pmatrix} , \tag{4.122}$$

which, by applying Eqs. (4.41)–(4.46), becomes

$$
\mathsf{IS}_{\mathrm{Rayl}} = \alpha^6 \cdot \left| \frac{\tilde{n}^2 - 1}{\tilde{n}^2 + 2} \right|^2
$$
$$
\times \begin{pmatrix}
+(1 + \cos^2 \vartheta)/2 & -(1 - \cos^2 \vartheta)/2 & 0 & 0 \\
-(1 - \cos^2 \vartheta)/2 & +(1 + \cos^2 \vartheta)/2 & 0 & 0 \\
0 & 0 & \cos \vartheta & 0 \\
0 & 0 & 0 & \cos \vartheta
\end{pmatrix} . \quad (4.123)
$$

4.10.2
Degree of Polarization

For unpolarized incident radiation, we know that from Eq. (4.103)

$$
P_{\mathrm{lin},0°,\mathrm{sca,unp,Rayl}} = -\frac{S_{\mathrm{Rayl},12}}{S_{\mathrm{Rayl},11}} . \quad (4.124)
$$

Thus, we find

$$
P_{\mathrm{lin},0°,\mathrm{sca,unp,Rayl}} = \frac{1 - \cos^2 \vartheta}{1 + \cos^2 \vartheta} . \quad (4.125)
$$

Even if unpolarized light impinges on a very small particle, the scattered radiation is always polarized to some extent. In cloud-free and haze-free atmospheres, the scattered radiant field is dominated by Rayleigh scattering originating from air molecules. Under the single-scattering approximation, the sky light is unpolarized ($P_{\mathrm{lin,sca,Rayl}} = 0$) if an observer is looking directly towards or away from the Sun (scattering angle ϑ is 0 or 180°), and 100% polarized ($P_{\mathrm{lin,sca,Rayl}} = 1$) when viewing the sky at a 90° angle from the Sun, see Figure 4.10a.

4.10.3
Rayleigh Phase Function for Unpolarized Incident Radiation: $\mathcal{P}_{\mathrm{unp,Rayl}}$

If we apply Eq. (4.54) to unpolarized incident radiation impinging on a very small spherical particle, we get

$$
f_{\mathrm{unp,Rayl}} = S_{\mathrm{Rayl},11} = \alpha^6 \cdot \left| \frac{\tilde{n}^2 - 1}{\tilde{n}^2 + 2} \right|^2 \cdot \frac{1}{2} \cdot (1 + \cos^2 \vartheta) . \quad (4.126)
$$

Normalizing $f_{\mathrm{unp,Rayl}}$ in the preceding equation gives rise to the Rayleigh phase function $\mathcal{P}_{\mathrm{unp,Rayl}}$, that is,

$$
\mathcal{P}_{\mathrm{unp,Rayl}}(\vartheta) = \frac{3}{4} \cdot (1 + \cos^2 \vartheta) . \quad (4.127)
$$

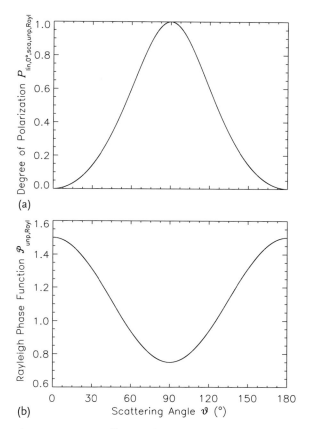

Figure 4.10 Degree of linear polarization $P_{\text{lin},0°,\text{sca,unp,Rayl}}$ given by Eq. (4.125) (a) and Rayleigh phase function $\mathcal{P}_{\text{unp,Rayl}}$ given by Eq. (4.127) (b), both as functions of scattering angle (ϑ).

The Rayleigh phase function is a symmetric function illustrated in Figure 4.10b. Note that the asymmetry factor associated with the Rayleigh phase function is zero. The degree of linear polarization by very small spherical particles, also a symmetrical function, is plotted in Figure 4.10a.

A comparison of the Rayleigh phase function with that of larger particles is shown in Figure 4.11. Compared are a spherical aerosol particle of 1 μm diameter and a spherical cloud droplet of 12 μm in diameter at wavelength $\lambda = 0.532$ μm and with unpolarized incident radiation. The forward scattering ($\vartheta = 0°$) becomes increasingly dominant for larger particles, and the oscillations of the phase function become more pronounced with increasing particle diameter, see Figure 4.8.

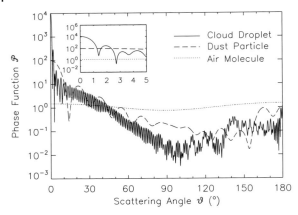

Figure 4.11 Phase function dependence on scattering angle for very small air molecules (Rayleigh), spherical dust particle of 1 μm diameter (refractive index $\tilde{n} = 1.53 + 0.008i$), and a spherical cloud droplet of 12 μm diameter (refractive index $\tilde{n} = 1.33 + 1.674 \times 10^{-8}i$), wavelength $\lambda = 0.532$ μm. The Rayleigh curve was simulated using Eq. (4.127).

4.10.4
Scattering Cross Section and Efficiency Factor

The Rayleigh scattering cross section can be calculated, applying Eq. (4.63), as

$$
C_{\text{sca,Rayl}} = \frac{1}{k^2} \iint\limits_{4\pi} f_{\text{unp,Rayl}}(\vartheta)\, d^2\Omega
$$

$$
= \pi r^2 \frac{8}{3} \cdot \alpha^4 \left| \frac{\tilde{n}^2 - 1}{\tilde{n}^2 + 2} \right|^2 . \tag{4.128}
$$

The corresponding scattering efficiency factor is given by

$$
Q_{\text{sca,Rayl}} = \frac{8}{3}\, \alpha^4 \cdot \left| \frac{\tilde{n}^2 - 1}{\tilde{n}^2 + 2} \right|^2 . \tag{4.129}
$$

4.10.5
Extinction and Absorption Cross Sections and Efficiency Factors

In Eqs. (4.120) and (4.121), the nonzero elements of the amplitude scattering matrix are the same in the forward direction and are given by

$$
A_{\text{Rayl}}(\vartheta = 0°) = A_{\text{Rayl,11}}(\vartheta = 0°) = A_{\text{Rayl,22}}(\vartheta = 0°)
$$

$$
= -i \cdot \alpha^3 \cdot \left(\frac{\tilde{n}^2 - 1}{\tilde{n}^2 + 2} \right) . \tag{4.130}
$$

If the preceding equation is applied to the optical theorem, the resultant extinction cross section contains the contribution from absorption but contains none from scattering because the expression does not consider the radiation reaction on the dipole (van de Hulst, 1957). Bohren and Huffman (1983) offered an insightful explanation of the Rayleigh extinction cross section inaccuracy when directly computed from the optical theorem. To include the radiation reaction, Eq. (4.130) needs to be modified into the following form (van de Hulst, 1957):

$$A_{\text{Rayl}}(\vartheta = 0°) = -i \cdot \alpha^3 \cdot \left(\frac{\tilde{n}^2 - 1}{\tilde{n}^2 + 2} \right) + \frac{2}{3} \cdot \alpha^6 \cdot \left(\frac{\tilde{n}^2 - 1}{\tilde{n}^2 + 2} \right)^2 . \tag{4.131}$$

Thus, the correct extinction efficiency factor is given by

$$Q_{\text{ext,Rayl}} = \frac{4}{\alpha^2} \cdot \Re e \left\{ A_{\text{Rayl}}(\vartheta = 0°) \right\}$$

$$= 4\alpha \cdot \Im m \left\{ \frac{(\tilde{n}^2 - 1)}{(\tilde{n}^2 + 2)} \right\} + \frac{8}{3} \alpha^4 \cdot \Re e \left\{ \left(\frac{\tilde{n}^2 - 1}{\tilde{n}^2 + 2} \right)^2 \right\} . \tag{4.132}$$

Using the expression

$$\Re e \left\{ \left(\frac{\tilde{n}^2 - 1}{\tilde{n}^2 + 2} \right)^2 \right\} - \left| \frac{\tilde{n}^2 - 1}{\tilde{n}^2 + 2} \right|^2 = -2 \left[\Im m \left\{ \frac{\tilde{n}^2 - 1}{\tilde{n}^2 + 2} \right\} \right]^2 , \tag{4.133}$$

the absorption efficiency can be calculated as

$$Q_{\text{abs,Rayl}} = 4\alpha \cdot \Im m \left\{ \frac{\tilde{n}^2 - 1}{\tilde{n}^2 + 2} \right\} - \frac{16}{3} \alpha^4 \cdot \left[\Im m \left\{ \frac{\tilde{n}^2 - 1}{\tilde{n}^2 + 2} \right\} \right]^2 . \tag{4.134}$$

For a sufficiently small α, we obtain

$$Q_{\text{abs,Rayl}} \approx 4\alpha \cdot \Im m \left\{ \frac{\tilde{n}^2 - 1}{\tilde{n}^2 + 2} \right\} . \tag{4.135}$$

4.10.6
Rayleigh Scattering as an Approximation of Lorenz–Mie Theory

Rayleigh scattering is an approximation of the Lorenz–Mie theory for small size parameters. Following van de Hulst (1957), to retain accuracy on the order of $O(\alpha^6)$, the nonzero elements of the Lorenz–Mie amplitude scattering matrix can be written as

$$A_{\text{Rayl,11}}(\vartheta) = -i \cdot \alpha^3 \cdot \left(\frac{\tilde{n}^2 - 1}{\tilde{n}^2 + 2} \right) \cdot \cos \vartheta - \frac{i \cdot \alpha^5}{3} \cdot \left(\frac{\tilde{n}^2 - 1}{\tilde{n}^2 + 2} \right)$$

$$\times \left[\frac{9}{5} \cdot \left(\frac{\tilde{n}^2 - 2}{\tilde{n}^2 + 2} \right) \cdot \cos \vartheta + \frac{1}{10} \cdot (\tilde{n}^2 + 2) \right.$$

$$\left. + \frac{1}{2} \cdot \left(\frac{\tilde{n}^2 + 2}{2\tilde{n}^2 + 3} \right) \cdot \cos 2\vartheta \right]$$

$$+ \frac{2\alpha^6}{3} \cdot \left(\frac{\tilde{n}^2 - 1}{\tilde{n}^2 + 2} \right)^2 \cdot \cos \vartheta$$

$$+ O(\alpha^7) , \tag{4.136}$$

and

$$
A_{\text{Rayl,22}}(\vartheta) = -i \cdot \alpha^3 \cdot \left(\frac{\tilde{n}^2 - 1}{\tilde{n}^2 + 2} \right) - \frac{i \cdot \alpha^5}{3} \cdot \left(\frac{\tilde{n}^2 - 1}{\tilde{n}^2 + 2} \right)
$$
$$
\times \left[\frac{9}{5} \cdot \left(\frac{\tilde{n}^2 - 2}{\tilde{n}^2 + 2} \right) + \frac{1}{10} \cdot (\tilde{n}^2 + 2) \cdot \cos \vartheta \right.
$$
$$
\left. + \frac{1}{2} \cdot \left(\frac{\tilde{n}^2 + 2}{2\tilde{n}^2 + 3} \right) \cdot \cos \vartheta \right]
$$
$$
+ \frac{2\alpha^6}{3} \cdot \left(\frac{\tilde{n}^2 - 1}{\tilde{n}^2 + 2} \right)^2
$$
$$
+ O(\alpha^7) . \tag{4.137}
$$

From the optical theorem, the extinction efficiency factor is given by

$$
Q_{\text{ext,Rayl}} = \frac{4}{\alpha^2} \mathfrak{Re} \left\{ A_{11,\text{Rayl}}(\vartheta = 0°) \right\} = \frac{4}{\alpha^2} \mathfrak{Re} \left\{ A_{22,\text{Rayl}}(\vartheta = 0°) \right\}
$$
$$
= 4\alpha \cdot \mathfrak{Im} \left\{ \frac{(\tilde{n}^2 - 1)}{(\tilde{n}^2 + 2)} \cdot \left[1 + \frac{\alpha^2}{3} \cdot \left(\frac{9}{5} \cdot \frac{\tilde{n}^2 - 1}{\tilde{n}^2 + 2} + \frac{1}{10} \cdot (\tilde{n}^2 + 2) \right. \right. \right.
$$
$$
\left. \left. \left. + \frac{1}{2} \cdot \frac{\tilde{n}^2 + 2}{2\tilde{n}^2 + 3} \right) \right] \right\}
$$
$$
+ \frac{8}{3} \alpha^4 \cdot \mathfrak{Re} \left\{ \left(\frac{\tilde{n}^2 - 1}{\tilde{n}^2 + 2} \right)^2 \right\} , \tag{4.138}
$$

and the scattering efficiency factor is given by

$$
Q_{\text{sca,Rayl}} = \frac{1}{\pi \cdot r^2} \cdot \frac{1}{k^2} \iint_{4\pi} \frac{|A_{11,\text{Rayl}}(\vartheta)|^2 + |A_{22,\text{Rayl}}(\vartheta)|^2}{2} \, d^2\Omega
$$
$$
= \frac{1}{\pi \cdot r^2} \cdot \frac{1}{k^2} \iint_{4\pi} \alpha^6 \cdot \left[\left| \frac{\tilde{n}^2 - 1}{\tilde{n}^2 + 2} \right|^2 \cdot \frac{1}{2} \cdot (1 + \cos^2 \vartheta) + O(\alpha^8) \right] d^2\Omega
$$
$$
= \frac{8}{3} \alpha^4 \cdot \left| \frac{\tilde{n}^2 - 1}{\tilde{n}^2 + 2} \right|^2 . \tag{4.139}
$$

We obtain the absorption efficiency factor for very small spheres from $Q_{\text{abs,Rayl}} = Q_{\text{ext,Rayl}} - Q_{\text{sca,Rayl}}$, which yields

$$
Q_{\text{abs,Rayl}} = 4\alpha \cdot \mathfrak{Im} \left\{ \frac{\tilde{n}^2 - 1}{\tilde{n}^2 + 2} \right\} \cdot \left[1 - \frac{4}{3} \alpha^3 \cdot \mathfrak{Im} \left\{ \frac{\tilde{n}^2 - 1}{\tilde{n}^2 + 2} \right\}^2 \right] . \tag{4.140}
$$

For a sufficiently small α, we obtain

$$
Q_{\text{abs,Rayl}} \approx 4\alpha \cdot \mathfrak{Im} \left\{ \frac{\tilde{n}^2 - 1}{\tilde{n}^2 + 2} \right\} . \tag{4.141}
$$

It is evident that the result of Eq. (4.141) is consistent with that of Eq. (4.135), although two different approaches are used to derive the results.

4.10.7
Rayleigh Scattering in the Atmosphere

For Rayleigh scattering in the atmosphere, the polarizability, a, is given by the Lorentz–Lorenz formula (Born and Wolf, 2003)

$$a = \frac{3}{n_{\text{mol}}} \cdot \left(\frac{\tilde{n}^2 - 1}{\tilde{n}^2 + 2} \right), \tag{4.142}$$

where n_{mol} is the number concentration of air molecules. A different form of Eq. (4.142), expressed in a general form in terms of the electric permittivity, is known as the Clausius–Mossotti formula, a relationship used in numerical techniques, for example, the discrete dipole approximation (Purcell and Pennypacker, 1973) for calculating light scattering.

With the polarizability, a, defined in Eq. (4.142), the corresponding Mueller matrix, extinction cross section, and scattering cross section can be computed. Absorption is linearly related to the size parameter, that is,

$$Q_{\text{abs,Rayl}} \sim \alpha, \tag{4.143}$$

see Eq. (4.141); whereas, the scattering efficiency factor for Rayleigh scattering is

$$Q_{\text{sca,Rayl}} \sim \alpha^4, \tag{4.144}$$

see Eq. (4.139). For a sufficiently small α and a nonzero imaginary part of the complex refractive index $\tilde{n}$, it follows that

$$Q_{\text{sca,Rayl}} \ll Q_{\text{abs,Rayl}} \approx Q_{\text{ext,Rayl}} \tag{4.145}$$

when all terms with α^4 are very small and α^2 in Eq. (4.132) becomes small. For a sufficiently small α, we can neglect scattering ($Q_{\text{sca,Rayl}} \sim \alpha^4$; $Q_{\text{abs,Rayl}} \sim \alpha$) and focus on absorption. Thus, the single-scattering albedo for very small spherical particles is approximated by

$$\tilde{\omega}_{\text{Rayl}} = \frac{Q_{\text{sca,Rayl}}}{Q_{\text{ext,Rayl}}} \sim \alpha^3. \tag{4.146}$$

$\tilde{\omega}_{\text{Rayl}}$ corresponds to α^3 for slightly absorbing sufficiently small particles; however, for nonabsorbing particles, $\tilde{\omega}_{\text{Rayl}} \equiv 1$ regardless of the size of α. For the Rayleigh scattering efficiency factor, from Eq. (4.139),

$$Q_{\text{sca,Rayl}} \sim \alpha^4 \sim \left(\frac{r}{\lambda} \right)^4, \tag{4.147}$$

which appears to be a reasonable approximation. In Figure 4.12a, the α^4 dependence is plotted and the exact results are shown in parallel with the α^4 approximation. The agreement is better than 3% for small values of $\alpha < 0.5$.

For very small air molecules and $r = \text{constant}$, Eq. (4.147) gives $Q_{\text{sca,Rayl}} \sim \lambda^{-4}$, and the small wavelength (blue) radiation appears more strongly scattered. This explains the blue color of a clear sky, which only contains air molecules. A low

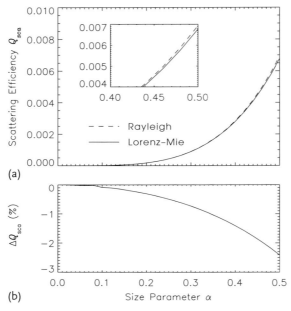

(a)

(b)

Figure 4.12 Exact results from the Lorenz–Mie theory for the scattering efficiency factor $Q_{s,Mie}$ compared with the approximate values of $Q_{sca,Rayl}$ given in Eq. (4.147) for very small spherical particles as a function of the size parameter $\alpha = 2\pi \cdot r/\lambda$ (a). (b) Percentage difference $\Delta Q_{sca} = (Q_{sca,Mie} - Q_{sca,Rayl})/Q_{sca,Mie}$ as a function of the size parameter α.

Sun during sunset appears red because the blue color has been largely removed from the direct radiation beam and red remains. Large particles suspended in the atmosphere affect the radiation scattering, and the blue color of the sky appears different. The presence of large haze particles produces a reddening in the sky due to the increased extinction of small wavelengths. For clouds with even larger particles, the scattering becomes wavelength-independent and, thus, clouds appear white.

The scattering cross section of the Rayleigh scattering for spherical particles $[Q_{sca,Rayl} \sim (r/\lambda)^4]$ becomes

$$C_{sca,Rayl} = \pi \cdot r^2 \cdot Q_{sca,Rayl} \sim \frac{r^6}{\lambda^4} \,. \tag{4.148}$$

An approximation formula by Nicolet (1984) is often used for the molecular scattering cross sections in units of cm^2, but not for efficiency factors

$$C_{sca,Rayl,N} = Q_{sca,Rayl,N} \cdot \pi \cdot r^2 \approx \frac{4.02 \times 10^{-28}}{\lambda^{4+x_{Rayl}}} \,, \tag{4.149}$$

where the wavelength λ is given in μm. The function x_{Rayl} is given by

$$x_{Rayl} = \begin{cases} 0.04 & \text{for} \quad \lambda > 0.55 \ \mu m \\ 0.389 \cdot \lambda + 0.094\,26/\lambda - 0.3228 & \text{for} \quad \lambda \leq 0.55 \ \mu m \end{cases} \,. \tag{4.150}$$

4.11
Scattering by Nonspherical Individual Particles

4.11.1
Analytical Approaches

To simulate the scattering properties of nonspherical particles, the most commonly used techniques include analytical, numerical, and approximate methods. We will only survey analytical approaches, and the discussion in this section is gleaned from the survey by Mishchenko et al. (2000), Chapter 2.

Analytical approaches are based on solving Maxwell's equations. We need to separate the variables into a set of variables conforming a coordinate system in which the equations for the resultant variables are solvable by analytical means. The incident EM field and the field inside the scatterer are expanded into eigenfunctions that are regular inside the scatterer; the scattered field outside the scatterer is expanded into eigenfunctions that reduce the outgoing waves at infinity. These series are generally double series and degeneration to a single series only occurs for spheres and infinite cylinders. Because of the need for continuity of the tangential electric and magnetic field components at a particular surface, the unknown expansion coefficients of the internal and scattered fields are determined from the known expansion coefficients of the incident electric field.

The separation of variables only works for the specific shapes listed in Table 4.1, together with respective references to be consulted for detailed descriptions.

Table 4.1 Particle shapes and respective references. This overview is partially compiled from the summary by Mishchenko et al. (2000).

Shape	Reference
Homogeneous, isotropic sphere	Love (1899)
	Lorenz (1890)
	Mie (1908)
	Debye (1909)
Concentric, core-mantle sphere	Aden and Kerker (1951)
Concentric, multilayered sphere	Wait (1963)
	Mikulski and Murphy (1963)
	Bhandari (1985)
Radially inhomogeneous spheres	Wyatt (1962)
Optically active spheres	Bohren (1974)
Homogeneous, infinite circular cylinders	Wait (1955)
Optically active cylinders	Bohren (1978)
Infinite, elliptical cylinders	Kim and Yeh (1991)
Homogeneous, isotropic spheroids	Oguchi (1973)
	Asano and Yamamoto (1975)
	Asano and Yamamoto (1995)

4.11.2
Mueller Matrix

For nonspherical particles, each of the 16 elements of the Mueller matrix can be nonzero, and may depend on the orientation of the particle with respect to the incident EM wave. If the Mueller elements are all nonzero, nonspherical scattering is indicated. If nonspherical particles are assumed to be randomly oriented with an equal number of mirror-image orientations, the Mueller matrix degenerates to the Lorenz–Mie structure given in Eq. (4.38).

An example of the elements of the Mueller matrix is given in Figure 4.13, adapted from Bi et al. (2010). Figure 4.13 shows the nonzero phase matrix elements associated with Pinatubo ash particles from measurements (Volten et al., 2001) and theoretical simulations based on the spherical and nonspherical models for the particle shape. In the nonspherical model, dust particles are assumed to be randomly distorted hexahedra. From the figure, it is evident that the nonspherical model is in much better agreement with the measurements, particularly in the case of $-P_{12}/P_{11}$.

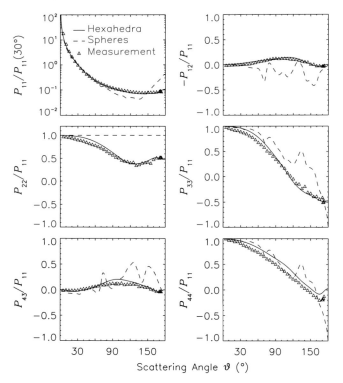

Figure 4.13 Comparison of simulated results of hexahedra with measurements for Pinatubo aerosol particles at a wavelength of 0.633 μm, from Bi et al. (2010). The measurements are from Volten et al. (2001). Reproduced with permission of the Optical Society of America (OSA) © OSA.

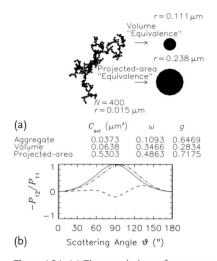

(a)

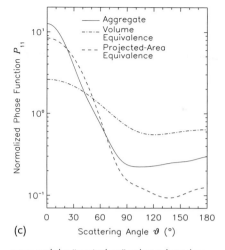

Figure 4.14 (a) The morphology of a soot aggregate and "equivalent" spheres of the same volume and projected area, (b) $-P_{12}/P_{11}$ computed for randomly oriented soot aggregates and the "equivalent" spheres based on volume and projected area, (c) The phase funtions corresponding to the $-P_{12}/P_{11}$ in (b). Courtesy of C. Liu.

Airborne soot is a unique aerosol species and has an important climate effect. Soot particles usually exist as polydisperse aggregates (Figure 4.14a) of similarly-sized small elements referred to as monomers (Liu et al., 2006; Sorensen and Roberts, 1997). Substantial errors can be incurred in scattering computation if these aggregates are approximated as "equivalent" spheres. For example, Figure 4.14b shows the comparison of $-P_{12}/P_{11}$ between randomly oriented soot aggregates against volume and projected-area "equivalent" spheres, whereas Figure 4.14c shows the corresponding phase funtions. The differences between the spherical and nonspherical results are obvious, particularly for the phase function.

4.11.3
Phase Function

Cirrus clouds are composed of ice crystals with various habits, including droxtals, bullet rosettes, aggregates, hollow columns, and hexagonal plates and columns. As an example of the scattering properties for these different ice crystal shapes, Figure 4.15 shows the phase function of ice crystals as computed using an improved geometric-optics method (IGOM) (Yang and Liou, 1996). The method is based on the ray-tracing technique for computing the near-field on the particle's surface and an exact integral relation between the near-field and the corresponding far-field.

The most striking difference between spherical and nonspherical scattering is the presence of halo peaks (compare Figures 4.8 and 4.15). The angles of scattering peaks corresponding to halos are observed at 22° and 46° for hexagonal shapes. Another important feature is the enhanced sideward scattering observed for non-

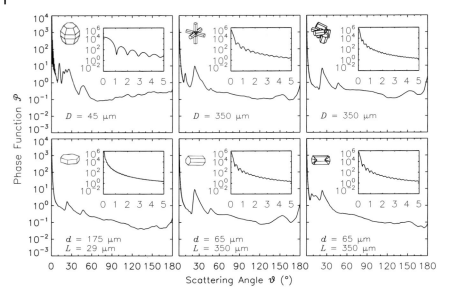

Figure 4.15 Nonspherical phase functions for differently shaped ice crystals, calculated on the basis of computations by Yang et al. (2000) at a wavelength of 0.65 μm. D represents the maximum dimension of droxtals, bullet rosettes, and aggregates (upper three panels). d is the cross sectional semiwidth of the hexagonal-shaped particles, L represents the height of the plates and columns (lower three panels).

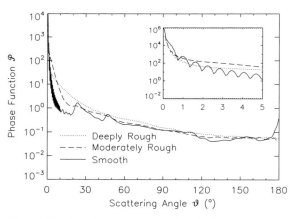

Figure 4.16 Phase functions at a wavelength of $\lambda = 0.65$ μm and $D = 500$ μm for the randomly oriented aggregates of plates with three different surface roughness conditions. Courtesy of Y. Xie.

spherical particles that does not indicate rainbow formation at the scattering angles of 138° and 129°.

Phase functions at a wavelength of 0.65 μm for randomly oriented aggregates of plates are shown in Figure 4.16. The ice crystal scattering properties are computed using the IGOM. To account for ice crystal surface roughness, we assume the slopes of the roughened facets on the particle surface to be randomly sampled on the basis of the Gaussian distribution and the Box–Muller method discussed by Yang et al. (2008). As can be seen in Figure 4.16, the presence of roughness can smooth many of the special features of the hexagonal plate nonspherical scattering function; in particular, it suppresses the halo formation.

4.11.4
Integrated Optical Properties

Let us have a look at the spectral variations of the extinction cross section, asymmetry factor, and single-scattering albedo for spherical and nonspherical particles; C_{ext}, g, and $\tilde{\omega}$. For this purpose, in this Section 4.11.4, we show two figures, Fig-

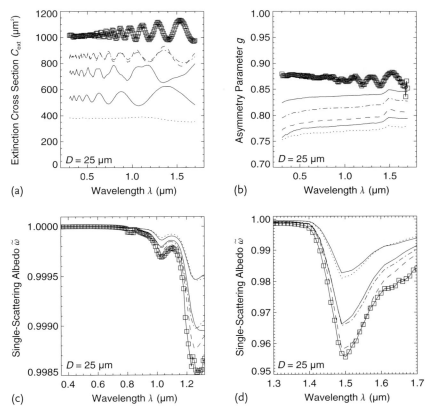

Figure 4.17 Individual ice crystals: Extinction cross section C_{ext} (a), asymmetry factor g (b) and single-scattering albedo $\tilde{\omega}$ (c, d) as a function of wavelength λ for ice crystals with a fixed maximum dimension of $D = 25$ μm. The curve notation is as: Solid lines with open squares for spheres; dashed lines for columns; dash-dot lines for hollow columns; dash-dot-dot-dot lines for plates; solid lines for bullet rosettes; dotted lines for aggregates. Adapted from Wendisch et al. (2005) with permission of the American Geophysical Union (AGU) © AGU.

ures 4.17 and 4.18, and summarize the discussion given in Wendisch et al. (2005). For more details, the reader is referred to the original paper by Wendisch et al. (2005).

Spectra of C_{ext} (Figure 4.17a), g (Figure 4.17b), and $\tilde{\omega}$ (Figure 4.17c,d) calculated assuming different crystal habits are compared in Figure 4.17. Spheres exhibit the largest values of C_{ext} and g because they have the largest cross sectional area for a given maximum particle dimension D. On the other hand, the extinction cross section and asymmetry parameter of aggregates are smallest. Spheres show the strongest absorption effects compared to all nonspherical shape assumptions, see Figure 4.17c,d.

Figure 4.18 shows the same data as Figure 4.17, but as a function of maximum particle dimension. The extinction coefficient increases and the single-scattering albedo decreases with increasing D. For the asymmetry parameter, the general trend is to increase with the maximum particle dimension.

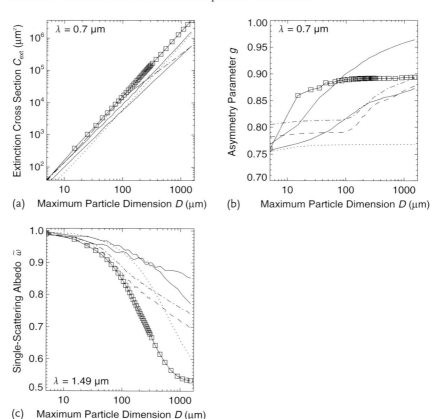

Figure 4.18 Individual ice crystals: Extinction cross section C_{ext} (a), asymmetry factor g (b) and single-scattering albedo $\tilde{\omega}$ (c) as functions of the maximum particle dimension D for a fixed wavelength of $\lambda = 0.7\,\mu m$ (a, b) and within an ice absorption band at $\lambda = 1.49\,\mu m$ (c). The curve notation is the same as in Figure 4.17. Adapted from Wendisch et al. (2005) with permission of the American Geophysical Union © AGU.

4.12
Geometric-Optics Method for Light Scattering by Large Particles

When the characteristic dimension of a scattering particle is much larger than the incident wavelength, the scattering of a plane EM wave by the particle can be solved from the basic principles of geometric optics in terms of Snel's law, the Fresnel formulas, and the Fraunhofer diffraction theory. Snel's law is often called Snell's law in the literature; it was named in honor of Willebrord Snel van Royen, also known as Willebrord van Roijen Snell or Snellius (Adams and Kattawar, 1997).

In the geometric-optics method, the incident radiation can be regarded as a bundle of localized rays. Each of the rays propagates along a rectilinear path until it impinges on a particle surface by which reflection and refraction occur. The reflected and transmitted/refracted rays also propagate along their rectilinear paths until the next reflection-refraction event occurs at a particle surface. In addition to reflection and refraction, incident electromagnetic energy can be redistributed in directions deviating from the incident direction due to diffraction inherently stemming from

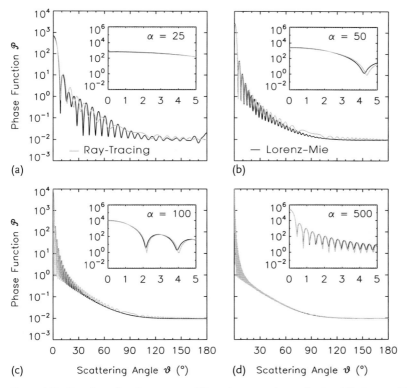

Figure 4.19 The phase function computed from the ray-tracing method and the Lorenz–Mie theory for individual liquid water spheres at the 10 μm wavelength for four size parameters $\alpha =$ 25 (a), $\alpha = 50$ (b), $\alpha = 100$ (c) and $\alpha = 500$ (d). Adapted from Yang and Liou (2009).

the phase interference of an incomplete wavefront caused by the particle's blocking the incident wave.

The geometric-optics method is unlikely to be applicable to particles with small size parameters ($\alpha < 20$) because either "A pencil of light of length l can exist only if its width at its base is large compared to $\sqrt{\lambda \cdot l}$," or "a pencil of width of the order of $p \cdot \lambda$ can be an independent existence over a length of the order of $p^2 \cdot \lambda$" (van de Hulst, 1957). Furthermore, van de Hulst also stated, "For a particle with a size of 20 or more times the wavelength, it is possible to distinguish fairly sharply between the rays incident on the particle and rays passing along the particle. Among the former, it is possible to distinguish rays hitting various parts of the particle's surface."

To illustrate the dependence of the accuracy of the geometric-optics method on the size parameter, Figure 4.19 shows the phase function calculated from the Lorenz–Mie theory and from the conventional geometric optics method in terms of the ray-tracing technique for four size parameters, $\alpha = 25, 50, 100,$ and 500. The refractive index for the calculation is $\tilde{n} = 1.218 + i \cdot 0.0508$, and the value of the refractive index for water at a wavelength of 10 µm. Figure 4.20 is identical to Fig-

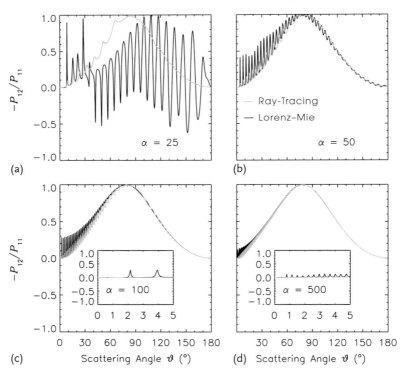

Figure 4.20 The same as Figure 4.19 but instead of the phase function the ratio $-P_{12}/P_{11}$ is plotted, as computed from the ray-tracing method and the Lorenz–Mie theory for individual liquid water spheres at the 10 µm wavelength for four size parameters $\alpha = 25$ (a), $\alpha = 50$ (b), $\alpha = 100$ (c) and $\alpha = 500$ (d). Adapted from Yang and Liou (2009).

ure 4.19, except that the ratio $-P_{12}/P_{11}$ is plotted instead of the phase function. It is evident from the two figures that the accuracy of the geometric optics is systematically improved with the increase in size parameter. For size parameters of 100 and 500, the ray-tracing solutions essentially converge to the corresponding exact results.

4.12.1
Directional Changes Due to Reflection and Transmission (Refraction) at a Plane Interface: Snel's Law

Figure 4.21 illustrates the incident (subscript "inc"), reflected (subscript "ref"), and transmitted/refracted (subscript "tra") rays within two media whose indices of refraction are $\tilde{n}_1$ and $\tilde{n}_2$. If the position vector of a point at the interface is $\vec{r}_p$, then the spatial components of the incident, reflected, and transmitted/refracted electric field vectors can be expressed as, see Eq. (4.6),

$$\vec{E}_{inc}(\vec{r}) = \vec{E}_{0,inc} \cdot \exp[i \cdot k_0 \cdot \tilde{n}_2 \, \hat{e}_{inc} \cdot (\vec{r} - \vec{r}_p)] \,, \tag{4.151}$$

$$\vec{E}_{ref}(\vec{r}) = \vec{E}_{0,ref} \cdot \exp[i \cdot k_0 \cdot \tilde{n}_2 \, \hat{e}_{ref} \cdot (\vec{r} - \vec{r}_p)] \,, \tag{4.152}$$

$$\vec{E}_{tra}(\vec{r}) = \vec{E}_{0,tra} \cdot \exp[i \cdot k_0 \cdot \tilde{n}_1 \, \hat{e}_{tra} \cdot (\vec{r} - \vec{r}_p)] \,. \tag{4.153}$$

In Eqs. (4.151)–(4.153), $k_0 = 2\pi/\lambda_0$ with λ_0 the wavelength in a vacuum, and $\vec{E}_{0,inc}$ is the incident electric field vector at the incident point $\vec{r}_p$. At an arbitrary location at the surface $\vec{r} - \vec{r}_p = d\hat{f}$, where d is arbitrary and $\hat{f}$ is a unit vector parallel to the interface of the media, the phases of the incident, reflected, and transmit-

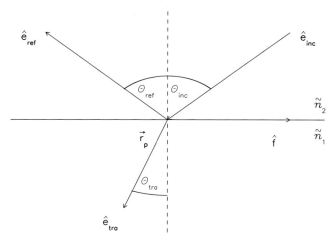

Figure 4.21 Geometric configuration of incident, reflected and transmitted/refracted rays. $\vec{r}_p$ represents the position vector of the incident point.

ted/refracted waves must be continuous, and we have

$$\tilde{n}_2 \, (\hat{e}_{\text{inc}} \cdot \hat{f}) \, d = \tilde{n}_2 \, (\hat{e}_{\text{ref}} \cdot \hat{f}) \, d = \tilde{n}_1 \, (\hat{e}_{\text{tra}} \cdot \hat{f}) \, d \, . \tag{4.154}$$

Because d in Eq. (4.154) is arbitrary, we have

$$\hat{e}_{\text{inc}} \cdot \hat{f} = \hat{e}_{\text{ref}} \cdot \hat{f} \, , \tag{4.155}$$

$$\tilde{n}_2 \, \hat{e}_{\text{inc}} \cdot \hat{f} = \tilde{n}_1 \, \hat{e}_{\text{tra}} \cdot \hat{f} \, . \tag{4.156}$$

The preceding equations reduce to a more explicit form, that is,

$$\sin \Theta_{\text{inc}} = \sin \Theta_{\text{ref}} \, , \tag{4.157}$$

$$\tilde{n}_2 \cdot \sin \Theta_{\text{inc}} = \tilde{n}_1 \cdot \sin \Theta_{\text{tra}} \, , \tag{4.158}$$

with Θ_{inc} as the incident angle of the incident ray in medium 2 and Θ_{tra} as the refraction angle of the transmitted/refracted beam in medium 1. Both angles are measured with respect to the normal of the interface. In general, $\tilde{n}_1$ and $\tilde{n}_2$ represent the complex refractive indices in media 1 and 2; $\tilde{n} = \tilde{n}_{\text{re}} + \text{i} \cdot \tilde{n}_{\text{im}}$.

Equations (4.157) and (4.158) constitute Snel's law for reflection and refraction or transmission. When the media are absorptive, the angles involved in Snel's law are complex quantities and do not have straightforward geometric meanings. When a plane wave is refracted from a nonabsorptive medium into an absorptive medium, $\tilde{n}_2$ and $\tilde{n}_1$ in Eqs. (4.157)–(4.158) are real and complex, respectively, and the transmitted/refracted wave is inhomogeneous (Bohren and Huffman, 1983; Born and Wolf, 2003) as the planes of constant phase are not parallel to those of constant amplitude. In such cases, effective refractive indices are necessary for the ray-tracing calculation (Yang and Liou, 2009). For simplicity, here we do not consider absorbing media, that is, we assume that both $\tilde{n}_1$ and $\tilde{n}_2$ are real.

Snel's law illustrates that when $\tilde{n}_2 < \tilde{n}_1$, $\sin \Theta_{\text{inc}} > \sin \Theta_{\text{tra}}$ and $\Theta_{\text{inc}} > \Theta_{\text{tra}}$. Consequently, a ray will always be bent towards the normal of the optically thicker medium or larger $\tilde{n}_{\text{re}}$. The change of ray direction is the result of a slower phase speed in the optically thicker medium. It is evident from Eqs. (4.151)–(4.153) that the wave constants in media 1 and 2 are

$$\tilde{n}_1 \cdot k_0 = \frac{2\pi}{\lambda_0/\tilde{n}_1} \, , \tag{4.159}$$

and

$$\tilde{n}_2 \cdot k_0 = \frac{2\pi}{\lambda_0/\tilde{n}_2} \, . \tag{4.160}$$

The wavelength of the incident wave is

$$\lambda_{\text{inc}} = \frac{\lambda_0}{\tilde{n}_2} \, ; \tag{4.161}$$

whereas, that of the transmitted/refracted wave is

$$\lambda_{\text{tra}} = \frac{\lambda_0}{\tilde{n}_1} , \tag{4.162}$$

and $\lambda_{\text{tra}} < \lambda_{\text{inc}}$ if $\tilde{n}_2 < \tilde{n}_1$. Furthermore, because the frequency does not change in the reflection and transmission/refraction processes, we must have

$$\frac{c_{\text{inc}}}{\lambda_{\text{inc}}} = \frac{c_{\text{tra}}}{\lambda_{\text{tra}}} . \tag{4.163}$$

Thus, $c_{\text{inc}} > c_{\text{tra}}$, if $\lambda_{\text{tra}} < \lambda_{\text{inc}}$.

The photo in Figure 4.22 illustrates a phenomenon that appears contrary to our expectation, as the image of the rod is refracted away from the normal to the surface in the optically thicker water. However, in Figure 4.23, the real rays are indicated by the solid arrows, and the dashed line reveals that our eyes do not always see the true picture.

Figure 4.22 Photo of a metal rod (tip of a spoon) dipped into water.

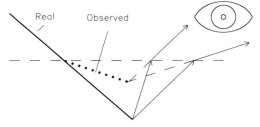

Figure 4.23 Refraction on a water surface: What reality and our eyes tell us.

Three possible configurations exist for Θ_{inc} and the refractive indices:

Case 1: $\Theta_{inc,nor} = 0°$

A ray hits the interface in normal direction. From Snel's law, we get

$$\sin \Theta_{inc,nor} = \sin \Theta_{tra,nor} = 0 , \tag{4.164}$$

with no change in direction for the transmitted/refracted ray, or

$$\Theta_{inc,nor} = \Theta_{tra,nor} = 0 . \tag{4.165}$$

Case 2: $\tilde{n}_2 < \tilde{n}_1$

For example, consider air ($\tilde{n}_2$) and water ($\tilde{n}_1$). From Snel's law, we obtain

$$\sin \Theta_{inc} > \sin \Theta_{tra} , \tag{4.166}$$

indicating the ray will bend towards the local normal direction or

$$\Theta_{inc} > \Theta_{tra} , \tag{4.167}$$

as illustrated in Figure 4.21.

Case 3: $\tilde{n}_2 > \tilde{n}_1$

For example, glass fiber ($\tilde{n}_2$) in air ($\tilde{n}_1$) and the transmission of radiation is from the fiber to the air. From Snel's law, we have

$$\sin \Theta_{inc} < \sin \Theta_{tra} . \tag{4.168}$$

The preceding equation illustrates that the ray will bend away from the local normal direction, that is,

$$\Theta_{inc} < \Theta_{tra} . \tag{4.169}$$

For $\Theta_{tra,T} = 90°$ ($\sin \Theta_{tra,T} = 1$), we get the critical incident angle $\Theta_{inc,T}$

$$\tilde{n}_2 \cdot \sin \Theta_{inc,T} = \tilde{n}_1 \cdot \sin \Theta_{tra,T} = \tilde{n}_1 , \tag{4.170}$$

which gives the critical incident angle of total reflection

$$\Theta_{inc,T} = \arcsin \left(\frac{\tilde{n}_1}{\tilde{n}_2} \right) = \arcsin(\tilde{m}) , \tag{4.171}$$

with the relative refractive index

$$\tilde{m} = \frac{\tilde{n}_1}{\tilde{n}_2} < 1 . \tag{4.172}$$

The term $\sin \Theta_{tra}$ always needs to be less than or equal to unity. If the incident angle is larger than the critical incident angle defined by Eq. (4.170), no refraction is possible and the incident ray is totally reflected with the occurrence of neither refraction nor transmission. $\Theta_{\text{inc,T}}$ is called the critical incident angle of total reflection and defines the threshold of total internal reflection within a denser medium. In the visible wavelength region, $\tilde{n}_2 \approx 1.33$ is for water and $\tilde{n}_1 \approx 1$ is for air. Hence, $\Theta_{\text{inc,T}} = 49°$ for a water surface below air.

A useful equation can be derived on the basis of Snel's law which connects the angles of transmitted/refracted and incident rays with the relative refractive index $\tilde{m}$, see Eq. (4.172). The derivation is

$$\sin \Theta_{\text{tra}} = \frac{\sin \Theta_{\text{inc}}}{\tilde{m}} \; ,$$

$$\sin^2 \Theta_{\text{tra}} = 1 - \cos^2 \Theta_{\text{tra}} = \frac{\sin^2 \Theta_{\text{inc}}}{\tilde{m}^2} \; , \tag{4.173}$$

which leads to

$$\cos \Theta_{\text{tra}} = \sqrt{1 - \left(\frac{\sin \Theta_{\text{inc}}}{\tilde{m}} \right)^2}$$

$$= \sqrt{1 - \left(\frac{\tilde{n}_2}{\tilde{n}_1} \right)^2 \cdot \sin^2 \Theta_{\text{inc}}} \; . \tag{4.174}$$

Equation (4.174) proves to be useful to replace $\cos \Theta_{\text{tra}}$ in the forthcoming Eqs. (4.212) and (4.213). For normal incidence ($\Theta_{\text{inc,nor}} = 0°$, $\sin \Theta_{\text{inc,nor}} = 0$), we obtain

$$\cos \Theta_{\text{tra,nor}} = \sqrt{1 - 0} = 1 \; . \tag{4.175}$$

4.12.2
The $\tilde{n}^2$ Law

If two media are nonabsorptive and the reflection at the interface of the media can be neglected, the ratio of the radiance in medium 2 to its counterpart in medium 1, that is, $I_{\text{inc}}/I_{\text{tra}}$, is proportional to $(\tilde{n}_2/\tilde{n}_1)^2$. This relationship is usually called the $\tilde{n}^2$ law.

To understand the $\tilde{n}^2$ law, let us consider a small area element, $\text{d}^2 A$, on the interface of the two media, regarding the refraction of radiation beam from medium 2 to 1. The radiant energy impinging on the area $\text{d}^2 A$ is assumed to be $\text{d}^5 E_{\text{rad,inc}}$; whereas, its counterpart for the refracted or transmitted beam is $\text{d}^5 E_{\text{rad,tra}}$. If the reflected energy can be neglected, it follows from the principle of energy conservation that

$$\text{d}^5 E_{\text{rad,inc}} = \text{d}^5 E_{\text{rad,tra}} \; . \tag{4.176}$$

Taking the derivative from both sides of Snel's law, see Eq. (4.158), we obtain

$$d(\sin \Theta_{inc} \cdot \tilde{n}_2) = d(\sin \Theta_{tra} \cdot \tilde{n}_1) \,, \tag{4.177}$$

$$\tilde{n}_2 \cdot \cos \Theta_{inc} \, d\Theta_{inc} = \tilde{n}_1 \cdot \cos \Theta_{tra} \, d\Theta_{tra} \,, \tag{4.178}$$

which gives

$$\frac{d\Theta_{tra}}{d\Theta_{inc}} = \frac{\tilde{n}_2 \cdot \cos \Theta_{inc}}{\tilde{n}_1 \cdot \cos \Theta_{tra}} \,. \tag{4.179}$$

According to the definition of radiance, see Eq. (3.30), and by using of Eqs. (2.37) and (4.179), we have

$$
\begin{aligned}
\frac{I_{inc}}{I_{tra}} &= \frac{d^5 E_{rad,inc} \Big/ \big(d^2 A \cdot \cos \Theta_{inc} \cdot \sin \Theta_{inc} \, d\Theta_{inc} \, d\varphi \, dt\big)}{d^5 E_{rad,tra} \Big/ \big(d^2 A \cdot \cos \Theta_{tra} \cdot \sin \Theta_{tra} \, d\Theta_{tra} \, d\varphi \, dt\big)} \\
&= \frac{\cos \Theta_{tra} \cdot \sin \Theta_{tra}}{\cos \Theta_{inc} \cdot \sin \Theta_{inc}} \left(\frac{d\Theta_{tra}}{d\Theta_{inc}}\right) \\
&= \frac{\sin \Theta_{tra} \cdot \tilde{n}_2}{\sin \Theta_{inc} \cdot \tilde{n}_1} \\
&= \left(\frac{\tilde{n}_2}{\tilde{n}_1}\right)^2 \,,
\end{aligned}
\tag{4.180}
$$

where $d\varphi$ is a small angle element along the direction normal to the incident plane, that is, the plane containing the incident and refracted beams. It is evident from the preceding equation that the quantity $I/\tilde{n}^2$ is invariant in transmitting from one transparent, that is, nonabsorptive, medium to another transparent medium if the reflection at the interface of the media can be neglected.

4.12.3
Fresnel Formulas for Reflection and Transmission

The Fresnel formulas give the amplitudes of the electric fields associated with the reflected and refracted waves. For simplicity, we only consider nonferromagnetic media, that is, their magnetic permeabilities are the same as that of vacuum ($\kappa = \kappa_0$). With this simplification, the refractive index of a nonferromagnetic medium is given by using Eq. (4.4) with $\kappa = \kappa_0$:

$$\tilde{n} = \sqrt{\frac{\epsilon \cdot \kappa_0}{\epsilon_0 \cdot \kappa_0}} = \sqrt{\frac{\epsilon}{\epsilon_0}} = \frac{c}{c_m} \,, \tag{4.181}$$

where c_m is defined by Eq. (3.15) and c is the speed of light in a vacuum, see Eq. (3.17). Consider a temporally harmonic EM wave in the form of

$$\begin{pmatrix} \vec{E} \\ \vec{H} \end{pmatrix} = \begin{pmatrix} \vec{E}_0 \\ \vec{H}_0 \end{pmatrix} \cdot \exp(i \cdot k_0 \cdot \tilde{n} \cdot \hat{e}_3 \cdot \vec{r} - i \cdot \omega_c \cdot t) \,, \tag{4.182}$$

where $k_0 = 2\pi/\lambda_0$ in units of m^{-1} is the modified wavenumber in a vacuum, ω_c is the circular frequency, and $\hat{\mathbf{e}}_3$ is the unit vector pointing to the direction of wave propagation. The curl pair of Maxwell's equations, see Eqs. (3.8) and (3.10), can be written for nonferromagnetic media ($\kappa = \kappa_0$) as

$$i \cdot k_0 \cdot \tilde{n} \cdot \hat{\mathbf{e}}_3 \times \vec{\mathbf{H}} = \epsilon \left(-i \cdot \omega_c\right) \vec{\mathbf{E}} \,, \tag{4.183}$$

$$i \cdot k_0 \cdot \tilde{n} \cdot \hat{\mathbf{e}}_3 \times \vec{\mathbf{E}} = -\kappa_0 \left(-i \cdot \omega_c\right) \vec{\mathbf{H}} \,. \tag{4.184}$$

From the dispersion relation given by Eq. (3.21), we have

$$\pm \frac{\omega_c}{k_0} = c = \frac{1}{\sqrt{\kappa_0 \cdot \epsilon_0}} \,. \tag{4.185}$$

From Eqs. (4.181) and (4.185), we obtain the following relations between the electric and magnetic field vectors:

$$\vec{\mathbf{E}} = -\frac{1}{\tilde{n}} \cdot \left(\sqrt{\frac{\epsilon_0}{\kappa_0}}\right)^{-1} \hat{\mathbf{e}}_3 \times \vec{\mathbf{H}} \,, \tag{4.186}$$

$$\vec{\mathbf{H}} = \tilde{n} \cdot \left(\sqrt{\frac{\epsilon_0}{\kappa_0}}\right) \hat{\mathbf{e}}_3 \times \vec{\mathbf{E}} \,. \tag{4.187}$$

Referring to Figure 4.21, we define the perpendicular component of the electric field vector to point out of the paper in the direction of $\hat{\mathbf{e}}_\perp$. The parallel components of the electric field vectors along the directions are given by

$$\hat{\mathbf{e}}_{\|\mathrm{inc}} = \hat{\mathbf{e}}_{\mathrm{inc}} \times \hat{\mathbf{e}}_\perp \,, \tag{4.188}$$

$$\hat{\mathbf{e}}_{\|\mathrm{ref}} = \hat{\mathbf{e}}_{\mathrm{ref}} \times \hat{\mathbf{e}}_\perp \,, \tag{4.189}$$

$$\hat{\mathbf{e}}_{\|\mathrm{tra}} = \hat{\mathbf{e}}_{\mathrm{tra}} \times \hat{\mathbf{e}}_\perp \,. \tag{4.190}$$

On the basis of Eq. (4.8), the electric field vectors associated with the incident, reflected, and refracted waves can be expressed by

$$\vec{\mathbf{E}}_{\mathrm{inc}} = E_{\|\mathrm{inc}} \cdot \hat{\mathbf{e}}_{\|\mathrm{inc}} + E_{\perp\mathrm{inc}} \cdot \hat{\mathbf{e}}_{\perp\mathrm{inc}} \,, \tag{4.191}$$

$$\vec{\mathbf{E}}_{\mathrm{ref}} = E_{\|\mathrm{ref}} \cdot \hat{\mathbf{e}}_{\|\mathrm{ref}} + E_{\perp\mathrm{ref}} \cdot \hat{\mathbf{e}}_{\perp\mathrm{ref}} \,, \tag{4.192}$$

$$\vec{\mathbf{E}}_{\mathrm{tra}} = E_{\|\mathrm{tra}} \cdot \hat{\mathbf{e}}_{\|\mathrm{tra}} + E_{\perp\mathrm{tra}} \cdot \hat{\mathbf{e}}_{\perp\mathrm{tra}} \,. \tag{4.193}$$

The magnetic vectors associated with Eqs. (4.191)–(4.193) can be calculated using Eq. (4.187).

The boundary conditions for the propagation of electromagnetic waves involving two media require that the tangential components of the electric and magnetic amplitude vectors are continuous across the interface of the media. Thus, for the perpendicular components of the electric amplitude vector, we have

$$E_{0,\perp\mathrm{inc}} + E_{0,\perp\mathrm{ref}} = E_{0,\perp\mathrm{tra}} \,. \tag{4.194}$$

The equation for the corresponding magnetic field vectors can be derived on the basis of Eq. (4.187), and

$$\tilde{n}_2 \cdot \cos \Theta_{\text{inc}} \cdot E_{0,\perp \text{inc}} - \tilde{n}_2 \cdot \cos \Theta_{\text{ref}} \cdot E_{0,\perp \text{ref}} = \tilde{n}_1 \cdot \cos \Theta_{\text{tra}} \cdot E_{0,\perp \text{tra}} \ . \quad (4.195)$$

The solutions to Eqs. (4.194) and (4.195) for the perpendicular amplitudes of the electric field vector are

$$t_\perp = \frac{E_{0,\perp \text{tra}}}{E_{0,\perp \text{inc}}} = \frac{2\tilde{n}_2 \cdot \cos \Theta_{\text{inc}}}{\tilde{n}_2 \cdot \cos \Theta_{\text{inc}} + \tilde{n}_1 \cdot \cos \Theta_{\text{tra}}}$$

$$= \frac{2 \cos \Theta_{\text{inc}}}{\cos \Theta_{\text{inc}} + \tilde{m} \cdot \cos \Theta_{\text{tra}}} \ , \quad (4.196)$$

and

$$r_\perp = \frac{E_{0,\perp \text{ref}}}{E_{0,\perp \text{inc}}} = \frac{\tilde{n}_2 \cdot \cos \Theta_{\text{inc}} - \tilde{n}_1 \cdot \cos \Theta_{\text{tra}}}{\tilde{n}_2 \cdot \cos \Theta_{\text{inc}} + \tilde{n}_1 \cdot \cos \Theta_{\text{tra}}}$$

$$= \frac{\cos \Theta_{\text{inc}} - \tilde{m} \cdot \cos \Theta_{\text{tra}}}{\cos \Theta_{\text{inc}} + \tilde{m} \cdot \cos \Theta_{\text{tra}}} \ , \quad (4.197)$$

where we have used $\Theta_{\text{inc}} = \Theta_{\text{ref}}$ and $\tilde{m} = \tilde{n}_1/\tilde{n}_2$. Similarly, for the parallel component, we have

$$\cos \Theta_{\text{inc}} \cdot E_{0,\|\text{inc}} - \cos \Theta_{\text{ref}} \cdot E_{0,\|\text{ref}} = \cos \Theta_{\text{tra}} \cdot E_{0,\|\text{tra}} \ , \quad (4.198)$$

and

$$\tilde{n}_2 \cdot E_{0,\|\text{inc}} + \tilde{n}_2 \cdot E_{0,\|\text{ref}} = \tilde{n}_1 \cdot E_{0,\|\text{tra}} \ . \quad (4.199)$$

The solutions to Eqs. (4.198) and (4.199) are

$$t_\| = \frac{E_{0,\|\text{tra}}}{E_{0,\|\text{inc}}} = \frac{2\tilde{n}_2 \cdot \cos \Theta_{\text{inc}}}{\tilde{n}_1 \cdot \cos \Theta_{\text{inc}} + \tilde{n}_2 \cdot \cos \Theta_{\text{tra}}}$$

$$= \frac{2 \cos \Theta_{\text{inc}}}{\tilde{m} \cdot \cos \Theta_{\text{inc}} + \cos \Theta_{\text{tra}}} \ , \quad (4.200)$$

and

$$r_\| = \frac{E_{0,\|\text{ref}}}{E_{0,\|\text{inc}}} = \frac{\tilde{n}_1 \cdot \cos \Theta_{\text{inc}} - \tilde{n}_2 \cdot \cos \Theta_{\text{tra}}}{\tilde{n}_1 \cdot \cos \Theta_{\text{inc}} + \tilde{n}_2 \cdot \cos \Theta_{\text{tra}}}$$

$$= \frac{\tilde{m} \cdot \cos \Theta_{\text{inc}} - \cos \Theta_{\text{tra}}}{\tilde{m} \cdot \cos \Theta_{\text{inc}} + \cos \Theta_{\text{tra}}} \ . \quad (4.201)$$

Equations (4.196), (4.197), (4.200), and (4.201) are known as the Fresnel formulas; $t_\|$, $t_\perp$, $r_\|$, and $r_\perp$ are the Fresnel coefficients for transmission and reflection, respectively.

4.12.4
Radiant Energy Changes for Transmission (Plane Interface)

General Transmissivity
The dimensionless transmissivity quantifies the relative changes of the radiant energy (fluxes) during transmission through a medium and represents the ratio of transmitted to incident radiant energy fluxes. To calculate the transmission of radiant energy from one medium to another medium, let us consider a differential area element, $d^2 A$, at the interface of the two media. The differential cross section of an incident beam impinging on $d^2 A$ is

$$d^2 A_{\text{inc}} = d^2 A \cdot \cos \Theta_{\text{inc}} . \tag{4.202}$$

The differential cross section of the transmitted beam associated with the incident beam is

$$d^2 A_{\text{tra}} = d^2 A \cdot \cos \Theta_{\text{tra}} . \tag{4.203}$$

We look at Eqs. (3.28), (3.29), and (4.14) for the definitions of radiant energy flux and irradiance. Thus, taking into account the principle of energy conservation, we define the dimensionless transmissivity, assuming a nonferromagnetic medium, as

$$
\begin{aligned}
\mathcal{T}_{\|} &= \frac{\Phi_{\|\text{tra}}}{\Phi_{\|\text{inc}}} = \frac{F_{\|\text{tra}}}{F_{\|\text{inc}}} \frac{d^2 A_{\text{tra}}}{d^2 A_{\text{inc}}} \\
&= \frac{\sqrt{\epsilon_1/\kappa_0}}{\sqrt{\epsilon_2/\kappa_0}} \cdot \frac{|E_{0,\|\text{tra}}|^2}{|E_{0,\|\text{inc}}|^2} \cdot \left(\frac{d^2 A \cdot \cos \Theta_{\text{tra}}}{d^2 A \cdot \cos \Theta_{\text{inc}}} \right) \\
&= \tilde{m} \cdot \left(\frac{\cos \Theta_{\text{tra}}}{\cos \Theta_{\text{inc}}} \right) \cdot |t_{\|}|^2 \\
&= \tilde{m} \cdot \left(\frac{\cos \Theta_{\text{tra}}}{\cos \Theta_{\text{inc}}} \right) \cdot \left| \frac{2 \cos \Theta_{\text{inc}}}{\tilde{m} \cdot \cos \Theta_{\text{inc}} + \cos \Theta_{\text{tra}}} \right|^2 \\
&= \frac{\tilde{n}_1}{\tilde{n}_2} \cdot \left(\frac{\cos \Theta_{\text{tra}}}{\cos \Theta_{\text{inc}}} \right) \cdot \left| \frac{2 \tilde{n}_1 \cdot \tilde{n}_2 \cdot \cos \Theta_{\text{inc}}}{\tilde{n}_1^2 \cdot \cos \Theta_{\text{inc}} + \tilde{n}_2 \cdot q} \right|^2 ,
\end{aligned}
\tag{4.204}
$$

and

$$
\begin{aligned}
\mathcal{T}_{\perp} &= \frac{\Phi_{\perp\text{tra}}}{\Phi_{\perp\text{inc}}} = \frac{F_{\perp\text{tra}}}{F_{\perp\text{inc}}} \frac{d^2 A_{\text{tra}}}{d^2 A_{\text{inc}}} \\
&= \frac{\sqrt{\epsilon_1/\kappa_0}}{\sqrt{\epsilon_2/\kappa_0}} \cdot \left(\frac{d^2 A \cdot \cos \Theta_{\text{tra}}}{d^2 A \cdot \cos \Theta_{\text{inc}}} \right) \cdot \frac{|E_{0,\perp\text{tra}}|^2}{|E_{0,\perp\text{inc}}|^2} \\
&= \tilde{m} \cdot \left(\frac{\cos \Theta_{\text{tra}}}{\cos \Theta_{\text{inc}}} \right) \cdot |t_{\perp}|^2 \\
&= \tilde{m} \cdot \left(\frac{\cos \Theta_{\text{tra}}}{\cos \Theta_{\text{inc}}} \right) \cdot \left| \frac{2 \cos \Theta_{\text{inc}}}{\cos \Theta_{\text{inc}} + \tilde{m} \cdot \cos \Theta_{\text{tra}}} \right|^2 \\
&= \frac{\tilde{n}_1}{\tilde{n}_2} \cdot \left(\frac{\cos \Theta_{\text{tra}}}{\cos \Theta_{\text{inc}}} \right) \cdot \left| \frac{2 \tilde{n}_2 \cdot \cos \Theta_{\text{inc}}}{\tilde{n}_2 \cdot \cos \Theta_{\text{inc}} + q} \right|^2 ,
\end{aligned}
\tag{4.205}
$$

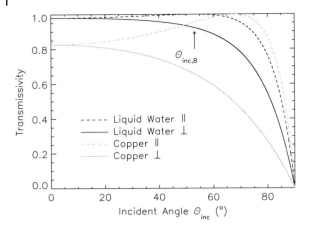

Figure 4.24 Transmissivity of liquid water as a function of angle of incident radiation Θ_{inc} as calculated from Eqs. (4.204) and (4.205). The Brewster angle for water is indicated by $\Theta_{inc,B}$.

with

$$q = \sqrt{\tilde{n}_1^2 - \tilde{n}_2^2 \cdot \sin^2 \Theta_{inc}} \ . \tag{4.206}$$

In deriving Eqs. (4.204) and (4.205), we utilized Eq. (4.4), which holds for nonferromagnetic media ($\kappa_1 = \kappa_2 = \kappa_0$) in the form of

$$\tilde{m} = \frac{\tilde{n}_1}{\tilde{n}_2} = \sqrt{\frac{\epsilon_1 \cdot \kappa_0}{\epsilon_2 \cdot \kappa_0}} = \sqrt{\frac{\epsilon_1}{\epsilon_2}} \ . \tag{4.207}$$

Thus, we have both $\mathcal{T}_\parallel$ and $\mathcal{T}_\perp$ as functions of the angle of the incident ray Θ_{inc} and of $\tilde{m}$. The formulas are illustrated in Figure 4.24.

Because each electric wave can be decomposed into a parallel and a perpendicular portion, the total transmissivity, $\mathcal{T}_{tot}$, can be derived by an appropriate average of $\mathcal{T}_\parallel$ and $\mathcal{T}_\perp$. If the incident radiation is unpolarized, then

$$\mathcal{T}_{unp,tot} = \frac{1}{2} \cdot (\mathcal{T}_\perp + \mathcal{T}_\parallel) \ . \tag{4.208}$$

Transmissivity at Normal Incidence
The transmissivity at normal incidence, $\mathcal{T}_{nor}$, with

$$\Theta_{inc,nor} = 0° \ ,$$
$$\cos \Theta_{inc,nor} = \cos \Theta_{tra,nor} = 1 \ ,$$
$$\sin \Theta_{inc,nor} = 0 \ ,$$
$$q_{nor} = \tilde{n}_1 \ , \tag{4.209}$$

is given by

$$\mathcal{T}_{\parallel nor} = \tilde{m} \cdot \left| \frac{2}{1 + \tilde{m}} \right|^2 = \frac{\tilde{n}_1}{\tilde{n}_2} \cdot \left| \frac{2\tilde{n}_2}{\tilde{n}_1 + \tilde{n}_2} \right|^2 \ . \tag{4.210}$$

By applying Eq. (4.175), we can obtain the same result for the perpendicular component, that is,

$$\mathcal{T}_{\perp \mathrm{nor}} = \tilde{m} \cdot \left| \frac{2}{1 + \tilde{m}} \right|^2 = \mathcal{T}_{\| \mathrm{nor}} = \mathcal{T}_{\mathrm{unp,tot,nor}} . \tag{4.211}$$

4.12.5
Radiant Energy Changes for Reflection (Plane Interface)

4.12.5.1 General Reflectivity

The reflected radiant energy flux is described by the reflectivity, $\mathcal{R}$, which is the ratio of reflected to incident radiant energy flux; whereas, the reflection takes place on a smooth interface. Similar to the case for transmissivity, the reflectivity, $\mathcal{R}$, is quantified by $\mathcal{R}_\|$ and $\mathcal{R}_\perp$, being the reflectivity parallel and perpendicular to the incident plane, using $\cos \Theta_{\mathrm{ref}} = \cos \Theta_{\mathrm{inc}}$:

$$
\begin{aligned}
\mathcal{R}_\| &= \frac{\Phi_{\| \mathrm{ref}}}{\Phi_{\| \mathrm{inc}}} = \frac{F_{\| \mathrm{ref}}}{F_{\| \mathrm{inc}}} \frac{\mathrm{d}^2 A_{\mathrm{ref}}}{\mathrm{d}^2 A_{\mathrm{inc}}} \\
&= \frac{\sqrt{\epsilon_2/\kappa_0}}{\sqrt{\epsilon_2/\kappa_0}} \cdot \frac{|E_{0,\| \mathrm{ref}}|^2}{|E_{0,\| \mathrm{inc}}|^2} \cdot \left(\frac{\mathrm{d}^2 A \cdot \cos \Theta_{\mathrm{ref}}}{\mathrm{d}^2 A \cdot \cos \Theta_{\mathrm{inc}}} \right) \\
&= \frac{|E_{0,\| \mathrm{ref}}|^2}{|E_{0,\| \mathrm{inc}}|^2} = |r_\||^2 \\
&= \left| \frac{\tilde{m} \cdot \cos \Theta_{\mathrm{inc}} - \cos \Theta_{\mathrm{tra}}}{\tilde{m} \cdot \cos \Theta_{\mathrm{inc}} + \cos \Theta_{\mathrm{tra}}} \right|^2 \\
&= \left| \frac{\tilde{n}_1^2 \cdot \cos \Theta_{\mathrm{inc}} - \tilde{n}_2 \cdot q}{\tilde{n}_1^2 \cdot \cos \Theta_{\mathrm{inc}} + \tilde{n}_2 \cdot q} \right|^2 ,
\end{aligned}
\tag{4.212}
$$

and

$$
\begin{aligned}
\mathcal{R}_\perp &= \frac{\Phi_{\perp \mathrm{ref}}}{\Phi_{\perp \mathrm{inc}}} = \frac{F_{\perp \mathrm{ref}}}{F_{\perp \mathrm{inc}}} \frac{\mathrm{d}^2 A_{\mathrm{ref}}}{\mathrm{d}^2 A_{\mathrm{inc}}} \\
&= \frac{\sqrt{\epsilon_2/\kappa_0}}{\sqrt{\epsilon_2/\kappa_0}} \cdot \frac{|E_{0,\perp \mathrm{ref}}|^2}{|E_{0,\perp \mathrm{inc}}|^2} \cdot \left(\frac{\mathrm{d}^2 A \cdot \cos \Theta_{\mathrm{ref}}}{\mathrm{d}^2 A \cdot \cos \Theta_{\mathrm{inc}}} \right) \\
&= \frac{|E_{0,\perp \mathrm{ref}}|^2}{|E_{0,\perp \mathrm{inc}}|^2} = |r_\perp|^2 \\
&= \left| \frac{\cos \Theta_{\mathrm{inc}} - \tilde{m} \cdot \cos \Theta_{\mathrm{tra}}}{\cos \Theta_{\mathrm{inc}} + \tilde{m} \cdot \cos \Theta_{\mathrm{tra}}} \right|^2 \\
&= \left| \frac{\tilde{n}_2 \cdot \cos \Theta_{\mathrm{inc}} - q}{\tilde{n}_2 \cdot \cos \Theta_{\mathrm{inc}} + q} \right|^2 ,
\end{aligned}
\tag{4.213}
$$

with q from Eq. (4.206). Similar to Eq. (4.203), we have also used the following equation,

$$\mathrm{d}^2 A_{\mathrm{ref}} = \mathrm{d}^2 A \cdot \cos \Theta_{\mathrm{ref}} , \tag{4.214}$$

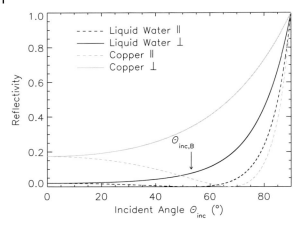

Figure 4.25 Specular reflectivity of liquid water as a function of the angle of incident radiation Θ_{inc} as calculated from Eqs. (4.212) and (4.213). The Brewster for water is indicated by $\Theta_{inc,B}$.

to derive Eqs. (4.212) and (4.213). The right sides of Eqs. (4.212) and (4.213) are functions of the angle of incident and the refractive indices. They were calculated using Eq. (4.174). Thus, we have both $\mathcal{R}_\parallel$ and $\mathcal{R}_\perp$ as functions of the angle of the incident ray Θ_{inc} and of $\tilde{m}$. Two terms exist for reflectivity because of polarization: $\mathcal{R}_\parallel$ defines reflectivity when the electric field vector is parallel to the plane of the incident electric wave, and $\mathcal{R}_\perp$ defines reflectivity when the electric field vector is perpendicular to the plane of the incident electric wave.

Figure 4.25 shows $\mathcal{R}_\parallel$ and $\mathcal{R}_\perp$ of liquid water and copper plane surfaces as functions of the incident angle, Θ_{inc}, at a wavelength of $\lambda = 0.5$ μm. The reflectivity is seen to be quite small near the normal incidence angle (nadir: $\Theta_{inc,nor} = 0°$) and to increase sharply as $\Theta_{inc} \to 90°$. Thus, a smooth water surface represents a poor reflector near nadir (i.e., normal, or $\Theta_{inc} \to 0°$) incidence, and a sharp reflector for glancing incidence ($\Theta_{inc} \to 90°$). In general, the reflectivity for vertical polarization ($\mathcal{R}_\parallel$) is much smaller than that of horizontal ($\mathcal{R}_\perp$) polarization ($\mathcal{R}_\parallel < \mathcal{R}_\perp$).

Similar to the total transmissivity, see Eq. (4.208), the total reflectivity, $\mathcal{R}_{tot}$, can be derived by an appropriate averaging of $\mathcal{R}_\parallel$ and $\mathcal{R}_\perp$. If the incident radiation is unpolarized, then

$$\mathcal{R}_{unp,tot} = \frac{1}{2} \cdot (\mathcal{R}_\perp + \mathcal{R}_\parallel) \, . \tag{4.215}$$

However, the reflected radiation is often polarized, even if the incident radiation is unpolarized.

Reflectivity at Normal Incidence

To obtain the reflectivity at normal incidence, $\mathcal{R}_{\parallel nor}$, we use Eqs. (4.209) and

$$\cos \Theta_{ref,nor} = 1 \, .$$

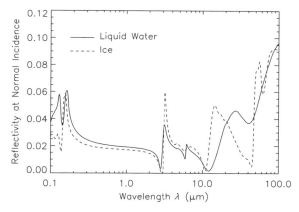

Figure 4.26 Reflectivity at normal incidence for water and ice surfaces as simulated from Eq. (4.218) as a function of wavelength. The data for the refractive index of liquid water are from Segelstein (1981), those for ice come from Warren and Brandt (2008), redrawn after Petty (2006).

Based on Eqs. (4.212) and (4.213), this gives

$$\mathcal{R}_{\parallel\mathrm{nor}} = \left| \frac{1 - \tilde{m}}{1 + \tilde{m}} \right|^2 = \left| \frac{\tilde{n}_1 - \tilde{n}_2}{\tilde{n}_1 + \tilde{n}_2} \right|^2 , \qquad (4.216)$$

and the reflectivity for the perpendicular component reduces to

$$\mathcal{R}_{\perp\mathrm{nor}} = \left| \frac{1 - \tilde{m}}{1 + \tilde{m}} \right|^2 = \mathcal{R}_{\parallel\mathrm{nor}} , \qquad (4.217)$$

and

$$\mathcal{R}_{\mathrm{unp,tot,nor}} = \frac{1}{2} \cdot (\mathcal{R}_{\perp\mathrm{nor}} + \mathcal{R}_{\parallel\mathrm{nor}}) = \left| \frac{1 - \tilde{m}}{1 + \tilde{m}} \right|^2 . \qquad (4.218)$$

For the normal incidence, no polarization distinction is needed. It is easy to show that $\mathcal{T}_{\mathrm{tot,nor}}$, see Eq. (4.211), and $\mathcal{R}_{\mathrm{tot,nor}}$ sum up to give unity, that is,

$$\mathcal{T}_{\mathrm{unp,tot,nor}} + \mathcal{R}_{\mathrm{unp,tot,nor}} = \tilde{m} \cdot \left| \frac{2}{1 + \tilde{m}} \right|^2 + \left| \frac{1 - \tilde{m}}{1 + \tilde{m}} \right|^2 = 1 , \qquad (4.219)$$

which corresponds to Eq. (3.57) without absorption.

The reflectivities of water and ice surfaces $\mathcal{R}_{\mathrm{nor}}$, with air above, are shown in Figure 4.26 for a normal incidence case.

Brewster Incident Angle

The Brewster incident angle or "polarizing angle," $\Theta_{\mathrm{inc,B}}$, is defined such that $\mathcal{R}_{\parallel B} = 0$, and the reflected electric field has only a perpendicular component with respect to the incident plane. We use Eqs. (4.212) for $\Theta_{\mathrm{ref}} \to \Theta_{\mathrm{ref,B}}$, $\Theta_{\mathrm{inc}} \to \Theta_{\mathrm{inc,B}}$, and $\Theta_{\text{tra}} \rightarrow \Theta_{\text{tra,B}}$:

$$\mathcal{R}_{\|B} = \left| \frac{\tilde{m} \cdot \cos \Theta_{\text{inc,B}} - \cos \Theta_{\text{tra,B}}}{\tilde{m} \cdot \cos \Theta_{\text{inc,B}} + \cos \Theta_{\text{tra,B}}} \right|^2 = 0 . \tag{4.220}$$

From Eq. (4.220), we get

$$\tilde{m} \cdot \cos \Theta_{\text{inc,B}} = \cos \Theta_{\text{tra,B}} . \tag{4.221}$$

We apply Eq. (4.174) to eliminate the $\cos \Theta_{\text{tra,B}}$-dependence, that is,

$$\cos \Theta_{\text{tra,B}} = \sqrt{1 - \left(\frac{\sin \Theta_{\text{inc,B}}}{\tilde{m}} \right)^2} . \tag{4.222}$$

Substitute Eq. (4.222) into Eq. (4.221) to yield the Brewster incident angle $\Theta_{\text{inc,B}}$:

$$\Theta_{\text{inc,B}} = \arcsin \sqrt{\frac{\tilde{m}^2}{\tilde{m}^2 + 1}} . \tag{4.223}$$

For an air–water interface in the visible spectral region, $\Theta_{\text{inc,B}} = 53°$.

4.12.6
Ray-Tracing Technique

To explain the ray-tracing technique, let us consider the scattering of collimated radiation by an ensemble of horizontally oriented hexagonal ice needles that randomly rotate around their axes. If the radiation source is from the local zenith direction and the scattered radiation is observed on a vertical plane (see Figure 4.27a) by neglecting the effect of particle end faces, we deal with a 2D scattering problem. This configuration of the radiation source and scattering particles can be used to explain the formation of a $22°$ moonlight halo associated with a cirrus cloud when the moon is directly overhead. Figure 4.27b illustrates various orders of reflections and refractions originating from a specific ray. The unit vectors $\hat{Z}_{\text{inc},i}$ and $\hat{Z}_{\text{sca},i}$ ($i = 1, 2, 3, \ldots$) denote the ith-order incident and outgoing directions, respectively. The unit vectors $\hat{n}_i$ ($i = 1, 2, 3, \ldots$) are normal to the particle faces where the reflections and refraction occur. Note, $\hat{n}_1$ and $\hat{n}_i$ ($i = 2, 3, 4, \ldots$) points outward and inward, respectively, in order that $\hat{n}_i$ always faces the incident rays.

According to Snel's law and the geometry shown in Figure 4.27b, it can be shown that

$$\hat{Z}_{\text{sca},1} = \hat{Z}_{\text{inc},1} - 2 \left(\hat{Z}_{\text{inc},1} \cdot \hat{n}_1 \right) \hat{n}_1 , \tag{4.224}$$

$$\hat{Z}_{\text{inc},2} = \frac{1}{\tilde{n}_{\text{re}}} \left\{ \hat{Z}_{\text{inc},1} - \left(\hat{n}_1 \cdot \hat{Z}_{\text{inc},1} \right) \hat{n}_1 \right.$$
$$\left. - \left[(\tilde{n}_{\text{re}})^2 - 1 + \left(\hat{n}_1 \cdot \hat{Z}_{\text{inc},1} \right)^2 \right]^{1/2} \hat{n}_1 \right\} , \tag{4.225}$$

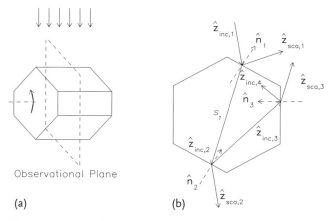

Figure 4.27 Schematic to introduce the Ray-Tracing Technique. (a) Particle's orientation with the observational plane normal to the symmetry axis of the particle; (b) ray paths.

and for $i = 2, 3, 4, \ldots$, that

$$
\hat{\mathbf{Z}}_{\text{sca},i} = \tilde{n}_{\text{re}} \left\{ \hat{\mathbf{Z}}_{\text{inc},i} - \left(\hat{\mathbf{n}}_i \cdot \hat{\mathbf{Z}}_{\text{inc},i} \right) \hat{\mathbf{n}}_i \right.
$$

$$
\left. - \left[(\tilde{n}_{\text{re}})^{-2} - 1 + \left(\hat{\mathbf{n}}_i \cdot \hat{\mathbf{Z}}_{\text{inc},i} \right)^2 \right]^{1/2} \hat{\mathbf{n}}_i \right\} ,
$$

(4.226)

$$
\hat{\mathbf{Z}}_{\text{inc},i+1} = \hat{\mathbf{Z}}_{\text{inc},i} - 2 \left(\hat{\mathbf{Z}}_{\text{inc},i} \cdot \hat{\mathbf{n}}_i \right) \hat{\mathbf{n}}_i .
$$

(4.227)

In the derivation of the preceding equations, we assume the direction of a refracted ray is determined by Snel's law via the real part of the refractive index if the particle is absorptive. An effective refractive index (Yang and Liou, 2009) should be used for an absorbing particle due to the inhomogeneous nature of refracted waves within the particle (Bohren and Huffman, 1983; Born and Wolf, 2003). However, the derivation is rather tedious and is not discussed here.

   If we assume that the incident radiation is unpolarized (note that natural light is unpolarized), the flux associated with various orders of scattered rays are given by

$$
\Phi_{\text{sca},1} = \frac{1}{2} \cdot \left[|\mathcal{R}_{\parallel,1}|^2 + |\mathcal{R}_{\perp,1}|^2 \right] \cdot F_{\text{inc}} \cdot \Delta A_i ,
$$

(4.228)

$$
\Phi_{\text{sca},2} = \frac{1}{2} \cdot \left[\left(1 - |\mathcal{R}_{\parallel,1}|^2 \right) \cdot \left(1 - |\mathcal{R}_{\parallel,2}|^2 \right) \right.
$$

$$
\left. + \left(1 - |\mathcal{R}_{\perp,1}|^2 \right) \cdot \left(1 - |\mathcal{R}_{\perp,2}|^2 \right) \right]
$$

$$
\times \exp(-4\pi \cdot \tilde{n}_{\text{im}} \cdot S_1 / \lambda) \cdot F_{\text{inc}} \cdot \Delta A_1 ,
$$

(4.229)

and

$$
\Phi_{\text{sca},i} = \frac{1}{2} \cdot \left[\left(1 - |\mathcal{R}_{\parallel,1}|^2 \right) \cdot |\mathcal{R}_{\parallel,2}|^2 \ldots |\mathcal{R}_{\parallel,i-1}|^2 \cdot \left(1 - |\mathcal{R}_{\parallel,i}|^2 \right) \right.
$$

$$
\left. + \left(1 - |\mathcal{R}_{\perp,1}|^2 \right) \cdot |\mathcal{R}_{\perp,2}|^2 \ldots |\mathcal{R}_{\perp,i-1}|^2 \cdot \left(1 - |\mathcal{R}_{\perp,i}|^2 \right) \right]
$$

$$
\times \exp[-4\pi \cdot \tilde{n}_{\text{im}} \cdot (S_1 + S_2 + \cdots + S_{i-1}) / \lambda] \cdot F_{\text{inc}} \cdot \Delta A_1 .
$$

(4.230)

In the preceding expressions, $\mathcal{R}_\parallel$ and $\mathcal{R}_\perp$ indicate Fresnel reflection coefficients for parallel and perpendicular components of the electric field, ΔA_1 indicates the cross section of the first-order incident ray, F_{inc} indicates the incident irradiance, and S_i denotes the distance D between the ith and $(i + 1)$th incidents.

If we consider a small angle region around the scattering angle ϑ, namely from $(\vartheta + \Delta\vartheta/2)$ to $(\vartheta - \Delta\vartheta/2)$, see Figure 4.27b, and add the contributions of all the outgoing rays emerging between $(\vartheta + \Delta\vartheta/2)$ and $(\vartheta - \Delta\vartheta/2)$, we obtain the flux density in the scattering direction ϑ, that is,

$$F_{\text{sca}}(\vartheta) = \frac{\sum_i \Delta\Phi_{\text{sca},i}}{2\pi \cdot R \cdot \Delta\vartheta} , \tag{4.231}$$

where R is the distance between the particle and an observational point in the far-field zone. From Eq. (4.231), the normalized form of the contributions of the externally reflected rays and transmitted to the phase function can be obtained as

$$\mathcal{P}_{\text{ref,tra}}(\vartheta) = \frac{1}{2\pi} \cdot \frac{F_{\text{sca}}}{\int_0^{2\pi} F_{\text{sca}}(\vartheta) \cdot R \, d\vartheta} . \tag{4.232}$$

Note, the normalization formalism for the 2D phase function is different from its 3D counterpart because the solid angle concept is not involved in the 2D case. The contribution of the externally reflected and transmitted rays to the extinction cross section is the projected area of the particle, that is,

$$C_{\text{ext,ref,tra}} = A_{\text{proj}} . \tag{4.233}$$

The absorption cross section is given by

$$C_{\text{abs,ref,tra}} = C_{\text{ext,ref,tra}} - \frac{\sum_i \Delta\Phi_{\text{sca},i}}{F_{\text{inc}}} , \tag{4.234}$$

where the summation is of all the outgoing or scattered rays.

4.12.7
Diffraction

In addition to refracted and reflected rays contributing to the scattered field, diffraction adds to the scattering of the incident radiation. A typical diffraction phenomenon, fundamentally due to an incomplete wave front, is associated with an incident EM wave passing through an opening in an opaque screen. In particle EM scattering, diffraction is caused by the particle's blocking the incident wave and producing an incomplete wave front. Let us consider the incident and scattering configuration, shown in Figure 4.28, for the scattering of light by a 2D hexagonal particle.

For simplicity, the electric field vector, $\vec{E}$, is assumed to be perpendicular to the scattering plane and is given by

$$\vec{E}(\vec{r}) = E(\vec{r}) \, \hat{e}_z , \tag{4.235}$$

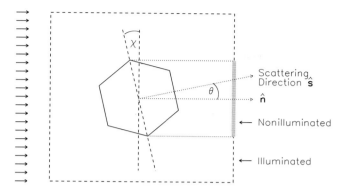

Figure 4.28 A rectangular path (dashed lines) enclosing a scattering particle, defined to account for the diffraction contribution to the scattered field. The rectangular path can be divided into illuminated and nonilluminated regions.

where the unit vector $\hat{\mathbf{e}}_z$ points out of the paper. Figure 4.28 illustrates the following relationship according to the theory of electrodynamics (Bates, 1975):

$$E(\vec{\mathbf{r}}) = \oint \hat{\mathbf{n}} \cdot \left[E_{\text{sca}}(\vec{\mathbf{r}}') \, \vec{\nabla}' \, G(\vec{\mathbf{r}}, \vec{\mathbf{r}}') - G(\vec{\mathbf{r}}, \vec{\mathbf{r}}') \, \vec{\nabla}' \, E_{\text{sca}}(\vec{\mathbf{r}}') \right] d\vec{\mathbf{r}}' , \qquad (4.236)$$

where the amplitude of the electric field on the left side of Eq. (4.236) is observed outside the rectangular path (dashed lines) enclosing the particle, and the electric field in the integrand is the scattered field. The integration is along the rectangular path, the unit vector $\hat{\mathbf{n}}$ is normal to the rectangular path, and the 2D Green's function G is given in terms of the zero-order Hankel function $H_0^{(1)}(|\vec{\mathbf{r}} - \vec{\mathbf{r}}'|)$,

$$G(\vec{\mathbf{r}}, \vec{\mathbf{r}}') = \frac{i}{4} \, H_0^{(1)}(|\vec{\mathbf{r}} - \vec{\mathbf{r}}'|) . \qquad (4.237)$$

By using the asymptotic behavior of the Hankel function in the far-field region with $k \cdot |\vec{\mathbf{r}} - \vec{\mathbf{r}}'| \to \infty$, we will obtain the scattered field as (Yang and Liou, 1995)

$$E_{\text{sca}}(\vec{\mathbf{r}}) = \sqrt{\frac{2}{\pi \cdot k \cdot R}} \cdot \exp\left[i \cdot \left(k \cdot R + \frac{3\pi}{4} \right) - i \cdot k \cdot x \right] f(\hat{\mathbf{s}}) \cdot E_{\text{inc}}(\vec{\mathbf{r}}) , \quad (4.238)$$

$$f(\hat{\mathbf{s}}) = \frac{i}{4} \cdot \frac{1}{|E_{\text{inc}}(\vec{\mathbf{r}})|} \oint \left[\frac{\partial E_{\text{sca}}(\vec{\mathbf{r}}')}{\partial n} + i \cdot k \cdot E_{\text{sca}}(\vec{\mathbf{r}}') \cdot (\hat{\mathbf{n}} \cdot \hat{\mathbf{s}}) \right]$$
$$\times \exp(-i \cdot k \cdot \hat{\mathbf{s}} \cdot \vec{\mathbf{r}}') \, dR' , \qquad (4.239)$$

where $\hat{\mathbf{s}}$ is a unit vector pointing along the scattering direction, and the incident wave is assumed to propagate in the direction of the x axis.

As shown in Figure 4.28, the rectangular path can be divided into illuminated and nonilluminated regions in the areas of unaccounted for reflection and refraction. The total electric field equals the incident field in the illuminated region and zero in the nonilluminated region. Because the total field is a linear combination of the incident and scattered fields, the scattered field along the rectangular path is given by

$$E_{sca}(\bar{r}) = \begin{cases} 0 \,, & \text{in the illuminated region} \\ -E_{inc}(\bar{r}) \,, & \text{in the nonilluminated region} \end{cases} . \tag{4.240}$$

Using the geometry shown in Figure 4.28 and Eqs. (4.239) and (4.240), we obtain the normalized form of the diffraction contribution to the phase function, that is,

$$\mathcal{P}_{dfr}(\vartheta) = \frac{2\pi \cdot (1 + \cos\vartheta)^2 \int_0^{\pi/6} [\sin(\alpha \sin\vartheta \cos\chi)]^2 \cdot (\sin\vartheta)^{-2} \, d\chi}{\int_0^{2\pi} (1 + \cos\vartheta)^2 \int_0^{\pi/6} [\sin(\alpha \sin\vartheta \cos\chi)]^2 \cdot (\sin\vartheta)^{-2} \, d\chi \, d\vartheta} \,, \tag{4.241}$$

where $\alpha = \pi \cdot D/\lambda$ is the size parameter, D is the maximum dimension of the 2D scattering particle, and χ is the angle indicating the orientation of the particle with respect to the incident radiation. In Figure 4.28, the diffraction contribution to the phase function can be shown to be the same as that in Eq. (4.241) when the electric field vector is parallel to the scattering plane.

In the conventional geometric-optics method, the total extinction cross section is twice the particle's projected area (onto a screen perpendicular to the incident direction) with half of the contribution originating from diffraction. Thus, if $\mathcal{P}_{ref,tra}$ and $\mathcal{P}_{dfr}$ indicate the normalized contributions by reflected and refracted rays and by diffraction, respectively, the total phase function is given by

$$\begin{aligned} \mathcal{P}(\vartheta) &= \frac{C_{ext}/2 - C_{abs}}{C_{ext}/2 - C_{abs} + C_{ext}/2} \cdot \mathcal{P}_{ref,tra}(\vartheta) \\ &\quad + \frac{C_{ext}/2}{C_{ext}/2 - C_{abs} + C_{ext}/2} \cdot \mathcal{P}_{dfr}(\vartheta) \\ &= \left(\frac{2\tilde{\omega} - 1}{2\tilde{\omega}}\right) \cdot \mathcal{P}_{ref,tra}(\vartheta) + \left(\frac{1}{2\tilde{\omega}}\right) \cdot \mathcal{P}_{dfr}(\vartheta) \,, \end{aligned} \tag{4.242}$$

where C_{ext} is the extinction cross section, C_{abs} is the absorption cross section, and $\tilde{\omega}$ is the single-scattering albedo.

Figure 4.29 shows the phase function of randomly oriented 2D hexagonal particles for unpolarized incident radiation. The index of refraction is 1.31, which at visible wavelengths is approximately equal to the real part of the refractive index of ice. Thus, the single-scattering albedo is one because the particles are not absorptive. Hence, diffraction and the reflected and transmitted/refracted rays contribute equally to the scattered field. Figure 4.29a shows the portion of the phase function contributed by the reflected and transmitted/refracted rays. For hexagonal particles, some incident rays are transmitted through a pair of parallel faces of the particle, and the transmitted rays are propagated in the incident direction. The contribution of these rays to the phase function is referred to as "delta-transmission" (Takano and Liou, 1989) and is represented in terms of the Dirac δ-function, an artifact of geometric optics. For simplicity, we do not deal with the delta-transmission contribution. This simplification gives rise to a sharp increase in the quantity $\mathcal{P}_{ref,tra}$ near

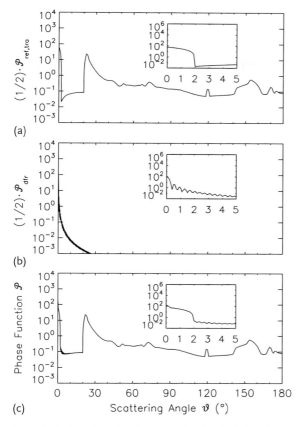

Figure 4.29 The phase function of randomly oriented 2D ice crystals with a refractive index of 1.31. (a) the contribution of reflected and transmitted/refracted rays to the phase function; (b) the contribution of diffraction to the phase function; (c) the total phase function.

the forward ($\vartheta = 0°$) direction. A pronounced peak at 22° is noted in the $\mathcal{P}_{ref,tra}$ curve and occurs from the contribution of the rays undergoing two sequential refractions without internal reflection. The scattering maximum at 154° is a direct result of rays undergoing several internal reflections. For the 3D case, Yang and Fu (2009) analyzed phase function contributions made by various orders of reflected and transmitted/refracted rays. Figure 4.29b shows the diffraction contribution to the phase function, and the diffracted energy appears to concentrate in a very narrow angular region around the forward direction. The diffraction contribution is essentially negligible for scattering angles larger than 10° for a size parameter of 500. As a result, the total phase function shown in Figure 4.29c is approximately the same as $\mathcal{P}_{ref,tra}$ at scattering angles larger than 10°. The angular distribution of diffracted energy depends on the size parameter; the larger the size parameter, the narrower the distribution of diffracted energy.

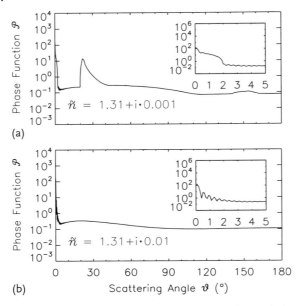

Figure 4.30 Phase functions for moderately (a) and strongly (b) absorptive randomly oriented 2D hexagonal particles.

Figure 4.30 illustrates the phase functions for moderately absorptive and strongly absorptive particles. The imaginary parts of the refractive index are 0.001 in Figure 4.30a and 0.1 in Figure 4.30b, and the corresponding single-scattering albedos are 0.5433 and 0.7046. For moderately absorptive particles, the phase function is quite smooth in the sidescattering and backscattering directions, and, although the 22° halo peak remains pronounced, the scattering maximum around 154° has been smoothed by the particle's absorption. For strongly absorptive particles, the phase function is essentially featureless, without any noticeable scattering peaks or maxima.

The previously discussed conventional geometric-optics method suffers from some shortcomings; the most prominent being the assumption that the extinction efficiency is two regardless of the size parameter. The angular distribution of the reflected and transmitted energy of nonabsorptive particles is independent of the size of the scattering particle and makes the conventional geometric-optics method inaccurate for small and moderate size parameters. Significant research has been conducted to improve the accuracy and applicability of the geometric-optics method. The principles of geometric optics have been applied to the calculation of the near-field, and rigorous electromagnetic relations have been employed to calculate the corresponding far-field (Muinonen, 1989; Yang and Liou, 1996, 1997). Furthermore, the edge effect (Jones, 1957a,b) and the above-/below-edge effects (Nussenzveig and Wiscombe, 1980, 1991), which are beyond the framework of the geometric-optics method, have been incorporated into calculations of the ex-

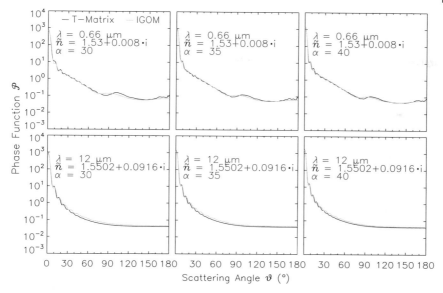

Figure 4.31 Comparison of the phase functions computed at two wavelengths from the T-matrix method and the IGOM for three size parameters, 30, 35, and 40. The size parameters are defined in terms of the radii of the equivalent volume spheres.

tinction and absorption efficiencies. After the improvements, the geometric-optics method (hereafter, the IGOM) is applicable to moderate and large size parameters with acceptable accuracy.

As an example, Figure 4.31 shows the phase functions of dust particles, assumed to be randomly oriented spheroids, computed from the exact T-matrix method (Mishchenko and Travis, 1994a) and the IGOM (Yang and Liou, 1997). The particle aspect ratio is assumed to be 1.7. As evidenced in Figure 4.31, the IGOM results closely agree with their T-matrix counterparts, particularly, at a dust particle absorptive infrared wavelength of 12 μm.

Figure 4.32 shows the extinction efficiency, single-scattering albedo, and asymmetry factor computed from the T-matrix method and the IGOM. The T-matrix method is a challenge to be applied to large size parameters, particularly with large aspect ratios, because of numerical instability and nonconvergence. The numerical limitation for large size parameters, however, is not present in the IGOM. Figure 4.32 indicates the transition from the T-matrix results to their IGOM counterparts to be continuous, thus, for practical applications, a combination of the two methods can cover the entire size parameter range.

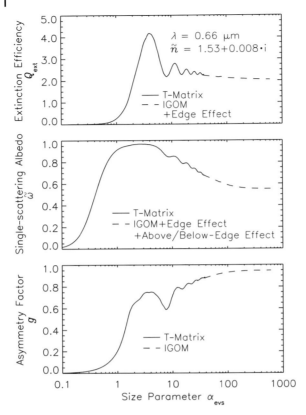

Figure 4.32 Comparison of the single-scattering properties of dust particles between the T-matrix solutions and their counterparts computed from a combination of IGOM, the edge effect, and the above/below-edge effect for the extinction efficiency and the single-scattering albedo, and the IGOM for the asymmetry factor. The particle shape is assumed to be a prolate spheroid with an aspect ratio of 1.8. The size parameter indicated in the x-axis is defined in terms of the equivalent-volume sphere.

4.13
Rainbow and Halo

Many atmospheric optical phenomena stem from the scattering of light (Greenler, 1990; Minnaert, 1993). The optical mechanisms of primary, secondary, and tertiary (third-order) rainbows are illustrated in Figure 4.33. To observe the primary and secondary rainbows, the observer and light source need to be on the same side of the scattering particles (rain drops), that is, the Sun is behind the observer while the rain is in front of the observer. The secondary rainbow is above the primary rainbow and has opposite color sequences. The tertiary rainbow is in the direction of the light source and is rarely observed in nature because of its weakness. To understand the formation of rainbows, let us consider the propagation of rays shown

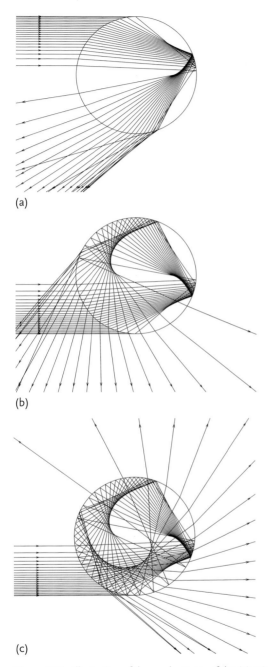

(a)

(b)

(c)

Figure 4.33 Illustration of the mechanisms of the (a) primary (first-order), (b) secondary (second-order), and (c) tertiary (third-order) rainbows. Courtesy of C. Liu

Table 4.2 The angular radii of the rainbows, data from Johnson (1954).

Color	Wavelength (nm)	Refractive index of liquid water	Angular radius of rainbows $(\pi - \Delta_{min})$		
			Primary	Secondary	Tertiary
Violet	404.7	1.3435	40° 36′	53° 36′	142° 08′
Green	546.1	1.3352	41° 46′	51° 38′	139° 10′
Yellow	577.0	1.3341	41° 58′	51° 18′	138° 42′
Red	656.3	1.3318	42° 18′	50° 40′	138° 00′

in Figure 4.33a. In this case, the deviation of a ray, Δ, is given by

$$\Delta = 2(\Theta_{inc} - \Theta_{tra}) + (\pi - 2\Theta_{tra}) , \tag{4.243}$$

where Θ_{inc} and Θ_{tra} are the incident and refractive angles of a ray impinging on the particle surface.

The minimum deviation, Δ_{min}, is the solution of the following equation:

$$\frac{d\Delta}{d\Theta_{inc}} = 0 . \tag{4.244}$$

Equation (4.244) and Snel's law lead to

$$\Theta_{inc,min} = \arccos \left[\sqrt{\frac{\tilde{n}_{lw}^2 - 1}{3}} \right] , \tag{4.245}$$

and

$$\Theta_{tra,min} = \arccos \left[\frac{2}{\tilde{n}_{lw}} \cdot \sqrt{\frac{\tilde{n}_{lw}^2 - 1}{3}} \right] , \tag{4.246}$$

where $\tilde{n}_{lw}$ is the refractive index of the liquid water droplets. At the minimum deviation, ray-bundling occurs, giving rise to a bright ring, the rainbow. The angular radius of the rainbow is determined by $(\pi - \Delta_{min})$. Table 4.2 lists the angular radii of the first three orders of rainbows. From Table 4.2, we can see that the color sequence of the primary rainbow is from red (upper) to violet (lower), and that the opposite is true for the secondary rainbow.

The geometric configuration of the primary rainbow is shown in Figure 4.34 as an illustration.

The mechanisms of halos are illustrated in Figure 4.35. Halos are produced by the bundling of rays refracted by ice crystals corresponding to the minimum deviation from the incident direction. When the rays are refracted from a side-to-side face of hexagonal ice crystals, the 22° halo is produced. The transmission of rays from a top face to a side face, or vice versa, causes the 46° halo. As evidenced from the phase functions shown in Figure 4.15, the 46° halo is much weaker as compared to the 22° halo.

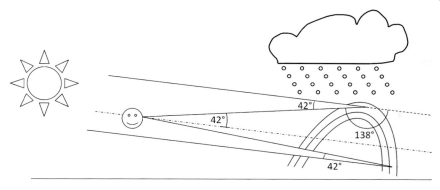

Figure 4.34 Schematic diagram of the geometric configuration of the primary (first-order) rainbow.

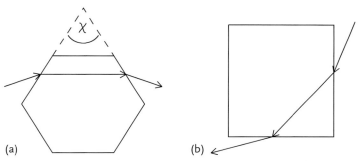

(a)

(b)

Figure 4.35 Illustration of the mechanisms of the (a) 22° and (b) 46° halos.

Table 4.3 The angular radii of the halos, data from Johnson (1954).

Color	Wavelength (nm)	Refractive index of ice	Angular radius of halos (Δ_{min}) 22° Halo	46° Halo
Violet	404.7	1.317	22° 22′	47° 16′
Yellow	577.0	1.310	21° 50′	45° 44′
Red	656.3	1.307	21° 34′	45° 06′

The angular radii of halos are the minimum deviation of the rays undergoing two sequential refractions through ice crystals and are determined by

$$\sin\left(\frac{\Delta_{min} + \chi}{2}\right) = \sin\left(\frac{\chi}{2}\right), \tag{4.247}$$

where χ is the prism angle. The scenarios shown in Figure 4.35 are for 60° and 90° prism angles. Table 4.3 lists the angular radii of the 22° and 46° halos.

The geometric constellation for halo formation is presented in Figure 4.36.

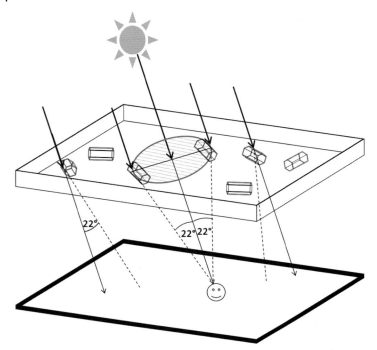

Figure 4.36 Schematic diagram of the geometric constellation of the 22° halo.

Problems

Problem 4.1 Refraction: Optical Fiber

An optical fiber consists of a glass cylinder (refractive index n_1) that is coated with a material (called cladding) with a different refractive index n_2. Assume there is no absorption within the fiber materials. The numerical aperture of the optical fiber is defined as the sine of the maximum angle of incidence of a ray that is transmitted through the fiber without loss.

a) Derive the formula for the numerical aperture dependent on the refractive indices n_1 and n_2 (setting $n_{air} = 1$). Give the corresponding maximum angle of incidence for which a fiber ($n_1 = 1.48$, $n_2 = 1.46$) is useful.

b) What happens to rays that enter the fiber at larger angles of incidence? What happens if a rat discovers a liking for cladding materials and bites holes into the cladding?

c) Assume that the fiber is illuminated with monochromatic red light with a wavelength of $\lambda = 666$ nm. What is the wavelength of the radiation inside the fiber and after it has exited the other end of the fiber?

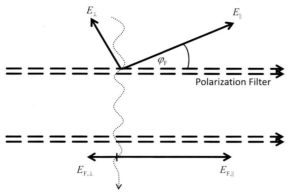

Figure 4.37 Geometry of a partly linearly polarized EM wave.

Problem 4.2 Polarization

The geometry of a partially linearly polarized EM wave passing a polarization filter is illustrated in Figure 4.37.

a) Give an equation for the irradiance $F_F(\varphi_F) = (1/2) \cdot \sqrt{\epsilon/\kappa} \cdot (E_{F,\perp}^2 + E_{F,\|}^2)$ measured behind the filter.

b) Use your result from (1) to calculate and plot the transmissivity of the polarization filter $\mathcal{T} = F_F/F_0$ dependent on the filter orientation (azimuth angle φ_F). Assume three different degrees of linear polarization $P_{\text{lin},0°} = 0/0.5/1$. Remember that the incident radiation F_0 is constant and defined by $F_0 = (1/2) \cdot \sqrt{\epsilon/\kappa} \cdot (E_\perp^2 + E_\|^2)$.

c) Using the general equation obtained in (1), show that two measurements, $F_F(\varphi_F)$ and $F_F(\varphi_F + 90°)$, at azimuth angles separated by $\Delta\varphi = 90°$, are sufficient to derive the total irradiance F_0 of the incident radiation.

Problem 4.3 Stokes Vector

Use the definition for the Stokes vector as given in the text, that is,

$$\vec{S} = \begin{pmatrix} F \\ Q \\ U \\ V \end{pmatrix} . \tag{4.248}$$

Consider the following electric wave, namely,

$$\vec{E} = E_\| \cdot \hat{e}_\| + E_\perp \cdot \hat{e}_\perp , \tag{4.249}$$

$$E_\| = E_{0,\|} \cdot \exp(i \cdot k \cdot z - i \cdot \omega_c \cdot t) , \tag{4.250}$$

$$E_\perp = E_{0,\perp} \cdot \exp(i \cdot k \cdot z - i \cdot \omega_c \cdot t + i \cdot \delta) , \tag{4.251}$$

where $E_\parallel$ and $E_\perp$ are the amplitudes of the parallel and perpendicular electric field vector components, $\hat{e}_\parallel$ and $\hat{e}_\perp$ are the unit vectors in parallel and perpendicular directions, and δ is the phase shift between the vertically and the horizontally polarized wave.

a) Derive an equation for each single component of the Stokes vector as defined above.
b) Give the Stokes vector $\vec{S}$ for:

1) $E_{0,\perp} = 0$,
2) $\delta = \pi/2$, $E_{0,\parallel} = E_{0,\perp}$,
3) $\delta = 0$, $E_{0,\parallel} = E_{0,\perp}$, and
4) $\delta = -\pi/2$, $E_{0,\parallel} = E_{0,\perp}$.

Identify how the radiation is polarized in each case and describe what the electric field vector does.

Problem 4.4 Asymmetry Factor

The dimensionless phase function $\mathcal{P}(\vartheta)$ represents the probability that a photon is scattered in a specific direction defined by the scattering angle ϑ. The asymmetry factor g describes the relative portion of radiative energy scattered in the forward versus the backward direction and is defined as the first moment of the phase function with respect to $\cos\vartheta$:

$$g = \frac{1}{4\pi} \iint\limits_{4\pi} \mathcal{P}(\vartheta) \cdot \cos\vartheta \, d^2\Omega \ . \tag{4.252}$$

With $\mu = \cos\vartheta$, we obtain

$$g = \frac{1}{2} \int\limits_{-1}^{1} \mathcal{P}(\mu) \cdot \mu \, d\mu \ . \tag{4.253}$$

a) What is the range of values that g may take? How can this be interpreted physically?
b) Derive the value of the phase function for the case of isotropic scattering (all directions equally likely, $\mathcal{P}_{\text{iso}}$=constant) using the normalization equation of $\mathcal{P}$.
c) Calculate g for isotropic scattering using your result from (b).

Problem 4.5 Rayleigh Phase Function

The Rayleigh phase function $\mathcal{P}_{\text{unp,Rayl}}$ describes the scattering of EM radiation by particles whose dimensions are small in comparison to the wavelength of the incident EM radiation whereby the incident radiation is assumed to be unpolarized.

The Rayleigh theory is an approximation for the exact Lorenz–Mie theory valid for spherical particles and applies, for example, to the scattering of visible EM radiation by air molecules or the scattering of microwave EM radiation by rain droplets. The Rayleigh phase function is given by

$$\mathcal{P}_{\text{unp,Rayl}}(\cos \vartheta) = \frac{3}{4} \cdot (1 + \cos^2 \vartheta) \,. \tag{4.254}$$

a) Check whether this phase function is properly normalized.
b) The nth moment of the phase function is defined by

$$\hat{m}_n = \frac{1}{2} \int_{-1}^{1} \mathcal{P}(\mu) \cdot \mu^n \, d\mu \,. \tag{4.255}$$

Derive a general equation for the moments of $\mathcal{P}_{\text{unp,Rayl}}$. Use the equation to calculate the first and second moment of $\mathcal{P}_{\text{unp,Rayl}}$.
c) Remember the first moment of the phase function is equal to the asymmetry factor $g = \hat{m}_1$. Confirm your result for g from above by plotting $\mathcal{P}_{\text{unp,Rayl}}$ as a function of the scattering angle ϑ. State why your plot agrees with the result from (b).
d) The degree of linear polarization of Rayleigh scattering is given by

$$\mathcal{P}_{\text{lin},0°,\text{sca,unp,Rayl}} = \frac{1 - \cos^2 \vartheta}{1 + \cos^2 \vartheta} \,. \tag{4.256}$$

For which scattering angle is the degree of linear polarization highest? What is the largest possible degree of polarization? What does that mean for the sky light and for photography (think of polarization filters)?

Problem 4.6 Henyey–Greenstein Phase Function

The Henyey–Greenstein phase function $\mathcal{P}_{\text{HG}}$ is often applied to parameterize phase functions of aerosol particles and is defined using the asymmetry factor g as

$$\mathcal{P}_{\text{HG}}(\cos \vartheta) = \frac{1 - g^2}{(1 + g^2 - 2g \cdot \cos \vartheta)^{3/2}} \,. \tag{4.257}$$

a) Show that $\mathcal{P}_{\text{HG}}$ is properly normalized.
b) For $g = 0.5$ and $g = 0.8$, plot $\mathcal{P}_{\text{HG}}$ as a function of ϑ and in a second plot as a function of $\cos \vartheta$, both with a logarithmic y axis. For which values of g (qualitatively) does $\mathcal{P}_{\text{HG}}$ get a sharp peak in the forward direction?
c) Confirm that the first moment of $\mathcal{P}_{\text{HG}}$ is indeed $\hat{m}_1 = g$ (use the short notation $\mu = \cos \vartheta$).

Problem 4.7 Lorenz–Mie Code

On the web site http://www.philiplaven.com/mieplot.htm (accessed November 2011), you will find a tool for Lorenz–Mie calculations. You can use it to simulate the scattering matrix elements for spherical particles of known refractive index. A small data base for liquid water, air, and various metals comes with the program. Download the program and familiarize yourself.

a) In the "Advanced" menu, set the refractive index of the sphere to that of water (use "IAPWS 5C Complex") and that of the surrounding medium to air. Apply a temperature of $20\,°C$. Plot the phase function for both parallel and perpendicular polarization (two different elements of the complex amplitude scattering matrix). Use the Sun as radiation source, and choose a wavelength of 500 nm. Use a size distribution with a mean radius of 1 μm and a standard deviation of 10%. You can use the screenshot function of your computer to obtain plots from this program, or export the data and plot with a different program.
b) Repeat (a) for a monodisperse size distribution (particle radius 1 μm). Explain why the results are different.
c) Make a polar plot for the same size (unpolarized). Use the View menu to switch to polar plots. Compare this with what you get for a 1 mm droplet. What is the feature at $139°$?

Problem 4.8 Reddening

Use the Lorenz–Mie tool introduced in Problem 4.7 with the same general settings. Plot the dimensionless extinction efficiency factors $Q_{ext,Mie}$ for nonabsorbing spheres (water) for a set of radii between 0.01 and 10 μm. Assuming that atmospheric aerosols have approximately the same refractive index as water, determine the range of aerosol radii that give rise to reddening when considering visible radiation (use a wavelength of 650 nm).

Problem 4.9 White Clouds

Use the Lorenz–Mie tool introduced in Problem 4.7 with the same general settings. Calculate the extinction efficiency factors $Q_{ext,Mie}$ for an aerosol particle (Aitken mode), a haze droplet (giant aerosol particle), and a cloud droplet as a function of wavelength (range 380–700 nm), and discuss why clouds are "white" and aerosols are not.

Problem 4.10 Fresnel Formulas

Please show that total transmissivity and reflectivity, described by the Fresnel formulas, add up to unity.

Problem 4.11 Halo Simulations

On the web site http://www.atoptics.co.uk/halo/halfeat.htm (accessed November 2011), you will find a tool called "HaloSim" which simulates halo phenomena based on the ray-tracing technique. Download the code, install and run. Make yourself familiar with the control panel. Vary the shape, orientation, aspect ratio (c/a) and concentration of the particles, such that:

a) A 22° halo,
b) Sun dogs,
c) An upper tangent arc, and
d) A parhelic circle

are visible in the simulations. Save the simulations as an image and mark the different halo features.

Problem 4.12 Halo Geometry

For a ray passing through a hexagonal ice crystal, as in Figure 4.35a, derive a formula for the scattering angle dependent on the angle of incidence. Use a plot program to illustrate this dependence and to show why the halo is seen at 22° from the Sun, although other scattering angles are possible as well. Note, the ice crystals can be randomly oriented in the atmosphere.

Problem 4.13 Single-Scattering Properties

This is a comprehension question (no calculation necessary). You learned that the optical properties of cloud particles are described by three quantities called single-scattering properties: the extinction cross section C_{ext}, the single-scattering albedo $\tilde{\omega}$, and the phase function, in this case characterized by the asymmetry factor g.

   Assume an optically thick cloud with a monodisperse droplet distribution ($r = 5\,\mu m$). For a wavelength of 1600 nm, the droplets are characterized by $C_{ext} = 2.0 \times 10^{-10}\,m^2$, $\tilde{\omega} = 1$, and $g = 0.85$. How are: (i) the upward irradiance at the top of the atmosphere $F_{TOA}^{\uparrow}$ reflected by the cloud layer; and, (ii) the downward irradiance at the surface $F_{BOA}^{\downarrow}$ transmitted through the cloud affected by the following changes of the single-scattering properties (assume in each case that all other parameters are kept constant):

a) Increase of C_{ext},
b) Decrease of $\tilde{\omega}$, and an
c) Increase of g.
d) Which changes of the cloud droplet size are related to (a)–(c)?

e) How do C_{ext}, $\tilde{\omega}$, and g change for a nonspherical ice crystal with a maximum geometric dimension equal to the diameter of the droplet given above? Briefly explain.

Hint: Consider the shape of the particles and assume that the real part of the refractive index of ice and liquid water to be equal at $\lambda = 1600$ nm.

5
Volumetric (Bulk) Optical Properties

In this chapter, we study the optical properties of a population of particles contained in a sufficiently small (from a practical perspective) volume element. The particles may be gas molecules, aerosol particles, cloud particles (liquid droplets and ice crystals), or precipitation particulates. The particles are characterized by different sizes, chemical compositions or refractive indices, and shapes. The volumetric optical properties, that is, the bulk optical properties of the sufficiently small volume containing many individual particles, are derived by averaging the optical properties of individual particles over the particle size distribution. Averaging gives rise to the volumetric (bulk) optical properties of an ensemble of small particles whose weights are taken into account.

The individual particles are assumed to be independent of each other in terms of their optical effects, provided that the particles are sufficiently separated, and multiple-scattering within the volume can be neglected. Furthermore, the scattered EM waves associated with the particles are incoherent.

5.1
Particle Size Distribution

5.1.1
Analytical Descriptions

The particle number size distribution (NSD) $n(D)$ is the number $\mathrm{d}N$ of particles within the maximum dimension interval $(D-\mathrm{d}D/2,\ D+\mathrm{d}D/2)$ of the small volume element and is given by

$$n(D) = \frac{\mathrm{d}N}{\mathrm{d}D} .$$ (5.1)

The units of $n(D)$ are m^{-4}, of $N(D)$ are m^{-3}, and of D are $\mu\mathrm{m}$. Alternative forms of the size distribution are used in the literature to clearly emphasize the mode structure or for other specific reasons, for example, in the form of

$$n(D) = \frac{\mathrm{d}N}{\mathrm{d}D} = \frac{1}{D} \cdot \frac{\mathrm{d}N}{\mathrm{d}\ln D} = \frac{1}{D \cdot \ln 10} \cdot \frac{\mathrm{d}N}{\mathrm{d}\log_{10} D} .$$ (5.2)

Theory of Atmospheric Radiative Transfer, First Edition. Manfred Wendisch and Ping Yang
© 2012 WILEY-VCH Verlag GmbH & Co. KGaA. Published 2012 by WILEY-VCH Verlag GmbH & Co. KGaA.

A description via either the particle surface, the particle cross section, or the particle volume is often used. For spherical particles, the surface and the volume size distributions are

$$n_{sur}(D) = \pi \cdot D^2 \cdot n(D) , \tag{5.3}$$

$$n_{vol}(D) = \frac{\pi}{6} \cdot D^3 \cdot n(D) . \tag{5.4}$$

The units are given as: $[n_{sur}(D)] = m^{-2}$; and, $[n_{vol}(D)] = m^{-1}$. The mass size distribution, $m(D)$, is obtained by multiplying $n_{vol}(D)$ by the density, ρ, of the relevant species

$$m(D) = \rho \cdot n_{vol}(D) = \frac{\pi}{6} \cdot \rho \cdot D^3 \cdot n(D) . \tag{5.5}$$

5.1.2
Integrated Microphysical Parameters

The total number of particles, N_{tot}, or particle number concentration, is given by

$$N_{tot} = \int_0^\infty n(D') \, dD' , \tag{5.6}$$

with N_{tot} in the units of m^{-3}. Integration of $m(D)$ yields the Liquid Water Content (LWC) for water droplets or the Ice Water Content (IWC) for ice crystals

$$LWC = \frac{\pi}{6} \cdot \varrho_{lw} \int_0^\infty n(D') \cdot D'^3 \, dD' , \tag{5.7}$$

$$IWC = \frac{\pi}{6} \cdot \varrho_{ice} \int_0^\infty n(D') \cdot D'^3 \, dD' , \tag{5.8}$$

where ϱ_{lw} and ϱ_{ice} represent the density of liquid water or ice, respectively. In the case of ice clouds, D' in Eq. (5.8) should be understood as the diameter of a volume-equivalent sphere of a nonspherical ice crystal. For liquid droplets, either the effective diameter, D_{eff}, or the effective radius, $r_{eff} = D_{eff}/2$, of the droplet size distribution is calculated from the ratio of the third to the second moment of the droplet size distribution (Hansen and Travis, 1974)

$$D_{eff} = 2 \cdot r_{eff} = \frac{\int_0^\infty n(D') \cdot D'^3 \, dD'}{\int_0^\infty n(D') \cdot D'^2 \, dD'} . \tag{5.9}$$

Typical values of these parameters for marine and continental clouds are given in Table 5.1. From the table, it is evident that the number concentration for continental clouds is much larger than its counterpart for marine clouds, although the LWC

Table 5.1 Typical values of N_{tot}, LWC, and D_{eff} for marine and continental clouds, from Miles et al. (2000).

Quantity	Marine clouds	Continental clouds
N_{tot} (cm^{-3})	74 ± 45	288 ± 159
LWC (g m^{-3})	0.18 ± 0.14	0.19 ± 0.21
D_{eff} (μm)	19.2 ± 4.7	10.8 ± 4.1

is similar in the two cases. Furthermore, D_{eff} is smaller for continental clouds than for marine clouds because cloud condensation nuclei are more abundant over land than over ocean.

For nonspherical particles, the effective diameter is calculated as

$$D_{eff} = \frac{3}{2} \cdot \frac{V_{int}}{A_{int}} , \tag{5.10}$$

where V_{int} and A_{int} represent the integrated volume and projected cross-sectional area on a plane perpendicular to the incident radiation of an ensemble of nonspherical particles.

5.1.3
Parameterizations

Different parametrization forms exist for size distribution. One example, often applied to aerosol particles, is the Junge distribution given by

$$\frac{dN}{d\log_{10} r} = N_0 \cdot r^{-b} , \tag{5.11}$$

where b is the Junge parameter. Here, we have $r = D/2$. The modified Gamma-distribution with three open parameters c, d, and e is often used for parameterizing the cloud droplet size distribution

$$\frac{dN}{d\log_{10} r} = N_0 \cdot r^{c+1} \cdot \exp(-d \cdot r^e) . \tag{5.12}$$

Another form of the Gamma-distribution given by Hansen (1971) is

$$\frac{dN}{dr} = N_0 \cdot r^{(1-3v_{eff})/v_{eff}} \cdot \exp\left(-\frac{r}{r_{eff} \cdot v_{eff}}\right), \tag{5.13}$$

where v_{eff} is the dimensionless effective variance and r_{eff} is the effective radius defined by Eq. (5.9). The logarithmic normal distribution of aerosol particles or clouds is given in the form of the number size distribution as

$$\frac{dN}{d\log_{10} D} = \sum_{i=1}^{2} \frac{N_{tot,i}}{\sqrt{2\pi} \cdot \log_{10} \sigma_i} \cdot \exp\left[-\frac{1}{2} \cdot \left(\frac{\log_{10} D - \log_{10} D_{n,i}}{\log_{10} \sigma_i}\right)^2\right]. \tag{5.14}$$

Table 5.2 Typical values of the mode parameters for fog droplet size distributions (logarithmic normal distribution).

Quantity (μm)	Small droplet mode $i = 1$	Large droplet mode (Drizzle) $i = 2$
$D_{n,i}$	—	23.5 ± 3.2
$D_{m,i}$	8.4 ± 1.8	29.2 ± 4.6
σ_i	1.51 ± 0.11	1.29 ± 0.06

The mass size distribution of the logarithmic normal distribution is

$$\frac{\mathrm{d}M}{\mathrm{d}\log_{10} D} = \sum_{i=1}^{2} \frac{\mathrm{LWC}_i}{\sqrt{2\pi} \cdot \log_{10}\sigma_i} \cdot \exp\left[-\frac{1}{2} \cdot \left(\frac{\log_{10} D - \log_{10} D_{m,i}}{\log_{10}\sigma_i}\right)^2\right].$$
(5.15)

Typical values of the mode parameters for fog droplet size distributions are given in Table 5.2 from Wendisch et al. (1998).

5.2
Volumetric (Bulk) Scattering, Absorption, and Extinction

The volumetric (bulk) optical properties of an ensemble of particles are obtained by the integration of the single-scattering properties, including the extinction, scattering, and absorption cross sections, C_{ext}, C_{sca}, and C_{abs}, in units of m^2; the single-scattering albedo $\tilde{\omega}$ (dimensionless); the asymmetry factor g (dimensionless); or the phase function p in units of sr^{-1}. The properties are weighted by the number size distribution $\mathrm{d}N/\mathrm{d}\log D$. The procedure corresponds to a number size distribution-weighted averaging of the single-scattering optical properties.

The spectral volumetric (bulk) extinction coefficient $\langle b_{ext}(\lambda)\rangle$ in units of m^{-1} is calculated by

$$\langle b_{ext}(\lambda)\rangle = \int_0^\infty C_{ext}(\lambda, D') \cdot \frac{\mathrm{d}N}{\mathrm{d}\log D}(D')\,\mathrm{d}\log D'.$$
(5.16)

In the preceding equation, the symbol $\langle \ldots \rangle$ is used to distinguish bulk optical properties from the single-particle properties symbolized without $\langle \ldots \rangle$. The volumetric (bulk) scattering coefficient $\langle b_{sca}(\lambda)\rangle$ in units of m^{-1} is calculated by

$$\langle b_{sca}(\lambda)\rangle = \int_0^\infty C_{sca}(\lambda, D') \cdot \frac{\mathrm{d}N}{\mathrm{d}\log D}(D')\,\mathrm{d}\log D'.$$
(5.17)

Correspondingly, the volumetric (bulk) absorption coefficient $\langle b_{abs}(\lambda) \rangle$, in units of m^{-1}, is calculated by

$$\langle b_{abs}(\lambda) \rangle = \int_0^\infty C_{abs}(\lambda, D') \cdot \frac{dN}{d\log D}(D')\, d\log D'$$

$$= \langle b_{ext} \rangle - \langle b_{sca} \rangle. \tag{5.18}$$

In a similar manner, the volumetric (bulk) single-scattering albedo, $\langle \tilde{\omega}(\lambda) \rangle$, and the volumetric (bulk) asymmetry factor, $\langle g(\lambda) \rangle$, both dimensionless, are obtained by

$$\langle \tilde{\omega}(\lambda) \rangle = \frac{\int_0^\infty \tilde{\omega}(\lambda, D') \cdot C_{ext}(\lambda, D') \cdot \frac{dN}{d\log D}(D')\, d\log D'}{\langle b_{ext}(\lambda) \rangle}, \tag{5.19}$$

and

$$\langle g(\lambda) \rangle = \frac{\int_0^\infty g(\lambda, D') \cdot C_{sca}(\lambda, D') \cdot \frac{dN}{d\log D}(D')\, d\log D'}{\langle b_{sca}(\lambda) \rangle}, \tag{5.20}$$

with $C_{sca} = \tilde{\omega} \cdot C_{ext}$. Additionally, the volumetric (bulk) phase function $\langle p(\lambda) \rangle$, in units of sr^{-1}, is calculated by a similar weighting procedure, that is,

$$\langle p(\lambda, \vartheta) \rangle = \frac{\int_0^\infty p(\lambda, \vartheta, D') \cdot C_{sca}(\lambda, D') \cdot \frac{dN}{d\log D}(D')\, d\log D'}{\langle b_{sca}(\lambda) \rangle}. \tag{5.21}$$

An example of calculation results based on cirrus particle size distribution measurements taken during the Cirrus Regional Study of Tropical Anvils and Cirrus Layers – Florida Area Cirrus Experiment (CRYSTAL-FACE) on 23 July 2002, is shown in Figure 5.1. Figure 5.1 and part of the discussion in this example are adapted from Wendisch et al. (2005). Profiles of the volumetric extinction coefficient, $\langle b_{ext}(\lambda) \rangle$, are shown in Figure 5.1a. The volumetric asymmetry factor, $\langle g(\lambda) \rangle$, is presented in Figure 5.1b, and the volumetric single-scattering albedo, $\langle \tilde{\omega}(\lambda) \rangle$, is depicted in Figure 5.1c. Different ice crystal habits are assumed in the simulations, see Figure 5.2. The curve notation is given in Table 5.3.

Figure 5.1a shows that $\langle b_{ext}(\lambda) \rangle$ is largest for spheres and smallest for aggregates. The influence of shape indicated by the spread between the curves is approximately 60%. Figure 5.1b,c show that there is almost no height dependence of both $\langle g(\lambda) \rangle$ and $\langle \tilde{\omega}(\lambda) \rangle$ throughout the cloud. $\langle g(\lambda) \rangle$ is largest for spheres and smallest for aggregates, which reproduces the findings of Figures 4.17b and 4.18b. $\langle \tilde{\omega}(\lambda) \rangle$ is anticorrelated with $\langle b_{ext}(\lambda) \rangle$. The spread of the the the curves, quantifying shape effects, is approximately 10%.

Figure 5.3 is an example of the volumetric (bulk) phase function for a particle population, $\langle p(\lambda) \rangle$, with different average particle sizes. The volumetric (bulk) phase functions, as calculated from the Lorenz–Mie theory and integrated using Eq. (5.21), are depicted. A gamma-distribution, Eq. (5.13), was applied to calculate the size distribution. Compared to the phase functions of individual particles, the volumetric (bulk) functions are smoother (see Figure 4.15).

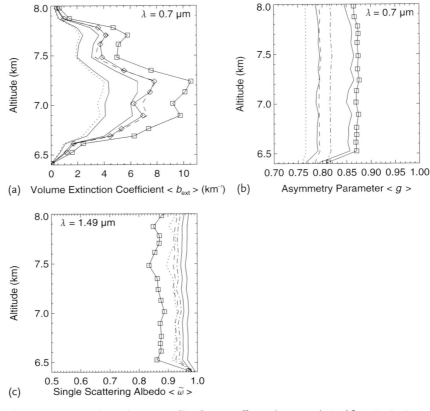

Figure 5.1 Ice crystal populations: Profile of the volumetric (bulk) extinction coefficient $\langle b_{ext}(\lambda) \rangle$ (a), asymmetry factor $\langle g(\lambda) \rangle$ (b), and single-scattering albedo $\langle \tilde{\omega}(\lambda) \rangle$ (c) at a wavelength of $\lambda = 0.7$ μm (a, b) and in the ice absorption band at $\lambda = 1.49$ μm (c). The curve notation is the same as in Figure 4.17. Additionally, the volumetric (bulk) extinction coefficient $b_{ext,vis}$ as derived from in situ instruments is included (solid lines with open diamonds in (a)). The size distribution data were taken on 23 July 2002 (cirrus of moderate optical thickness). Adapted from Wendisch et al. (2005) with permission of the American Geophysical Union © AGU.

Table 5.3 Curve notation for Figure 5.1.

Crystal habit	Indication
Spheres	Solid lines with open squares
Columns	Dashed lines
Hollow columns	Dash-dot lines
Plates	Dash-dot–dot-dot lines
Bullet rosettes	Solid lines
Aggregates	Dotted lines

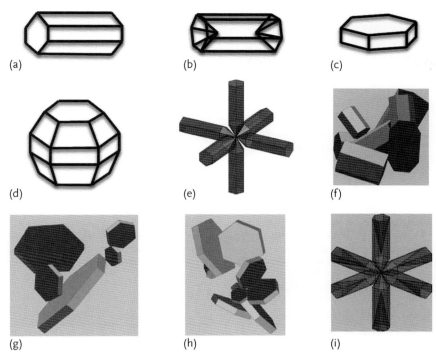

Figure 5.2 Crystal shapes assumed for the simulation results presented in Figure 5.1. (a) Solid column, (b) hollow column, (c) plate, (d) droxtal, (e) bullet rosette, (f) column aggregate, (g,h) plate aggregates, (i) hollow column rosette. Courtesy of Y. Xie and Z. Zhang.

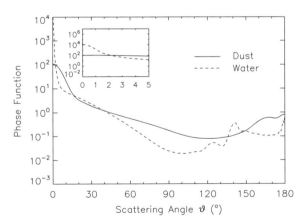

Figure 5.3 Comparison of spherical phase functions (volumetric) for dust particles and a water cloud with different average sizes of the size distribution. The size distribution in the form given by Eq. (5.13) was applied with an effective variance of 0.1 and an effective radius r_{eff} of 1 μm for the dust particle size distribution and 12 μm for the liquid water droplet. The wavelength is 0.532 μm.

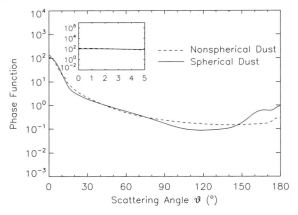

Figure 5.4 Comparison of spherical and nonspherical phase functions (volumetric) for dust particles. The size distribution in the form given by Eq. (5.13) was applied with the effective variance $\nu_{eff} = 0.1$ and an effective radius r_{eff} of 1 μm for the dust particle size distribution. The wavelength is 0.532 μm. Spheroids (with a ratio of long to small axis of 1.7) were assumed for the simulation of the nonspherical phase function of the dust size distribution.

Another example of the volumetric (bulk) phase functions for spherical and nonspherical dust particle size distributions is shown in Figure 5.4. As evidenced in the figure, the nonspherical dust particles scatter more in the sideward direction when compared to the phase function simulated by the Lorenz–Mie theory.

In the following chapters, we will omit using $\langle \ldots \rangle$ for the volumetric (bulk) optical properties.

Problems

Problem 5.1 Gamma Number Size Distribution

For ice crystal populations, a Gamma function is commonly used to represent the crystal number size distribution (NSD). A simplified special case of the Gamma NSD is the exponential NSD defined by the total particle concentration N_{tot} and the mean particle diameter $\overline{D}$, that is,

$$n(D) = \frac{dN}{dD} = \frac{N_{tot}}{\overline{D}} \cdot \exp\left(-\frac{1}{\overline{D}} \cdot D\right). \tag{5.22}$$

a) Show that $\overline{D}$ is the first moment of the NSD. First normalize the NSD by dividing by N_{tot}.
b) Use the definition of the effective diameter D_{eff} as the ratio of the third moment to the second moment of the NSD (keep in mind that this holds only for spherical particles) and find a relationship between $\overline{D}$ and D_{eff}.

c) With regard to the result of (b), which particle sizes contribute more to the absorption of the entire particle population?

Hint: The following integrals will help.

$$\int x^3 \cdot \exp(-x)\, dx = -\exp(-x) \cdot (x^3 + 3x^2 + 6x + 6),\qquad (5.23)$$

$$\int x^2 \cdot \exp(-x)\, dx = -\exp(-x) \cdot (x^2 + 2x + 2),\qquad (5.24)$$

$$\int x \cdot \exp(-x)\, dx = -\exp(-x) \cdot (x + 1).\qquad (5.25)$$

Problem 5.2 Ice Crystals

A cirrus cloud was observed with an ice water content of $IWC = 500\,\mathrm{mg\,m^{-3}}$ and a total crystal concentration of $N_{tot} = 1.0 \times 10^4\,\mathrm{l^{-1}}$. Assume a monodisperse crystal population with a mean projected area $\overline{A}_{proj} = 2.0 \times 10^{-9}\,\mathrm{m^2}$ for each ice crystal.

a) Calculate the effective diameter D_{eff} of the ice crystals.
b) The single-scattering properties of ice crystals are difficult to derive. Therefore, an approximation can be applied for radiative transfer simulations by replacing the nonspherical ice crystals by spheres. Note that the scattering properties of a sphere can be exactly computed from the Lorenz–Mie theory. Apply this method to the given cirrus cloud assuming spheres with

1) a volume equivalent to the ice crystals, and
2) assuming spheres of equivalent projected area.

Calculate the corresponding effective diameter D_{eff} in both cases.
c) Compare the resulting D_{eff} and state if the two methods underestimate or overestimate the absorption of the cirrus cloud. How must the spheres be constructed in order to avoid the differences in the absorption?

6
Radiative Transfer Equation

In Chapter 4, we discussed the optical properties of individual particles (illustrated schematically on the left side of Figure 6.1). These particles can be air molecules, spherical particles (e.g., aerosol particles, cloud droplets, precipitation droplets), or nonspherical particles (e.g., dust particles, ice crystals, large precipitation particulates). In Chapter 5, we considered spherical and nonspherical particles within a small cube and neglected the optical interaction between the particles (shown in the center of Figure 6.1). We used the particle size distribution to derive volumetric (bulk) scattering, absorption, and extinction properties of the particles. In this chapter, we will consider a number of small air cubes that form a vertical air column composed of air molecules, aerosol particles, and cloud or precipitation particulates (see the right side of Figure 6.1). Furthermore, the vertical variations of the optical properties within the column represent the counterparts of a plane-parallel atmosphere. We will account for interactions within a layer and between layers of a plane-parallel atmosphere; however, we will ignore emission and treat only multiple-scattering and absorption. The main objective of this chapter will be to derive and interpret the radiative transfer equation (RTE). Specifically, we

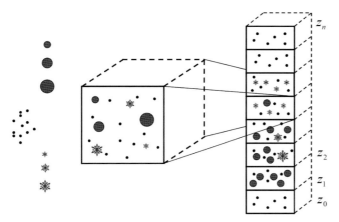

Figure 6.1 Introductory schematic diagram. Gas molecules, small black dots; aerosol particles and cloud droplets, gray spheres; and, ice particles, nonspherical crystals.

Theory of Atmospheric Radiative Transfer, First Edition. Manfred Wendisch and Ping Yang.
© 2012 WILEY-VCH Verlag GmbH & Co. KGaA. Published 2012 by WILEY-VCH Verlag GmbH & Co. KGaA.

will employ a phenomenological approach to deduce the RTE and, consequently, present the general solutions to the RTE.

6.1
Optical Thickness

We consider a plane-parallel atmosphere with the optical properties only allowed to vary in the vertical direction. For the extinction coefficient case, we have

$$b_{ext}(\lambda, \vec{r}) = b_{ext}(\lambda, z) . \tag{6.1}$$

We introduce the optical thickness as a vertical coordinate defined by

$$\tau(\lambda, z) = \int_{z}^{\infty} b_{ext}(\lambda, z') \, dz' . \tag{6.2}$$

τ is a function of altitude, z, above the surface, and we use a simple coordinate transformation. Note that τ does not depend on μ. At the surface, we have

$$\tau^{*}(\lambda) \equiv \tau(\lambda, z = 0) = \int_{0}^{\infty} b_{ext}(\lambda, z') \, dz' . \tag{6.3}$$

At the top of the atmosphere (TOA), $z \to \infty$ and $\tau \to 0$. Thus, τ varies between a value of τ^{*} at the surface and zero at the TOA. The optical thickness between the two levels z_1 and z_2 (with $z_1 < z_2$) is given by

$$\tau(\lambda, z_1) - \tau(\lambda, z_2) = \int_{z_1}^{z_2} b_{ext}(\lambda, z') \, dz' . \tag{6.4}$$

To quantify the radiation propagation within a medium, we can use optical thickness rather than geometric thickness to specify the radiation field variation within the medium.

6.2
Lambert–Bouguer Law

6.2.1
Differential and Exponential Forms

In the literature, the Lambert–Bouguer law is sometimes known as Beer's law. This law describes the extinction (attenuation) of spectral radiance, I_λ, along a slant path of length s through a medium without accounting for the contributions of multiple-

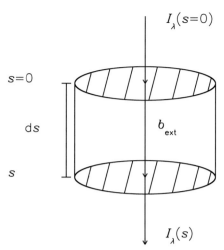

$I_\lambda(s=0)$

$s=0$

ds

b_{ext}

s

$I_\lambda(s)$

Figure 6.2 Illustration of the Lambert–Bouguer law.

scattering and emission, see Figure 6.2. Within a phenomenological framework, the change of spectral radiance I_λ is given by

$$d I_\lambda = -b_{\text{ext}}(\lambda, s) \, ds \cdot I_\lambda , \tag{6.5}$$

where $b_{\text{ext}}(\lambda, s) = b_{\text{sca}}(\lambda, s) + b_{\text{abs}}(\lambda, s)$ is the volumetric (bulk) extinction coefficient in units of m^{-1}. Equation (6.5) is an expression of the Lambert–Bouguer law in differential form. Integration of Eq. (6.5) yields an exponential decrease in the radiance, that is,

$$I_\lambda(s) = I_\lambda(s = 0) \cdot \exp\left[-\int_0^s b_{\text{ext}}(\lambda, s') \, ds' \right] . \tag{6.6}$$

The preceding equation represents the exponential form of the Lambert–Bouguer law. Equation (6.6) describes the attenuation of the radiance associated with a collimated radiation beam, see Section 3.2, without consideration of contributions by multiple-scattering and emission. To express the law in terms of τ, we consider the upward (positive μ, see Figure 3.6) radiation in the atmosphere and we have

$$dz = \cos\theta \, ds = \mu \, ds , \tag{6.7}$$

where $\mu = \cos\theta$. Because τ and z are measured downward and upward, respectively, from Eq. (6.2) we obtain

$$d\tau(\lambda, z) = -b_{\text{ext}}(\lambda, z) \, dz . \tag{6.8}$$

Incorporating Eq. (6.7) and Eq. (6.8) into the right side of Eq. (6.5) and integrating the resultant equation, we get

$$I_\lambda[\tau(\lambda, z)] = I_\lambda[\tau(\lambda, z')] \cdot \exp\left[-\frac{\tau(\lambda, z') - \tau(\lambda, z)}{\mu} \right]; \quad \text{for} \quad z > z' . \tag{6.9}$$

6.2.2
Application to Direct Solar Irradiance $S_{dir,\lambda}$

Direct solar radiation, distinguished from the diffuse component resulting from multiple-scattering events in the atmosphere, can be treated as collimated radiation because the Sun and the Earth are separated by a mean distance on the order of 1.5×10^{11} m. Thus, we can apply the Lambert–Bouguer law to solve the transfer of direct solar radiation. To do so, we first define the differential slant path distance, ds, in relation to the differential vertical distance, dz, see Figure 6.3, for downward (negative μ, see Figure 3.6) direct solar radiation

$$ds = -\frac{dz}{\cos \theta_0} = -\frac{dz}{\mu_0} \, , \tag{6.10}$$

where θ_0 is the solar zenith angle and the negative sign indicates the direct solar radiation propagates downward.

Using Eqs. (6.10) and (6.5) and replacing the radiance, I_λ, by the direct solar irradiance, $S_{dir,\lambda}$, we have

$$\frac{d S_{dir,\lambda}}{S_{dir,\lambda}} = -b_{ext}(\lambda, z)\left(-\frac{dz}{\mu_0}\right) = b_{ext}(\lambda, z)\frac{dz}{\mu_0} \, . \tag{6.11}$$

From this relationship, we may replace the slant path distance, s, by the geometric altitude, z, and subsequently the optical thickness, τ, may be used as the vertical

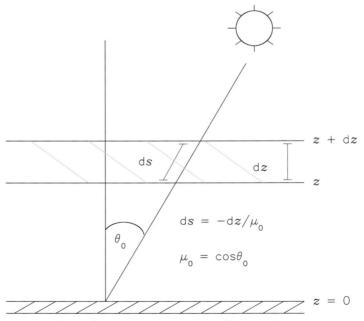

Figure 6.3 Slant and vertical paths in a plane-parallel atmosphere.

coordinate instead of the geometric height, z. We obtain the irradiance change from Eqs. (6.11) and (6.8), by

$$dS_{\text{dir},\lambda} = -\frac{d\tau(\lambda, z)}{\mu_0} \cdot S_{\text{dir},\lambda} . \tag{6.12}$$

Consequently, the Lambert–Bouguer law as a function of τ instead of slant path s reads

$$S_{\text{dir},\lambda}[\tau(\lambda, z)] = S_{\text{dir},\lambda,\text{TOA}} \cdot \exp\left[-\frac{\tau(\lambda, z)}{\mu_0}\right], \tag{6.13}$$

where

$$S_{\text{dir},\lambda,\text{TOA}} = S_{\text{dir},\lambda}(\tau = 0) . \tag{6.14}$$

The quantity $S_{\text{dir},\lambda,\text{TOA}}$ represents the direct solar spectral irradiance at the TOA. Equation (6.13) fully describes the attenuation (i.e., extinction) of direct solar irradiance within the atmosphere. The direct solar radiance incident at the TOA, $S_{\text{dir},\lambda,\text{TOA}}$, is exponentially attenuated with respect to the spectral optical thickness τ. For a low Sun (large θ_0 and small μ_0), the attenuation is quite substantial due to a long ray path-length.

6.3
General Formulation of the RTE

6.3.1
Spectral Photon Density Function

A photon is considered an infinitesimally small physical entity (a package of energy) with zero rest mass and with energy $E_{\text{phot}} = h \cdot \nu$, see Eq. (3.26). Additionally, $\nu = c/\lambda = c \cdot \tilde{\nu}$ with frequency ν and wavenumber $\tilde{\nu}$ defined as $\tilde{\nu} = 1/\lambda$. The units are $[\nu] = \text{s}^{-1}$, $[h] = \text{J s}$, and $[E_{\text{phot}}] = \text{J}$. The radiant energy propagates into the direction $\hat{s}$ and undergoes interactions with matter, either by scattering or by absorption-emission. Between the interactions, the EM radiation propagates at the speed of light, c, and does not change its frequency, ν. In this text, we will only consider elastic scattering, that is, the induced or scattered radiation due to the interaction between the incident radiation and matter has the same frequency as its incident counterpart.

We define the spectral photon density function $\xi_\lambda = \xi_\lambda(\vec{r}, \hat{s}, t)$ by

$$d^6 n_{\text{phot}} = \xi_\lambda(\vec{r}, \hat{s}, t) \, d^3 V \, d^2 \Omega \, d\lambda , \tag{6.15}$$

with the differential number of photons $d^6 n_{\text{phot}}$ (dimensionless, a function of $\vec{r}$, $\hat{s}$, and t) contained within a six-dimensional (6D) infinitesimal element specified by the volume element $d^3 V$ (m^3) centered at $\vec{r}$; the solid angle interval $d^2 \Omega$ (sr) around the propagation direction $\hat{s}$; and, the wavelength interval $d\lambda$ (μm) in the interval $[\lambda, \lambda + d\lambda]$. From Eq. (6.15), it becomes straightforward to arrive at the spectral photon density function given by

$$\xi_\lambda(\vec{r}, \hat{s}, t) = \frac{d^6 n_{phot}}{d^3 V \, d^2 \Omega \, d\lambda} . \tag{6.16}$$

The spectral photon density function in the wavelength domain, ξ_λ, is given in units of $m^{-3} \, sr^{-1} \, \mu m^{-1}$, and its counterpart in the frequency domain, ξ_ν, is in units of $m^{-3} \, sr^{-1} \, Hz^{-1}$. ξ_λ is the photon number concentration in a 6D volume.

From Eq. (3.40), the radiant energy propagating within the solid angle element $d^2 \Omega$ and within $d^3 V$ can be expressed in terms of radiance as

$$d^6 E_{rad} = \frac{I_\lambda(\vec{r}, \hat{s}, t)}{c} d^3 V \, d^2 \Omega \, d\lambda . \tag{6.17}$$

We can also express the preceding radiant energy using the differential number of photons $d^6 n_{phot}$ contained within a 6D infinitesimal element by

$$\begin{aligned} d^6 E_{rad} &= d^6 n_{phot} \cdot E_{phot} \\ &= d^6 n_{phot} \cdot h \cdot \frac{c}{\lambda} \\ &= h \cdot \frac{c}{\lambda} \cdot \xi_\lambda(\vec{r}, \hat{s}, t) \, d^3 V \, d^2 \Omega \, d\lambda . \end{aligned} \tag{6.18}$$

Comparing Eqs. (6.17) and (6.18), we obtain

$$\frac{I_\lambda(\vec{r}, \hat{s}, t)}{c} = h \cdot \frac{c}{\lambda} \cdot \xi_\lambda(\vec{r}, \hat{s}, t) , \tag{6.19}$$

or

$$\xi_\lambda(\vec{r}, \hat{s}, t) = I_\lambda(\vec{r}, \hat{s}, t) \cdot \frac{\lambda}{h \cdot c^2} . \tag{6.20}$$

In the frequency domain, the relation is

$$\xi_\nu(\vec{r}, \hat{s}, t) = I_\nu(\vec{r}, \hat{s}, t) \cdot \frac{1}{c \cdot h \cdot \nu} . \tag{6.21}$$

If the contributions due to multiple-scattering and emission are neglected, the spectral photon density variation follows the Lambert–Bouguer law in the form of

$$\xi_\lambda(s) = \xi_\lambda(s = 0) \cdot \exp\left[-\int_0^s b_{ext}(\lambda, s') \, ds' \right] , \tag{6.22}$$

where $b_{ext}(\lambda, s)$ is the volumetric (bulk) extinction coefficient in units of m^{-1}. Equation (6.22) follows by introducing $I_\lambda = h \cdot c^2/\lambda \cdot \xi_\lambda$, see Eq. (6.19), into Eq. (6.6). Pomraning (1973) showed that if a medium is homogeneous, that is, b_{ext} is independent of the path-length s, the mean free path-length $\bar{s}_{\text{ext}}$ is

$$
\begin{aligned}
\bar{s}_{\text{ext}} &= \frac{\int_0^\infty s' \cdot \xi_\lambda(s')\,\mathrm{d}s'}{\int_0^\infty \xi_\lambda(s')\,\mathrm{d}s'} \\
&= \frac{\int_0^\infty s' \cdot \xi_\lambda(s'=0) \cdot \exp\left[-b_{\text{ext}}(\lambda) \cdot s'\right]\,\mathrm{d}s'}{\int_0^\infty \xi_\lambda(s'=0) \cdot \exp\left[-b_{\text{ext}}(\lambda) \cdot s'\right]\,\mathrm{d}s'} \\
&= \frac{1}{b_{\text{ext}}(\lambda)} \; .
\end{aligned}
\tag{6.23}
$$

From a statistical perspective, $1/\bar{s}_{\text{ext}}$ is the probability that a photon will be lost within a unit distance along the propagation direction. Furthermore, we can define the mean free path-lengths, $\bar{s}_{\text{sca}}$ and $\bar{s}_{\text{abs}}$, associated with scattering and absorption in the form

$$
\bar{s}_{\text{sca}} = \frac{1}{b_{\text{sca}}(\lambda)} \; ,
\tag{6.24}
$$

and

$$
\bar{s}_{\text{abs}} = \frac{1}{b_{\text{abs}}(\lambda)} \; .
\tag{6.25}
$$

Thus, the relation $b_{\text{ext}}(\lambda) = b_{\text{sca}}(\lambda) + b_{\text{abs}}(\lambda)$ leads to

$$
\frac{1}{\bar{s}_{\text{ext}}} = \frac{1}{\bar{s}_{\text{sca}}} + \frac{1}{\bar{s}_{\text{abs}}} \; .
\tag{6.26}
$$

As explained by Pomraning (1973), the physical meaning of Eq. (6.26) is that the probability of photon lost in the extinction process is the sum of the probabilities of photon scattering and absorbtion.

6.3.2
Radiative Transfer Equation in Scattering Media

Traditionally, the radiative transfer equation is derived from phenomenological theory. A rigorous derivation of this basic equation from the fundamental EM theory was not available until the work by Mishchenko et al. (2002, 2006), Mishchenko (2008), and Tsang and Kong (2001), who proved that the usual phenomenological theory of radiative transfer is indeed a corollary of Maxwell's equations. However, the linkage between the radiative transfer equation and the Maxwell equations requires advanced knowledge in EM theory and a complicated mathematical derivation and will not be addressed in this text.

Following Pomraning (1973) and Zdunkowski et al. (2007), we adopted a phenomenological approach analogous to hydrodynamic analysis to introduce the equation of radiative transfer. In hydrodynamics, the Eulerian and Lagrangian methods can be applied to the study of fluid motion. In the former approach, fluid motion is observed at fixed spatial locations through which the fluid passes; whereas, in the latter approach, fluid motion was analyzed following a certain fluid

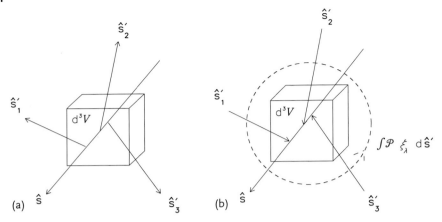

Figure 6.4 Schematic view of out-scattering (a) and in-scattering (b) processes. Partly adapted from Zdunkowski et al. (2007). Courtesy of A. Bott and T. Trautmann.

parcel that moves as a function of space and time. In the following discussion, we employ the Eulerian method. Specifically, we consider the variation of the number of photons within a time-independent 6D volume element $d^3 V d^2 \Omega \, d\lambda$, where

$$d^3 V = dx \, dy \, dz \,, \tag{6.27}$$

$$d^2 \Omega = \sin \theta \, d\theta \, d\varphi \,. \tag{6.28}$$

We denote the number of the photons, $d^6 n_{\text{phot}}$, within the preceding 6D volume element using Eq. (6.15). Because $d^3 V d^2 \Omega \, d\lambda$ is independent of time, the temporal variation of the photon number in the 6D volume is quantified by

$$
\begin{aligned}
d^6 \dot{n}_{\text{phot}} = \frac{\partial d^6 n_{\text{phot}}}{\partial t} &= \frac{\partial}{\partial t} \left[\xi_\lambda (\mathbf{r}, \hat{\mathbf{s}}, t) \, d^3 V \, d^2 \Omega \, d\lambda \right] \\
&= \dot{\xi}_\lambda (\mathbf{r}, \hat{\mathbf{s}}, t) \, d^3 V \, d^2 \Omega \, d\lambda \,.
\end{aligned}
\tag{6.29}
$$

The variation in the number of photons specified in Eq. (6.29) can be separated into the components associated with five processes: (a) photon convergence associated with the streaming of photons through the 6D volume element; (b) photon loss due to the absorption of photons by the medium; (c) photon loss due to scattering, that is, some photons are scattered out of the 6D volume; (d) photon gain due to scattering, that is, some photons are scattered into the 6D volume element from outside due to the multiple-scattering events in the medium; and, (e) photon gains due to emission. The loss and gain of photons due to scattering are schematically illustrated by Figure 6.4a,b, respectively. In the following, we will discuss, in detail, the variation in the number of photons associated with (a) to (e).

Convergence due to streaming: $(d^6 \dot{n}_{\text{phot}})_a$ Here, the term convergence is used to denote both convergence and divergence, depending on whether a quantity of concern is positive or negative. To calculate $(d^6 \dot{n}_{\text{phot}})_a$, let us consider the contribution due to photon streaming in the vertical z-direction (upward) through the surface area element $dxdy$:

$$
\begin{aligned}
\left(d^6 \dot{n}_{phot}\right)_{a,z} &= (\text{Streaming-In} - \text{Streaming-Out})_z \\
&= \left\{ c \cdot s_3 \cdot \xi_\lambda \left[\vec{r}(x, y, z), \hat{s}, t \right] \right. \\
&\quad \left. - c \cdot s_3 \cdot \xi_\lambda \left[\vec{r}(x, y, z + dz), \hat{s}, t \right] \right\} dx \, dy \, d^2\Omega \, d\lambda \\
&= -\frac{\partial}{\partial z} \left[c \cdot s_3 \cdot \xi_\lambda(\vec{r}, \hat{s}, t) \right] dx \, dy \, dz \, d^2\Omega \, d\lambda \\
&= -\frac{\partial}{\partial z} \left[c \cdot s_3 \cdot \xi_\lambda(\vec{r}, \hat{s}, t) \right] d^3 V \, d^2\Omega \, d\lambda \, ,
\end{aligned} \tag{6.30}
$$

where $c \cdot s_3$ is the speed of photon streaming in the vertical direction. If we include the contributions due to photon streaming in all three directions (x, y, z), we have

$$
\begin{aligned}
\left(d^6 \dot{n}_{phot}\right)_a &= \left(d^6 \dot{n}_{phot}\right)_{a,x} + \left(d^6 \dot{n}_{phot}\right)_{a,y} + \left(d^6 \dot{n}_{phot}\right)_{a,z} \\
&= -\left\{ \frac{\partial}{\partial x} \left[c \cdot s_1 \cdot \xi_\lambda(\vec{r}, \hat{s}, t) \right] + \frac{\partial}{\partial y} \left[c \cdot s_2 \cdot \xi_\lambda(\vec{r}, \hat{s}, t) \right] \right. \\
&\quad \left. + \frac{\partial}{\partial z} \left[c \cdot s_3 \cdot \xi_\lambda(\vec{r}, \hat{s}, t) \right] \right\} d^3 V \, d^2\Omega \, d\lambda \, .
\end{aligned} \tag{6.31}
$$

Using the following velocity vector, that is,

$$
c \cdot \hat{s} = c \cdot \begin{pmatrix} s_1 \\ s_2 \\ s_3 \end{pmatrix} = \begin{pmatrix} \dot{x} \\ \dot{y} \\ \dot{z} \end{pmatrix} \, , \tag{6.32}
$$

we get

$$
\begin{aligned}
\left(d^6 \dot{n}_{phot}\right)_a &= -\vec{\nabla} \cdot \left[\hat{s} \cdot c \cdot \xi_\lambda(\vec{r}, \hat{s}, t) \right] d^3 V \, d^2\Omega \, d\lambda \\
&= -c \cdot \hat{s} \cdot \left[\vec{\nabla} \xi_\lambda(\vec{r}, \hat{s}, t) \right] d^3 V \, d^2\Omega \, d\lambda \, .
\end{aligned} \tag{6.33}
$$

Loss due to absorption: $(d^6 \dot{n}_{phot})_b$ Analogous to Eq. (6.8), we define the dimensionless spectral optical thickness of absorption, τ_{abs}, by

$$
d\tau_{abs} = -b_{abs}(\lambda, \vec{r}) \, ds = -\frac{ds}{\bar{s}_{abs}} \, , \tag{6.34}
$$

where we assume the medium to be stationary; the absorption and scattering properties are time-independent. According to the physical meaning of $\bar{s}_{abs}$, see Eq. (6.25), the quantity defined in Eq. (6.34) indicates the probability that a photon is absorbed within the infinitesimal 6D volume element along the distance element ds. For simplicity, we omit the time dependence of the volume absorption coefficient from the following analysis. The negative sign in Eq. (6.34) indicates that photons are lost due to absorption. The differentiation of Eq. (6.34) with respect to time yields

$$\frac{d\tau_{abs}}{dt} = -b_{abs}(\lambda, \vec{r}) \frac{ds}{dt} = -b_{abs}(\lambda, \vec{r}) \cdot c \,. \tag{6.35}$$

The total rate of photon absorption is obtained by the product of the photon number and the probability that a photon will be absorbed, from Eqs. (6.15) and (6.35),

$$\left(d^6 \dot{n}_{phot}\right)_b = d^6 n_{phot} \frac{d\tau_{abs}}{dt}$$

$$= -\xi_\lambda(\vec{r}, \hat{s}, t) \cdot b_{abs}(\lambda, \vec{r}) \cdot c \, d^3 V \, d^2 \Omega \, d\lambda \,. \tag{6.36}$$

Loss due to out-scattering: $(d^6 \dot{n}_{phot})_c$ As part of the extinction process, the first-order scattering events within the 6D volume element tend to decrease the number of photons within $d^3 V d^2 \Omega d\lambda$ by alternating the propagation directions of the photons. Similar to the case of absorption, the probability that a photon is scattered within the 6D volume element along ds is

$$d\tau_{sca} = -b_{sca}(\lambda, \vec{r}) \, ds = -\frac{ds}{\bar{s}_{sca}} \,. \tag{6.37}$$

Thus,

$$\frac{d\tau_{sca}}{dt} = -b_{sca}(\lambda, \vec{r}) \frac{ds}{dt} = -b_{sca}(\lambda, \vec{r}) \cdot c \,. \tag{6.38}$$

According to the physical meaning of $\bar{s}_{sca}$, the decrease in the number of photons within the 6D volume element from first-order scattering is given by

$$\left(d^6 \dot{n}_{phot}\right)_c = d^6 n_{phot} \cdot \frac{d\tau_{sca}}{dt}$$

$$= -\xi_\lambda(\vec{r}, \hat{s}, t) \cdot b_{sca}(\lambda, \vec{r}) \cdot c \, d^3 V \, d^2 \Omega \, d\lambda \,. \tag{6.39}$$

Gain due to in-scattering: $(d^6 \dot{n}_{phot})_d$ Opposite to the process just discussed, scattering events alternate the propagation directions of some photons from $\hat{s}'$ to $\hat{s}$. This process tends to increase the number of photons within the 6D volume element. In order to collect all the photons scattered from all directions $\hat{s}'$ into the direction $\hat{s}$, we need to integrate over all directions of $\hat{s}'$ and weight with the probability that a photon is actually scattered into the direction $\hat{s}$. The probability that a photon is scattered from $\hat{s}'$ to $\hat{s}$ is given by the dimensionless phase function $\mathcal{P}(\lambda, \vec{r}, \hat{s}', \hat{s})$. Thus, we obtain

$$\left(d^6 \dot{n}_{phot}\right)_d = c \cdot b_{sca}(\lambda, \vec{r}) \int \frac{\mathcal{P}(\lambda, \vec{r}, \hat{s}', \hat{s})}{4\pi} \cdot \xi_\lambda(\vec{r}, \hat{s}', t) \, d\hat{s}' \, d^3 V \, d^2 \Omega \, d\lambda \,. \tag{6.40}$$

Gain due to emission: $(d^6 \dot{n}_{phot})_e$ The emission rate, photon production per second, is given by

$$\left(d^6 \dot{n}_{phot}\right)_e = j_{emi,\lambda}(\vec{r}, \hat{s}, t) \, d^3 V \, d^2 \Omega \, d\lambda \,, \tag{6.41}$$

where $j_{emi,\lambda}$ indicates the source coefficient of emission given in units of $m^{-3}\,sr^{-1}\,\mu m^{-1}\,s^{-1}$. Here, we assume the source coefficient to depend on direction $\hat{s}$.

6.3.3
Photon Budget Equation

The governing equation for the variation of the photon number within the 6D volume element is given by

$$d^6\dot{n}_{phot} = \frac{\partial \xi_\lambda(\vec{r},\hat{s},t)}{\partial t}\, d^3V\, d^2\Omega\, d\lambda$$
$$= \left(d^6\dot{n}_{phot}\right)_a + \left(d^6\dot{n}_{phot}\right)_b + \left(d^6\dot{n}_{phot}\right)_c$$
$$+ \left(d^6\dot{n}_{phot}\right)_d + \left(d^6\dot{n}_{phot}\right)_e . \tag{6.42}$$

In terms of the photon number density function, Eq. (6.42) can be rewritten as

$$\frac{1}{c}\cdot\frac{\partial \xi_\lambda(\vec{r},\hat{s},t)}{\partial t} = -\hat{s}\cdot\vec{\nabla}\xi_\lambda(\vec{r},\hat{s},t)$$
$$- [b_{abs}(\lambda,\vec{r}) + b_{sca}(\lambda,\vec{r})]\cdot \xi_\lambda(\vec{r},\hat{s},t)$$
$$+ b_{sca}(\lambda,\vec{r})\int \frac{\mathcal{P}(\lambda,\vec{r},\hat{s}',\hat{s})}{4\pi}\cdot \xi_\lambda(\vec{r},\hat{s}',t)\, d\hat{s}'$$
$$+ \frac{1}{c}\cdot j_{emi,\lambda}(\vec{r},\hat{s},t) . \tag{6.43}$$

In the preceding expression of Eq. (6.43), the 6D volume element $d^3Vd^2\Omega\,d\lambda$ has been canceled from all terms. Note that $b_{abs} + b_{sca}$ is the extinction coefficient b_{ext}.

6.3.4
3D Time-Dependent and Stationary RTE for Total Radiance

To determine the time-dependent 3D RTE in terms of spectral radiance, we substitute Eq. (6.20) into Eq. (6.43). After simplification, we obtain

$$\frac{1}{c}\frac{\partial I_\lambda(\vec{r},\hat{s},t)}{\partial t} + \hat{s}\cdot\vec{\nabla} I_\lambda(\vec{r},\hat{s},t) = -b_{ext}(\lambda,\vec{r})\cdot I_\lambda(\vec{r},\hat{s},t)$$
$$+ b_{sca}(\lambda,\vec{r})\int \frac{\mathcal{P}(\lambda,\vec{r},\hat{s}',\hat{s})}{4\pi}\cdot I_\lambda(\vec{r},\hat{s}',t)\, d\hat{s}' + J_{emi,\lambda}(\vec{r},\hat{s},t) , \tag{6.44}$$

which introduces the source function $J_{emi,\lambda}$ of emission as

$$J_{emi,\lambda}(\vec{r},\hat{s},t) = h\cdot\frac{c}{\lambda}\cdot j_{emi,\lambda}(\vec{r},\hat{s},t) . \tag{6.45}$$

Equation (6.44) is a scalar RTE applicable to spectral radiance. From the arguments in Sections 6.3.2 and 6.3.3, it is evident that the polarization state of the radiation field is not taken into consideration in the analysis of the photon budget. This is a weakness in the phenomenological approach for deriving the RTE.

To derive the 3D stationary RTE, we will neglect the time dependence (i.e., no local changes; $\partial/\partial t = 0$) and obtain the standard form of the 3D RTE as

$$\hat{s} \cdot \vec{\nabla} I_\lambda(\vec{r}, \hat{s}) = -b_{\text{ext}}(\lambda, \vec{r}) \cdot I_\lambda(\vec{r}, \hat{s})$$
$$+ b_{\text{sca}}(\lambda, \vec{r}) \int \frac{P(\lambda, \vec{r}, \hat{s}', \hat{s})}{4\pi} \cdot I_\lambda(\vec{r}, \hat{s}') \, d\hat{s}'$$
$$+ J_{\text{emi}, \lambda}(\vec{r}, \hat{s}) \,. \tag{6.46}$$

Equation (6.46) is interpreted as the gradient of the radiance in the direction $\hat{s}$ given by the attenuation of the radiation beam by scattering and absorption events, the contribution due to multiple-scattering, and the enhancement due to emission.

6.3.5
3D Stationary RTE for Diffuse Radiance

The RTE can be split into diffuse and direct parts because the radiance is usually split into direct and diffuse portions. We can derive the equation for the diffuse part of the RTE, and we can compute the direct portion in a straightforward manner.

Splitting of diffuse and direct radiance   In the atmosphere, the radiation field can be regarded as the sum of two components: the diffuse radiation due to the multiple-scattering in the atmosphere or the nonspecular reflection by the land or roughened oceanic surface, and the direct radiation that is a quasi-collimated radiation component associated with the unattenuated portion of the incident solar radiation. Quantitatively, the total radiance, I_λ, can be written as

$$I_\lambda(\vec{r}, \hat{s}) = I_{\text{diff}, \lambda}(\vec{r}, \hat{s}) + I_{\text{dir}, \lambda}(\vec{r}, \hat{s})$$
$$= I_{\text{diff}, \lambda}(\vec{r}, \hat{s}) + S_{\text{dir}, \lambda}(\vec{r}) \cdot \delta(\hat{s} - \hat{s}_0) \,. \tag{6.47}$$

In Eq. (6.47), the subscript zero indicates that the quantity is associated with the direct solar radiation, in which direction is specified in terms of solar zenith and azimuthal angles θ_0 and φ_0, that is, $\hat{s}_0 = \hat{s}_0(\theta_0, \varphi_0)$.

Diffuse-direct splitting of the gradient term   The following transform holds for any function, not only for I_λ, but also for $S_{\text{dir}, \lambda}$ and $I_{\text{diff}, \lambda}$; here, we demonstrate it for the total radiance I_λ

$$\frac{d I_\lambda}{ds} = \frac{\partial I_\lambda}{\partial x}\frac{dx}{ds} + \frac{\partial I_\lambda}{\partial y}\frac{dy}{ds} + \frac{\partial I_\lambda}{\partial z}\frac{dz}{ds}$$
$$= \frac{\partial I_\lambda}{\partial x} s_1 + \frac{\partial I_\lambda}{\partial y} s_2 + \frac{\partial I_\lambda}{\partial z} s_3$$
$$= \hat{s} \cdot \vec{\nabla} I_\lambda(\vec{r}, \hat{s}) \,. \tag{6.48}$$

Applying Eq. (6.48) to the total radiance given by Eq. (6.47) yields

$$\hat{s} \cdot \vec{\nabla} I_\lambda(\vec{r}, \hat{s}) = \frac{d I_\lambda}{ds} = \frac{d I_{\text{diff}, \lambda}}{ds} + \frac{d S_{\text{dir}, \lambda}(\vec{r})}{ds} \cdot \delta(\hat{s} - \hat{s}_0) \,. \tag{6.49}$$

The Lambert–Bouguer law in the form of Eq. (6.5) is applied to nonscattered direct solar radiation, $I_\lambda \to S_{\text{dir},\lambda}$ in Eq. (6.6), producing

$$\frac{d\,S_{\text{dir},\lambda}(\vec{\mathbf{r}})}{ds} = -b_{\text{ext}}(\lambda, \vec{\mathbf{r}}) \cdot S_{\text{dir},\lambda}(\vec{\mathbf{r}}) \,. \tag{6.50}$$

Furthermore, Eq. (6.48) is applied to the diffuse radiance $I_{\text{diff},\lambda}$,

$$\hat{\mathbf{s}} \cdot \vec{\nabla} I_{\text{diff},\lambda}(\vec{\mathbf{r}}, \hat{\mathbf{s}}) = \frac{d\,I_{\text{diff},\lambda}}{ds} \,, \tag{6.51}$$

and continuing from Eq. (6.49)

$$\hat{\mathbf{s}} \cdot \vec{\nabla} I_\lambda(\vec{\mathbf{r}}, \hat{\mathbf{s}}) = \hat{\mathbf{s}} \cdot \vec{\nabla} I_{\text{diff},\lambda}(\vec{\mathbf{r}}, \hat{\mathbf{s}}) - b_{\text{ext}}(\lambda, \vec{\mathbf{r}}) \cdot S_{\text{dir},\lambda}(\vec{\mathbf{r}}) \cdot \delta(\hat{\mathbf{s}} - \hat{\mathbf{s}}_0) \,. \tag{6.52}$$

Diffuse-direct splitting of the multiple-scattering term We look at the multiple-scattering term of Eq. (6.46) and introduce the splitting, Eq. (6.47), as

$$b_{\text{sca}}(\lambda, \vec{\mathbf{r}}) \int \frac{\mathcal{P}(\lambda, \vec{\mathbf{r}}, \hat{\mathbf{s}}', \hat{\mathbf{s}})}{4\pi} \cdot I_\lambda(\vec{\mathbf{r}}, \hat{\mathbf{s}}') \, d\hat{\mathbf{s}}'$$
$$= b_{\text{sca}}(\lambda, \vec{\mathbf{r}}) \int \frac{\mathcal{P}(\lambda, \vec{\mathbf{r}}, \hat{\mathbf{s}}', \hat{\mathbf{s}})}{4\pi} \cdot I_{\text{diff},\lambda}(\vec{\mathbf{r}}, \hat{\mathbf{s}}') \, d\hat{\mathbf{s}}'$$
$$+ b_{\text{sca}}(\lambda, \vec{\mathbf{r}}) \int \frac{\mathcal{P}(\lambda, \vec{\mathbf{r}}, \hat{\mathbf{s}}', \hat{\mathbf{s}})}{4\pi} \cdot S_{\text{dir},\lambda}(\vec{\mathbf{r}}) \cdot \delta(\hat{\mathbf{s}}' - \hat{\mathbf{s}}_0) \, d\hat{\mathbf{s}}' \,. \tag{6.53}$$

The second term on the right side of Eq. (6.53) represents the photons scattered out of the direct beam by the first-order scattering events. We apply the property of the Dirac δ-function, given by Eq. (2.29), to this term, that is,

$$\phi \to \mathcal{P} \,,$$
$$x \to \hat{\mathbf{s}}' \,,$$
$$x_0 \to \hat{\mathbf{s}}_0 \,. \tag{6.54}$$

With this transformation, it follows that

$$b_{\text{sca}}(\lambda, \vec{\mathbf{r}}) \int \frac{\mathcal{P}(\lambda, \vec{\mathbf{r}}, \hat{\mathbf{s}}', \hat{\mathbf{s}})}{4\pi} \cdot S_{\text{dir},\lambda}(\vec{\mathbf{r}}) \cdot \delta(\hat{\mathbf{s}}' - \hat{\mathbf{s}}_0) \, d\hat{\mathbf{s}}'$$
$$= b_{\text{sca}}(\lambda, \vec{\mathbf{r}}) \cdot \frac{\mathcal{P}(\lambda, \vec{\mathbf{r}}, \hat{\mathbf{s}}_0, \hat{\mathbf{s}})}{4\pi} \cdot S_{\text{dir},\lambda}(\vec{\mathbf{r}}) \,. \tag{6.55}$$

Thus, the multiple-scattering term of Eq. (6.46) is given by

$$b_{\text{sca}}(\lambda, \vec{\mathbf{r}}) \int \frac{\mathcal{P}(\lambda, \vec{\mathbf{r}}, \hat{\mathbf{s}}', \hat{\mathbf{s}})}{4\pi} \cdot I_\lambda(\vec{\mathbf{r}}, \hat{\mathbf{s}}') \, d\hat{\mathbf{s}}'$$
$$= b_{\text{sca}}(\lambda, \vec{\mathbf{r}}) \int \frac{\mathcal{P}(\lambda, \vec{\mathbf{r}}, \hat{\mathbf{s}}', \hat{\mathbf{s}})}{4\pi} \cdot I_{\text{diff},\lambda}(\vec{\mathbf{r}}, \hat{\mathbf{s}}') \, d\hat{\mathbf{s}}'$$
$$+ b_{\text{sca}}(\lambda, \vec{\mathbf{r}}) \cdot \frac{\mathcal{P}(\lambda, \vec{\mathbf{r}}, \hat{\mathbf{s}}_0, \hat{\mathbf{s}})}{4\pi} \cdot S_{\text{dir},\lambda}(\vec{\mathbf{r}}) \,. \tag{6.56}$$

Diffuse component of the RTE By substituting the direct and diffuse gradient, radiance, and multiple-scattering terms into the RTE, see Eq. (6.46), we obtain

$$
\begin{aligned}
\hat{s} \cdot \vec{\nabla} I_{\text{diff},\lambda}(\vec{r},\hat{s}) - b_{\text{ext}}(\lambda,\vec{r}) \cdot S_{\text{dir},\lambda}(\vec{r}) \cdot \delta(\hat{s}-\hat{s}_0) = \\
- b_{\text{ext}}(\lambda,\vec{r}) \cdot I_{\text{diff},\lambda}(\vec{r},\hat{s}) - b_{\text{ext}}(\lambda,\vec{r}) \cdot S_{\text{dir},\lambda}(\vec{r}) \cdot \delta(\hat{s}-\hat{s}_0) \\
+ b_{\text{sca}}(\lambda,\vec{r}) \int \frac{\mathcal{P}(\lambda,\vec{r},\hat{s}',\hat{s})}{4\pi} \cdot I_{\text{diff},\lambda}(\vec{r},\hat{s}')\, d\hat{s}' \\
+ b_{\text{sca}}(\lambda,\vec{r}) \cdot \frac{\mathcal{P}(\lambda,\vec{r},\hat{s}_0,\hat{s})}{4\pi} \cdot S_{\text{dir},\lambda}(\vec{r}) \\
+ J_{\text{emi},\lambda}(\vec{r},\hat{s}) \, .
\end{aligned}
\tag{6.57}
$$

Two terms in Eq. (6.57) cancel each other, and with a slight rearrangement we have for the diffuse component of the RTE

$$
\begin{aligned}
\hat{s} \cdot \vec{\nabla} I_{\text{diff},\lambda}(\vec{r},\hat{s}) = {}& -b_{\text{ext}}(\lambda,\vec{r}) \cdot I_{\text{diff},\lambda}(\vec{r},\hat{s}) \\
& + b_{\text{sca}}(\lambda,\vec{r}) \cdot \frac{\mathcal{P}(\lambda,\vec{r},\hat{s}_0,\hat{s})}{4\pi} \cdot S_{\text{dir},\lambda}(\vec{r}) \\
& + b_{\text{sca}}(\lambda,\vec{r}) \int \frac{\mathcal{P}(\lambda,\vec{r},\hat{s}',\hat{s})}{4\pi} \cdot I_{\text{diff},\lambda}(\vec{r},\hat{s}')\, d\hat{s}' \\
& + J_{\text{emi},\lambda}(\vec{r},\hat{s}) \, .
\end{aligned}
\tag{6.58}
$$

Note that Eq. (6.58) only includes diffuse radiation. The gradient of diffuse radiance along direction $\hat{s}$ is determined by extinction, first-order scattering of direct solar radiation, multiple-scattering, and emission.

6.4
1D RTE for a Horizontally Homogeneous Atmosphere

6.4.1
Independent Variables

For a horizontally homogeneous atmosphere, in the related previous equations we replace

$$
\vec{r} \to s \quad \text{or} \quad \tau \, ,
\tag{6.59}
$$

and change the notation for the angles

$$
\hat{s} \to (\mu,\varphi) \, ; \quad \hat{s}' \to (\mu',\varphi') \, .
\tag{6.60}
$$

For the downward direction of the Sun, we replace

$$
\hat{s}_0 \to (-\mu_0,\varphi_0) \, ,
\tag{6.61}
$$

and

$$\int \dots d\hat{s}' \rightarrow \int\limits_{0}^{2\pi} \int\limits_{-1}^{1} \dots d\mu' d\varphi' \,. \tag{6.62}$$

6.4.2
Standard Form of 1D RTE for Diffuse Radiance

To obtain the 1D RTE for diffuse radiance, we can rewrite Eq. (6.58) as a function of slant path s and apply Eq. (6.51), that is,

$$\frac{1}{b_{\text{ext}}(\lambda, s)} \frac{d I_{\text{diff},\lambda}(s, \mu, \varphi)}{ds} = -I_{\text{diff},\lambda}(s, \mu, \varphi)$$

$$+ \tilde{\omega}(\lambda, s) \cdot \frac{\mathcal{P}(\lambda, s, -\mu_0, \varphi_0, \mu, \varphi)}{4\pi} \cdot S_{\text{dir},\lambda}(s)$$

$$+ \tilde{\omega}(\lambda, s) \int\limits_{0}^{2\pi} \int\limits_{-1}^{1} \frac{\mathcal{P}(\lambda, s, \mu', \varphi', \mu, \varphi)}{4\pi} \cdot I_{\text{diff},\lambda}(s, \mu', \varphi') \, d\mu' \, d\varphi'$$

$$+ \frac{1}{b_{\text{ext}}(\lambda, s)} \cdot J_{\text{emi},\lambda}(s) \,, \tag{6.63}$$

which includes the definition of the single-scattering albedo

$$\tilde{\omega}(\lambda, s) = \frac{b_{\text{sca}}(\lambda, s)}{b_{\text{ext}}(\lambda, s)} \,. \tag{6.64}$$

To rewrite Eq. (6.63) as a function of the optical thickness τ, we consider the relationship between the extinction coefficient and optical thickness with the "plane-parallel atmosphere" assumption, see Eq. (6.8). We use Eq. (6.13) for direct solar irradiance and specify the source function in terms of Planck's function $B_\lambda[T(\tau)]$ as

$$J_{\text{emi},\lambda}(\tau) = b_{\text{abs}}(\lambda) \cdot B_\lambda[T(\tau)] \,, \tag{6.65}$$

where $b_{\text{abs}}(\lambda)/b_{\text{ext}}(\lambda) = 1 - \tilde{\omega}(\lambda)$ is the single-scattering co-albedo. Furthermore, we apply Eq. (6.7) in the form of $ds = dz/\mu$, yielding

$$b_{\text{ext}}(\lambda, s) ds = b_{\text{ext}}(\lambda, z) \frac{dz}{\mu} = -\frac{1}{\mu} d\tau \,. \tag{6.66}$$

The standard form of the 1D RTE for the diffuse radiance in a plane-parallel, horizontally homogeneous atmosphere with the vertical coordinate $\tau(\lambda, z)$ is then given by

$$\mu \frac{d I_{\text{diff},\lambda}(\tau,\mu,\varphi)}{d\tau} = I_{\text{diff},\lambda}(\tau,\mu,\varphi)$$

$$- \tilde{\omega}(\lambda,\tau) \cdot S_{\text{dir},\lambda,\text{TOA}} \cdot e^{-\tau/\mu_0} \cdot \frac{\mathcal{P}(\lambda,\tau,-\mu_0,\varphi_0,\mu,\varphi)}{4\pi}$$

$$- \tilde{\omega}(\lambda,\tau) \int\limits_{0}^{2\pi} \int\limits_{-1}^{1} I_{\text{diff},\lambda}(\tau,\mu',\varphi') \cdot \frac{\mathcal{P}(\lambda,\tau,\mu',\varphi',\mu,\varphi)}{4\pi} \, d\mu' \, d\varphi'$$

$$- [1 - \tilde{\omega}(\tau)] \cdot B_\lambda[T(\tau)] \,. \tag{6.67}$$

The concise 1D RTE version is

$$\mu \frac{d I_{\text{diff},\lambda}(\tau,\mu,\varphi)}{d\tau} = I_{\text{diff},\lambda} - \left\{ J_{\text{dir},\lambda} + J_{\text{diff},\lambda} + (1 - \tilde{\omega}) \cdot B_\lambda[T(\tau)] \right\} \,. \tag{6.68}$$

The RTE is an integro-differential equation for the diffuse spectral radiance $I_{\text{diff},\lambda}$ and contains a differential and an integral term. The equation is also known as the Schwarzschild–Emden differential equation and represents a general radiative transfer equation for a plane-parallel medium. The source terms on the right are rearranged and interpreted as follows:

a) First-order scattering of direct solar radiation to direction (μ,φ):

$$J_{\text{dir},\lambda}(\tau,\mu,\varphi) = \tilde{\omega}(\lambda,\tau) \cdot S_{\text{dir},\lambda,\text{TOA}} \cdot e^{-\tau/\mu_0} \cdot \frac{\mathcal{P}(\lambda,\tau,-\mu_0,\varphi_0,\mu,\varphi)}{4\pi} \,. \tag{6.69}$$

The incident solar radiation, $S_{\text{dir},\lambda,\text{TOA}}$, at the TOA is exponentially attenuated according to the Lambert–Bouguer law (larger extinction for larger optical thickness) until it is scattered into the direction (μ,φ) to enhance $I_{\text{diff},\lambda}(\tau,\mu,\varphi)$. The scattered portion is quantified by the phase function, $\mathcal{P}$, and the absorption is determined by the single-scattering albedo $\tilde{\omega}$.

b) Multiple-scattering:

$$J_{\text{diff},\lambda}(\tau,\mu,\varphi) = \tilde{\omega}(\lambda,\tau) \int\limits_{0}^{2\pi} \int\limits_{-1}^{1} I_{\text{diff},\lambda}(\tau,\mu',\varphi') \cdot \frac{\mathcal{P}(\lambda,\tau,\mu',\varphi',\mu,\varphi)}{4\pi} \, d\mu' \, d\varphi' \,. \tag{6.70}$$

This term includes the diffuse radiances scattered into the direction (μ,φ). Therefore, the incoming diffuse radiances are multiplied by the phase function and integrated over the solid angle. Again, the single-scattering albedo $\tilde{\omega}$ determines how much of the diffuse radiance is scattered in the extinction process.

c) Thermal emission:

$$[1 - \tilde{\omega}(\tau)] \cdot B_\lambda[T(\tau)] \,, \tag{6.71}$$

with Planck's function $B_\lambda[T(\tau)]$ given by Eq. (3.58). The contribution of this source term can usually be neglected in the solar spectral region (0.2–5 μm), except in some remote sensing applications. For example, both solar and terrestrial emission need to be considered in the 3.7 μm spectral band.

The upper boundary condition at the TOA ($\tau = 0$) warrants that no downward diffuse radiance enters the atmosphere, that is,

$$I_{\mathrm{diff},\lambda}^{\downarrow}(\tau = 0, -\mu, \varphi) = 0 \, . \tag{6.72}$$

To specify the lower boundary condition, we introduce the Bidirectional Reflectance Distribution Function (BRDF, dimensionless), $\gamma(\lambda, \mu_0, \varphi_0, \mu, \varphi)$, associated with the surface as

$$
\begin{aligned}
I_{\mathrm{diff},\lambda}^{\uparrow}(\tau = \tau^*, \mu, \varphi) \\
= \frac{1}{\pi} \int_0^{2\pi} \int_0^1 \gamma(\lambda, \mu', \varphi', \mu, \varphi) \cdot I_\lambda^{\downarrow}(\tau^*, -\mu', \varphi') \cdot \mu' \, d\mu' \, d\varphi' \\
+ \varepsilon(\lambda, \tau^*) \cdot B_\lambda[T(\tau^*)] \, ,
\end{aligned}
\tag{6.73}
$$

where $\varepsilon(\lambda, \tau^*)$ is the surface emissivity. According to Eqs. (6.47) and (6.13), the total (diffuse plus direct) downward radiance at the surface is

$$
\begin{aligned}
I_\lambda^{\downarrow}(\tau = \tau^*, -\mu, \varphi) = I_{\mathrm{diff},\lambda}^{\downarrow}(\tau = \tau^*, -\mu, \varphi) \\
+ S_{\mathrm{dir},\lambda,\mathrm{TOA}} \cdot e^{-\tau^*/\mu_0} \cdot \delta(\mu - \mu_0) \cdot \delta(\varphi - \varphi_0) \, .
\end{aligned}
\tag{6.74}
$$

Substituting Eq. (6.74) into Eq. (6.73) and considering the contribution due to emission, we obtain

$$
\begin{aligned}
I_{\mathrm{diff},\lambda}^{\uparrow}(\tau = \tau^*, \mu, \varphi) = \frac{\gamma(\lambda, \mu_0, \varphi_0, \mu, \varphi)}{\pi} \cdot \mu_0 \cdot S_{\mathrm{dir},\lambda,\mathrm{TOA}} \cdot e^{-\tau^*/\mu_0} \\
+ \frac{1}{\pi} \int_0^{2\pi} \int_0^1 \gamma(\lambda, \mu', \varphi', \mu, \varphi) \cdot I_{\mathrm{diff},\lambda}^{\downarrow}(\tau^*, -\mu', \varphi') \cdot \mu' \, d\mu' \, d\varphi' \\
+ \varepsilon(\lambda, \tau^*) \cdot B_\lambda[T(\tau^*)] \, .
\end{aligned}
\tag{6.75}
$$

The physical meaning of the first and second terms on the right side of Eq. (6.75) are schematically illustrated by Figures 6.5 and 6.6, respectively. Specifically, the first term represents the reflection of the attenuated direct solar radiation; whereas, the second term specifies the reflection of diffuse radiation reaching the surface.

The last term on the right side of Eq. (6.75) quantifies the thermal emission of the surface corresponding to the surface temperature $T(\tau^*)$.

The surface BRDF, $\gamma(\lambda, \mu_0, \varphi_0, \mu, \varphi)$, can be calculated from the angle-dependent spectral upward radiance $I_{\mathrm{diff},\lambda}^{\uparrow}(\tau^*, \mu, \varphi)$ associated with a collimated incident radiation field,

$$I_{\mathrm{inc},\lambda}^{\downarrow}(-\mu, \varphi, \tau^*) = F_{\mathrm{inc},\lambda}^{\downarrow}(\tau^*) \cdot \delta(\mu - \mu_0) \cdot \delta(\varphi - \varphi_0) \, , \tag{6.76}$$

as

$$\gamma(\lambda, \mu_0, \varphi_0, \mu, \varphi) = \frac{\pi \cdot I_{\mathrm{diff},\lambda}^{\uparrow}(\tau^*, \mu, \varphi)}{\mu_0 \cdot F_{\mathrm{inc},\lambda}^{\downarrow}(\tau^*)} \, . \tag{6.77}$$

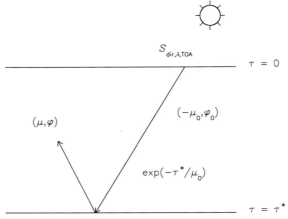

Figure 6.5 Surface reflection of direct solar radiation.

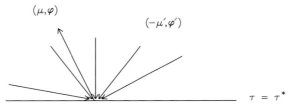

Figure 6.6 Surface reflection of diffuse radiation.

Furthermore, the planetary (or local) surface albedo is defined as

$$\gamma_{\mathrm{p}}(\lambda, \mu_0) = \frac{1}{\mu_0 \cdot F_{\lambda,\mathrm{inc}}^{\downarrow}(\tau^*)} \int_0^{2\pi} \int_0^1 I_{\mathrm{diff},\lambda}^{\uparrow}(\tau^*, \mu', \varphi') \cdot \mu' \, \mathrm{d}\mu' \, \mathrm{d}\varphi'$$

$$= \frac{1}{\pi} \int_0^{2\pi} \int_0^1 \gamma(\lambda, \mu_0, \varphi_0, \mu', \varphi') \cdot \mu' \, \mathrm{d}\mu' \, \mathrm{d}\varphi' \, . \tag{6.78}$$

We introduce the spherical (or, global) surface albedo as

$$\gamma_{\mathrm{g}}(\lambda) = \frac{\int_0^1 \mu_0 \cdot F_{\mathrm{inc},\lambda}^{\downarrow}(\tau^*) \cdot \gamma_{\mathrm{p}}(\lambda, \mu_0) \, \mathrm{d}\mu_0}{\int_0^1 \mu_0 \cdot F_{\mathrm{inc},\lambda}^{\downarrow}(\tau^*) \, \mathrm{d}\mu_0}$$

$$= 2 \int_0^1 \mu_0 \cdot \gamma_{\mathrm{p}}(\lambda, \mu_0) \, \mathrm{d}\mu_0 \, . \tag{6.79}$$

The planetary albedo is the μ-weighted average of the BRDF over reflection direction; whereas, the spherical albedo is its counterpart averaged over both the reflection and incident directions. Several examples of the spectral spherical surface albedo for typical surface types are given in Figure 6.7.

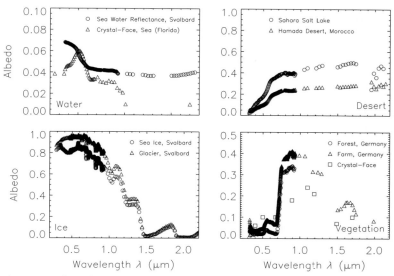

Figure 6.7 Spherical surface albedos for various surface conditions. Courtesy of E. Bierwirth.

6.4.3
Downward Diffuse Radiance

To derive the formal solution of the RTE in the case of downward radiation, we will rewrite the differential Schwarzschild–Emden Eq. (6.68) in the form of

$$
- \mu \, \frac{\mathrm{d} I_{\mathrm{diff},\lambda}^{\downarrow}(\tau, -\mu, \varphi)}{\mathrm{d}\tau} = I_{\mathrm{diff},\lambda}^{\downarrow}(\tau, -\mu, \varphi)
$$
$$
- \left\{ J_{\mathrm{dir},\lambda}(\tau, -\mu, \varphi) + J_{\mathrm{diff},\lambda}(\tau, -\mu, \varphi) + (1 - \tilde{\omega}) \cdot B_{\lambda}[T(\tau)] \right\}. \tag{6.80}
$$

Multiplying Eq. (6.80) with $\mathrm{e}^{\tau/\mu}$ and rearranging the terms, gives

$$
\mathrm{e}^{\tau/\mu} \cdot \mathrm{d} I_{\mathrm{diff},\lambda}^{\downarrow}(\tau, -\mu, \varphi) + I_{\mathrm{diff},\lambda}^{\downarrow}(\tau, -\mu, \varphi) \cdot \mathrm{e}^{\tau/\mu} \, \frac{\mathrm{d}\tau}{\mu}
$$
$$
= \left\{ J_{\mathrm{dir},\lambda}(\tau, -\mu, \varphi) + J_{\mathrm{diff},\lambda}(\tau, -\mu, \varphi) + (1 - \tilde{\omega}) \cdot B_{\lambda}[T(\tau)] \right\} \cdot \mathrm{e}^{\tau/\mu} \, \frac{\mathrm{d}\tau}{\mu}. \tag{6.81}
$$

The first two terms in Eq. (6.81) can be combined (via the chain rule) as

$$
\mathrm{e}^{\tau/\mu} \cdot \mathrm{d} I_{\mathrm{diff},\lambda}^{\downarrow}(\tau, -\mu, \varphi) + I_{\mathrm{diff},\lambda}^{\downarrow}(\tau, -\mu, \varphi) \cdot \mathrm{e}^{\tau/\mu} \, \frac{\mathrm{d}\tau}{\mu}
$$
$$
= \mathrm{d} \left[\mathrm{e}^{\tau/\mu} \cdot I_{\mathrm{diff},\lambda}^{\downarrow}(\tau, -\mu, \varphi) \right]. \tag{6.82}
$$

Thus, we obtain

$$
\int_0^\tau d\left[e^{\tau'/\mu} \cdot I_{\text{diff},\lambda}^{\downarrow}(\tau', -\mu, \varphi) \right]
$$

$$
= \int_0^\tau \left\{ J_{\text{dir},\lambda}(\tau', -\mu, \varphi) + J_{\text{diff},\lambda}(\tau', -\mu, \varphi) + (1 - \tilde{\omega}) \cdot B_\lambda[T(\tau')] \right\} \cdot e^{\tau'/\mu} \frac{d\tau'}{\mu} \cdot
$$

$$(6.83)$$

Equation (6.83) leads to

$$
e^{\tau/\mu} \cdot I_{\text{diff},\lambda}^{\downarrow}(\tau, -\mu, \varphi) - I_{\text{diff},\lambda}^{\downarrow}(\tau = 0, -\mu, \varphi)
$$

$$
= \int_0^\tau \left\{ J_{\text{dir},\lambda}(\tau', -\mu, \varphi) + J_{\text{diff},\lambda}(\tau', -\mu, \varphi) + (1 - \tilde{\omega}) \cdot B_\lambda[T(\tau')] \right\}
$$

$$
\times e^{\tau'/\mu} \frac{d\tau'}{\mu} \cdot
$$

$$(6.84)$$

According to the upper boundary condition, see Eq. (6.72), and substituting the detailed expressions for the source terms $J_{\text{dir},\lambda}$ and $J_{\text{diff},\lambda}$, see Eqs. (6.69) and (6.70), we obtain

$$
I_{\text{diff},\lambda}^{\downarrow}(\tau, -\mu, \varphi)
$$

$$
= \int_0^\tau S_{\text{dir},\lambda,\text{TOA}} \cdot e^{-\tau'/\mu_0} \cdot \tilde{\omega}(\lambda, \tau') \cdot \frac{P(\lambda, \tau', -\mu_0, \varphi_0, -\mu, \varphi)}{4\pi}
$$

$$
\times e^{-(\tau - \tau')/\mu} \frac{d\tau'}{\mu} + \int_0^\tau \tilde{\omega}(\lambda, \tau') \int_0^{2\pi} \int_{-1}^{1} I_{\text{diff},\lambda}(\tau', \mu', \varphi')
$$

$$
\times \frac{P(\lambda, \tau', \mu', \varphi', -\mu, \varphi)}{4\pi} d\mu' d\varphi' e^{-(\tau - \tau')/\mu} \frac{d\tau'}{\mu}
$$

$$
+ \int_0^\tau [1 - \tilde{\omega}(\lambda, \tau')] \cdot B_\lambda[T(\tau')] \cdot e^{-(\tau - \tau')/\mu} \frac{d\tau'}{\mu} \cdot
$$

$$(6.85)$$

The interpretation of the three terms on the right side in Eq. (6.85) is given in the following.

First-order scattering of direct solar radiation

$$
\int_0^\tau S_{\text{dir},\lambda,\text{TOA}} \cdot e^{-\tau'/\mu_0} \cdot \tilde{\omega}(\lambda, \tau') \cdot \frac{P(\lambda, \tau', -\mu_0, \varphi_0, -\mu, \varphi)}{4\pi} \cdot e^{-(\tau - \tau')/\mu} \frac{d\tau'}{\mu} \cdot \quad (6.86)
$$

Due to the integration from zero to τ, all first-order scattering events in the layer between zero and τ are summed, see Figure 6.8. $S_{\text{dir},\lambda,\text{TOA}} \cdot e^{-\tau'/\mu_0}$ represents

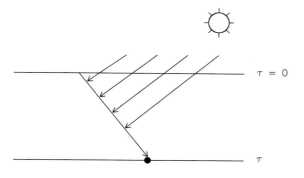

Figure 6.8 First-order scattering of direct solar radiation, downward radiance.

the decrease of solar radiation until it reaches the altitude τ', where the first-order scattering events take place. The phase function $\mathcal{P}(\lambda, \tau', -\mu_0, \varphi_0, -\mu, \varphi)$ regulates the amount of radiation arriving in the direction $(-\mu_0, \varphi_0)$ and scattering into direction $(-\mu, \varphi)$. A part of the radiation is absorbed and is defined by $\tilde{\omega}(\lambda, \tau')$. After the first-order scattering events, the scattered radiation is attenuated by a factor of $e^{-(\tau-\tau')/\mu}$.

Multiple-scattering

$$\int_0^\tau \tilde{\omega}(\lambda, \tau') \int_0^{2\pi} \int_{-1}^1 I_{\text{diff},\lambda}(\tau, \mu, \varphi) \cdot \frac{\mathcal{P}(\lambda, \tau', \mu', \varphi', -\mu, \varphi)}{4\pi} \, d\mu' \, d\varphi'$$

$$\times \, e^{-(\tau-\tau')/\mu} \, \frac{d\tau'}{\mu} \, . \tag{6.87}$$

On the basis of integration from zero to τ, all multiple-scattering events in the layer between zero and τ are summed, see Figure 6.9. All the diffuse radiation at altitude τ' is weighted with the respective phase function $\mathcal{P}(\lambda, \tau', \mu', \varphi', -\mu, \varphi)$ at altitude τ'. The inner integral gives the diffuse contribution in the direction (μ, φ) at altitude τ'. After the scattering event, the radiation is attenuated by a factor of $e^{-(\tau-\tau')/\mu}$.

Atmospheric emission

$$\int_0^\tau [1 - \tilde{\omega}(\lambda, \tau')] \cdot B_\lambda[T(\tau')] \cdot e^{-(\tau-\tau')/\mu} \, \frac{d\tau'}{\mu} \, . \tag{6.88}$$

Contributions of the thermal emission at the different altitudes τ' are decreased by the factor $e^{-(\tau-\tau')/\mu}$ and integrated over altitude. After the downward diffuse

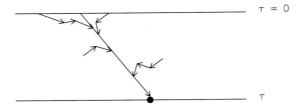

Figure 6.9 Multiple-scattering of diffuse solar radiation.

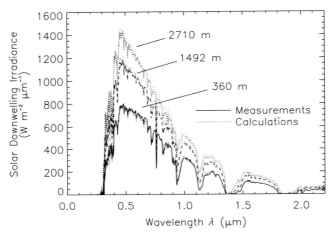

Figure 6.10 Downward spectral irradiance as measured from an aircraft during the Saharan Mineral Dust Experiment (SAMUM) in Morocco. The flight altitudes are given above ground level. The attenuation is caused by a layer of Saharan dust 3.5 km thick. Courtesy of E. Bierwirth.

radiance is obtained, the total (direct plus diffuse) downward spectral irradiance at altitude τ is given by

$$F_\lambda^\downarrow(\tau) = \mu_0 \cdot S_{\mathrm{dir},\lambda,\mathrm{TOA}} \cdot e^{-\tau/\mu_0} + \int_0^{2\pi} \int_0^1 I_{\mathrm{diff}}^\downarrow(\tau, -\mu, \varphi) \cdot \mu \, d\mu \, d\varphi \ . \tag{6.89}$$

An example of the downward irradiance as measured at different altitudes for the solar spectral region is shown in Figure 6.10.

6.4.4
Upward Radiance

To derive the formal solution for the upward radiance, we rewrite the Schwarz-schild–Emden Eq. (6.68) as

$$
\mu \frac{d I_{\mathrm{diff},\lambda}^{\uparrow}(\tau,\mu,\varphi)}{d\tau} = I_{\mathrm{diff},\lambda}^{\uparrow}(\tau,\mu,\varphi)
$$
$$
- \left\{ J_{\mathrm{dir},\lambda}(\tau,\mu,\varphi) + J_{\mathrm{diff},\lambda}(\tau,\mu,\varphi) + (1-\tilde{\omega})\cdot B_{\lambda}[T(\tau)] \right\}.
$$
(6.90)

We multiply Eq. (6.90) with $e^{-\tau/\mu}$ and rearrange the terms to obtain

$$
e^{-\tau/\mu}\cdot d I_{\mathrm{diff},\lambda}^{\uparrow}(\tau,\mu,\varphi) - I_{\mathrm{diff},\lambda}^{\uparrow}(\tau,\mu,\varphi)\cdot e^{-\tau/\mu}\frac{d\tau}{\mu}
$$
$$
= -\left\{ J_{\mathrm{dir},\lambda}(\tau,\mu,\varphi) + J_{\mathrm{diff},\lambda}(\tau,\mu,\varphi) + (1-\tilde{\omega})\cdot B_{\lambda}[T(\tau)] \right\}\cdot e^{-\tau/\mu}\frac{d\tau}{\mu} .
$$
(6.91)

The first two terms in Eq. (6.91) are combined, which gives

$$
e^{-\tau/\mu}\cdot d I_{\mathrm{diff},\lambda}^{\uparrow}(\tau,\mu,\varphi) - I_{\mathrm{diff},\lambda}^{\uparrow}(\tau,\mu,\varphi)\cdot e^{-\tau/\mu}\frac{d\tau}{\mu}
$$
$$
= d\left[e^{-\tau/\mu}\cdot I_{\mathrm{diff},\lambda}^{\uparrow}(\tau,\mu,\varphi) \right].
$$
(6.92)

Thus,

$$
\int_{\tau}^{\tau^*} d\left[e^{-\tau'/\mu}\cdot I_{\mathrm{diff},\lambda}^{\uparrow}(\tau',\mu,\varphi) \right]
$$
$$
= -\int_{\tau}^{\tau^*} \left\{ J_{\mathrm{dir},\lambda}(\tau',\mu,\varphi) + J_{\mathrm{diff},\lambda}(\tau',\mu,\varphi) + (1-\tilde{\omega})\cdot B_{\lambda}[T(\tau')] \right\}\cdot e^{-\tau'/\mu}\frac{d\tau'}{\mu} .
$$
(6.93)

The integral on the left side of Eq. (6.93) gives

$$
\int_{\tau}^{\tau^*} d\left[e^{-\tau'/\mu}\cdot I_{\mathrm{diff},\lambda}^{\uparrow}(\tau',\mu,\varphi) \right]
$$
$$
= e^{-\tau^*/\mu}\cdot I_{\mathrm{diff},\lambda}^{\uparrow}(\tau^*,\mu,\varphi) - e^{-\tau/\mu}\cdot I_{\mathrm{diff},\lambda}^{\uparrow}(\tau,\mu,\varphi) .
$$
(6.94)

By substituting Eq. (6.94) into Eq. (6.93) and simplifying the resultant expressions, we obtain

$$
I_{\text{diff},\lambda}^{\uparrow}(\tau,\mu,\varphi) = I_{\text{diff},\lambda}^{\uparrow}(\tau^{*},\mu,\varphi) \cdot e^{-(\tau^{*}-\tau)/\mu} + \int_{\tau}^{\tau^{*}} \{ J_{\text{dir},\lambda}(\tau',\mu,\varphi)
$$

$$
+ J_{\text{diff},\lambda}(\tau',\mu,\varphi) + (1 - \tilde{\omega}) \cdot B_{\lambda}[T(\tau')] \} \cdot e^{-(\tau'-\tau)/\mu} \, \frac{d\tau'}{\mu} . \tag{6.95}
$$

Using the lower boundary condition, see Eq. (6.75), and substituting the detailed expressions for the source terms $J_{\text{dir},\lambda}$, $J_{\text{diff},\lambda}$, and $J_{\text{emi},\lambda}$, see Eqs. (6.69)–(6.71), we obtain

$$
I_{\text{diff},\lambda}^{\uparrow}(\tau,\mu,\varphi)
$$

$$
= + \frac{\gamma(\lambda,\mu_0,\varphi_0,\mu,\varphi)}{\pi} \cdot S_{\text{dir},\lambda,\text{TOA}} \cdot \mu_0 \cdot e^{-\tau^{*}/\mu_0} \cdot e^{-(\tau^{*}-\tau)/\mu}
$$

$$
+ \int_0^{2\pi} \int_0^1 \frac{\gamma(\lambda,\mu',\varphi',\mu,\varphi)}{\pi} \cdot I_{\text{diff},\lambda}^{\downarrow}(\tau^{*},-\mu',\varphi') \cdot \mu' \, d\mu' \, d\varphi' \cdot e^{-(\tau^{*}-\tau)/\mu}
$$

$$
+ \int_{\tau}^{\tau^{*}} S_{\text{dir},\lambda,\text{TOA}} \cdot e^{-\tau'/\mu_0} \cdot \tilde{\omega}(\lambda,\tau') \cdot \frac{\mathcal{P}(\lambda,\tau',-\mu_0,\varphi_0,\mu,\varphi)}{4\pi} \cdot e^{-(\tau'-\tau)/\mu} \, \frac{d\tau'}{\mu}
$$

$$
+ \int_{\tau}^{\tau^{*}} \tilde{\omega}(\lambda,\tau') \int_0^{2\pi} \int_{-1}^1 I_{\text{diff},\lambda}(\tau',\mu',\varphi') \cdot \frac{\mathcal{P}(\lambda,\tau',\mu',\varphi',\mu,\varphi)}{4\pi} \, d\mu' \, d\varphi'
$$

$$
\times e^{-(\tau'-\tau)/\mu} \, \frac{d\tau'}{\mu}
$$

$$
+ \int_{\tau}^{\tau^{*}} [1 - \tilde{\omega}(\lambda,\tau')] \cdot B_{\lambda}[T(\tau')] \cdot e^{-(\tau'-\tau)/\mu} \, \frac{d\tau'}{\mu}
$$

$$
+ \varepsilon(\lambda,\tau^{*}) \cdot B_{\lambda}[T(\tau^{*})] \cdot e^{-(\tau^{*}-\tau)/\mu} . \tag{6.96}
$$

The six terms on the right side of Eq. (6.96) are discussed as follows.

Surface reflection of direct solar radiation

$$
S_{\text{dir},\lambda,\text{TOA}} \cdot e^{-\tau^{*}/\mu_0} \cdot \frac{\gamma(\lambda,\mu_0,\varphi_0,\mu,\varphi)}{\pi} \cdot \mu_0 \cdot e^{-(\tau^{*}-\tau)/\mu} . \tag{6.97}
$$

Three events occur: first, an exponential decrease of the direct solar radiation with $e^{-\tau^{*}/\mu_0}$; second, reflection occurs at the surface; and third, an exponential decrease takes place with $e^{-(\tau^{*}-\tau)/\mu}$, see Figure 6.11.

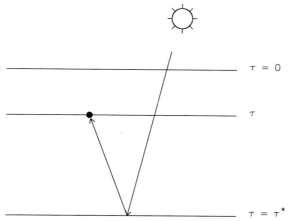

Figure 6.11 Surface reflection of direct solar radiation.

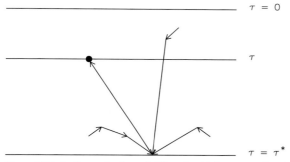

Figure 6.12 Surface reflection of diffuse (multiple-scattered) solar radiation.

Surface reflection of diffuse (due to multiple-scattering) solar radiation

$$\int_0^{2\pi} \int_0^1 \frac{\gamma(\lambda, \mu', \varphi', \mu, \varphi)}{\pi} \cdot I^{\downarrow}_{\text{diff}, \lambda}(\tau^*, -\mu', \varphi') \cdot \mu' \cdot d\mu' d\varphi' \cdot e^{-(\tau^* - \tau)/\mu} . \quad (6.98)$$

The diffuse radiance, $I^{\downarrow}_{\text{diff}, \lambda}$, reaching the surface from all directions $(-\mu', \varphi')$ in the upper hemisphere, is reflected according to the surface properties given by $\gamma(\lambda, \mu', \varphi', \mu, \varphi)$, see Figure 6.12. The reflected radiance is decreased by a factor of $e^{-(\tau^* - \tau)/\mu}$, but is no longer scattered on its way from the surface to the observational point. An integration over the upper hemisphere is performed because diffuse radiation may originate from all directions of the upper hemisphere.

First-order scattering of direct solar radiation

$$\int_\tau^{\tau^*} S_{\text{dir}, \lambda, \text{TOA}} \cdot e^{-\tau'/\mu_0} \cdot \tilde{\omega}(\lambda, \tau') \cdot \frac{\mathcal{P}(\lambda, \tau', -\mu_0, \varphi_0, \mu, \varphi)}{4\pi} \cdot e^{-(\tau' - \tau)/\mu} \frac{d\tau'}{\mu} . \quad (6.99)$$

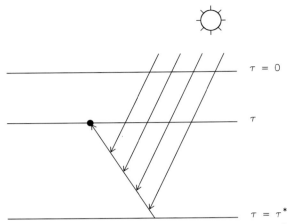

Figure 6.13 Primary scattering of direct solar radiation, upward radiance.

By integration from τ to τ^*, all primary scattering events in the layer between τ and τ^* are summed, see Figure 6.13. Before the first-order scattering event, $S_{\text{dir},\lambda,\text{TOA}} \cdot e^{-\tau'/\mu_0}$ represents the decrease of solar radiation until reaching altitude τ' where the first scattering events primarily take place. Here, the phase function $\mathcal{P}(\lambda, \tau', -\mu_0, \varphi_0, \mu, \varphi)$ regulates the amount of radiation arriving out of direction $(-\mu_0, \varphi_0)$ and scattered into direction (μ, φ). The part absorbed is defined by $\tilde{\omega}(\lambda, \tau')$. After the first-order scattering event, the scattered radiation is further attenuated by a factor of $e^{-(\tau'-\tau)/\mu}$.

Multiple-scattering of diffuse solar radiation

$$
\int_{\tau}^{\tau^*} \tilde{\omega}(\lambda, \tau') \int_{0}^{2\pi} \int_{-1}^{1} I_{\text{diff},\lambda}(\tau', \mu', \varphi') \cdot \frac{\mathcal{P}(\lambda, \tau', \mu', \varphi', \mu, \varphi)}{4\pi} \, d\mu' \, d\varphi' \cdot e^{-(\tau'-\tau)/\mu} \, \frac{d\tau'}{\mu} .
$$

(6.100)

On the basis of integration from τ to τ^*, all multiple-scattering events in the layer between τ and τ^* are summed, see Figure 6.14. All diffuse radiation at altitude τ' is weighted with the respective phase function $\mathcal{P}(\lambda, \tau', -\mu', \varphi', \mu, \varphi)$ at altitude τ'. The inner integral gives the diffuse contribution in direction (μ, φ) at altitude τ'. After the scattering event, the radiation is decreased by a factor of $e^{-(\tau'-\tau)/\mu}$.

Atmospheric emission

$$
\int_{\tau}^{\tau^*} [1 - \tilde{\omega}(\lambda, \tau')] \cdot B_\lambda[T(\tau')] \cdot e^{-(\tau'-\tau)/\mu} \, \frac{d\tau'}{\mu} .
$$

(6.101)

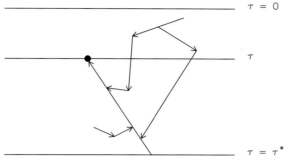

$\tau = 0$

τ

$\tau = \tau^*$

Figure 6.14 Multiple-scattering of diffuse solar radiation.

This term represents the contributions of the thermal emission at the different altitudes τ'. The contributions are decreased with the factor $e^{-(\tau'-\tau)/\mu}$ and are subsequently integrated over altitude.

Surface emission

$$\varepsilon(\lambda, \tau^*) \cdot B_\lambda[T(\tau^*)] \cdot e^{-(\tau^*-\tau)/\mu} \ . \tag{6.102}$$

This contribution results from the Planck emission of the surface ($\tau' = \tau^*$) and is decreased by a factor of $e^{-(\tau^*-\tau)/\mu}$.

Problems

Problem 6.1 Cloud Optics

Consider a cloud extending from 3000 to 3100 m altitude, with a liquid water content of LWC $= 0.2\,\mathrm{mg\,l^{-1}}$, and consisting of droplets with an effective radius of 12 μm.

a) Calculate the optical depth of the cloud.
b) The differential form of the Lambert–Bouguer law for direct solar irradiance S_{dir} is given by

$$\frac{\mathrm{d}\,S_{\mathrm{dir}}}{S_{\mathrm{dir}}} = -\frac{\mathrm{d}\tau}{\mu_0} \ . \tag{6.103}$$

Solve the differential equation and calculate the attenuation of direct solar radiation for:

1) the Sun in the zenith;
2) the Sun at 30° above the horizon.

c) Show that a reduction of the droplet size by half decreases the transmitted radiation if the LWC is kept constant.

Problem 6.2 Solar Constant

The solar constant F_k is defined as the broadband (wavelength-integrated) extraterrestrial irradiance $S_{dir,\lambda,TOA}$ at the mean distance between the Sun and the Earth, that is,

$$F_k = \int_{\lambda_1}^{\lambda_2} S_{dir,\lambda,TOA}\, d\lambda \ . \tag{6.104}$$

The solar spectrum spans the region from $\lambda_1 = 0.2\ \mu m$ to $\lambda_1 = 5\ \mu m$ wavelengths.

a) Use the NREL website http://www.nrel.gov/midc/solpos/solpos.html (accessed November 2011) to calculate and plot the diurnal pattern of the solar zenith angle θ_0 in Kiruna and Buenos Aires (find the latitudes and longitudes for these cities on the internet) for the following dates: 4 July, 21 December, 3 January, and 20 March. Use a plot program of your choice. Associate the terms "equinox," "solstice," "perihelion," and "aphelion" with the four given dates and explain what they mean (with the help of your plots).

b) For Buenos Aires, calculate the extraterrestrial solar irradiance $S_{dir,TOA}(\theta_0)$ as a function of the solar zenith angle θ_0 for local solar noon on three days: 1 October, 4 July, and 3 January.
Use the extraterrestrial direct solar irradiance for normal incidence F_k and multiply it with the cosine of the solar zenith angle to obtain $S_{dir,TOA}(\theta_0)$. The solar extraterrestrial irradiance is actually varying and is monitored by instruments like the total solar irradiance monitor (TIM). Use the latest values, which can be found at the following web address: http://lasp.colorado.edu/lisird/ (accessed November 2011) $\rightarrow$ Total Solar Irradiance $\rightarrow$ SORCE Total Solar Irradiance (TSI) $\rightarrow$ Adjust the plot.

c) Now, use the web program from (a), insert F_k from (b) and recalculate the extraterrestrial irradiance $S_{dir,TOA}(\theta_0)$ for Buenos Aires for all three days. Compare these with your result from (b). Explain the differences.

Problem 6.3 Application of the Lambert–Bouguer Law

Consider the following differential equation:

$$\mu_0 \frac{d\,S_{dir}}{d\tau} = -S_{dir}(\tau) \ . \tag{6.105}$$

This represents the differential form of the Lambert–Bouguer law applied to the direct solar radiation. $S_{dir,TOA}$ enters the atmosphere at the top of atmosphere (TOA). The variable τ is called the optical thickness, and represents a measure of the opacity of the atmosphere. At TOA, $\tau = 0$.

a) Solve this ordinary differential equation. Explain why $S_{dir}(\tau)/S_{dir,TOA}$ is smaller when the Sun is near the horizon than for an overhead Sun.

b) A further simplification is the use of the relative air mass M_0 (dimensionless), which is defined as

$$M_0 = \frac{1}{\mu_0} = \frac{1}{\cos \theta_0} \; . \tag{6.106}$$

Calculate the value of M_0 for which the Sun becomes invisible. Use the definition of the visibility threshold $S_{dir}(\tau)/S_{dir,TOA} = 0.02$. Assume three different scenarios:

1) Molecular atmosphere (Rayleigh): $\tau = 0.14$,
2) Dusty atmosphere: $\tau = 1$, and
3) Cloudy atmosphere: $\tau = 10$.

What are the corresponding solar zenith angles?

c) For the case of a molecular atmosphere, why and how (qualitatively) will your result of the maximum solar zenith angle from (2) differ from a "real" atmosphere? Name and discuss two reasons (think about sunset).

Problem 6.4 Layer Absorption

One of the consequences of the radiative transfer equation is a relationship between the divergence of the net irradiance vector, $\vec{F}_{net}$, and the actinic flux density in a scattering and absorbing atmosphere, that is,

$$\vec{\nabla} \cdot \vec{F}_{net} = b_{abs} \cdot F_{act} \; . \tag{6.107}$$

Design a strategy for airborne measurements of $\vec{F}_{net}$ and F_{act} to determine the mean volumetric absorption coefficient of a horizontally homogeneous aerosol layer. Which of the two radiative quantities needs to be measured precisely?

7
Numerical and Approximate Solution Techniques for the RTE

7.1
Legendre and Fourier Expansions

7.1.1
Expansion of Phase Function in Terms of Legendre Polynomials

In many numerical approaches to solving RTE, the phase function is expanded into a series of the Legendre polynomials, for example, see Nagel et al. (1978), or Zdunkowski et al. (2007). An infinite number of expansion terms are required to represent the original phase function precisely, but the expansion is limited to N terms for practical applications. Specifically, we can express the phase function as

$$\mathcal{P}(\cos \vartheta) = \sum_{n=0}^{\infty} C_n \cdot P_n(\cos \vartheta) \approx \sum_{n=0}^{N-1} C_n \cdot P_n(\cos \vartheta) \,. \tag{7.1}$$

Multiplying Eq. (7.1) with $P_m(\cos \vartheta)$, integrating the resultant expression over $\cos \vartheta \in [-1, 1]$, and using the orthogonality properties of the Legendre polynomials given by Eq. (2.44), we obtain

$$\int_{-1}^{1} \mathcal{P}(\cos \vartheta) \cdot P_m(\cos \vartheta) \mathrm{d} \cos \vartheta \approx \int_{-1}^{1} \sum_{n=0}^{N-1} C_n \cdot P_n(\cos \vartheta) \cdot P_m(\cos \vartheta) \, \mathrm{d} \cos \vartheta$$

$$= \sum_{n=0}^{N-1} C_n \cdot \int_{-1}^{1} P_n(\cos \vartheta) \cdot P_m(\cos \vartheta) \, \mathrm{d} \cos \vartheta$$

$$= \sum_{n=0}^{N-1} C_n \cdot \delta_{nm} \cdot \left(\frac{2}{2n + 1} \right)$$

$$= C_m \cdot \left(\frac{2}{2m + 1} \right) . \tag{7.2}$$

Theory of Atmospheric Radiative Transfer, First Edition. Manfred Wendisch and Ping Yang
© 2012 WILEY-VCH Verlag GmbH & Co. KGaA. Published 2012 by WILEY-VCH Verlag GmbH & Co. KGaA.

Thus, the expansion coefficients in Eq. (7.1) are found by

$$C_n = \frac{2n+1}{2} \int_{-1}^{1} \mathcal{P}(\cos \vartheta) \cdot P_n(\cos \vartheta) \, \mathrm{d} \cos \vartheta \; . \tag{7.3}$$

From the preceding equation, it can be proven that $C_0 = 1$ using $P_0 = 1$, see Eq. (2.45), and applying the normalization condition of the phase function, see Eqs. (4.60) and (2.37),

$$
\begin{aligned}
C_0 &= \frac{1}{2} \int_{-1}^{1} \mathcal{P}(\cos \vartheta) \cdot P_0(\cos \vartheta) \, \mathrm{d} \cos \vartheta \\
&= \frac{1}{2} \int_{-1}^{1} \mathcal{P}(\cos \vartheta) \, \mathrm{d} \cos \vartheta \\
&= \frac{1}{2} \int_{-1}^{1} \mathcal{P}(\mu) \, \mathrm{d}\mu \\
&= 1 \; .
\end{aligned}
\tag{7.4}
$$

For the Legendre coefficient C_1, the following relation holds:

$$
\begin{aligned}
C_1 &= \frac{3}{2} \int_{-1}^{1} \mathcal{P}(\cos \vartheta) \cdot P_1(\cos \vartheta) \, \mathrm{d} \cos \vartheta \\
&= \frac{3}{2} \int_{-1}^{1} \mathcal{P}(\cos \vartheta) \cdot \cos \vartheta \, \mathrm{d} \cos \vartheta \\
&= 3g \; ,
\end{aligned}
\tag{7.5}
$$

where g is the asymmetry factor of the phase function, see Eq. (4.64).

As an example of the phase function expansion, we consider the Henyey–Greenstein phase function $\mathcal{P}_{HG}(\cos \vartheta)$, see Eq. (4.257), which is widely used in various problems concerning radiative transfer. To represent $\mathcal{P}_{HG}(\cos \vartheta)$ in terms of the Legendre polynomials, we begin with the generating function (Arfken and Weber, 2005) for the Legendre polynomials given by

$$\frac{1}{(1 - 2x \cdot g + g^2)^{1/2}} = \sum_{n=0}^{\infty} g^n \cdot P_n(x) \; , \quad |g| < 1 \; . \tag{7.6}$$

Taking the derivative of the preceding equation with respect to g yields

$$-\frac{1}{2} \cdot \frac{(-2x + 2g)}{(1 - 2x \cdot g + g^2)^{3/2}} = \sum_{n=0}^{\infty} n \cdot g^{n-1} \cdot P_n(x) \; . \tag{7.7}$$

By multiplying Eq. (7.7) with $2g$ and by adding the resultant expression and Eq. (7.6), we get

$$\frac{(1 - 2x \cdot g + g^2) + (2x \cdot g - 2g^2)}{(1 - 2x \cdot g + g^2)^{3/2}} = \sum_{n=0}^{\infty} (2n + 1) \cdot g^n \cdot P_n(x) \,. \tag{7.8}$$

If we let $x = \cos \vartheta$ and simplify the left side of Eq. (7.8), we obtain

$$\frac{1 - g^2}{(1 - 2g \cdot \cos \vartheta + g^2)^{3/2}} = \sum_{n=0}^{\infty} (2n + 1) \cdot g^n \cdot P_n(\cos \vartheta) \,. \tag{7.9}$$

The left side of Eq. (7.9) is the Henyey–Greenstein phase function $\mathcal{P}_{HG}(\cos \vartheta)$, see Eq. (4.257). Thus, the expansion coefficients for this phase function are given by comparison with Eq. (7.1), that is,

$$C_{HG,n} = (2n + 1) \cdot g^n \,. \tag{7.10}$$

Another widely used phase function in radiative transfer simulations, particularly in the case of clear-sky atmospheres, is the Rayleigh phase function, see Eq. (4.127), which can be written in the form

$$
\begin{aligned}
\mathcal{P}_{unp,Rayl}(\vartheta) &= \frac{3}{4} \cdot (1 + \cos^2 \vartheta) \\
&= 1 + \frac{3}{4} \cdot \cos^2 \vartheta - \frac{1}{4} \\
&= 1 + \frac{1}{2} \cdot \frac{1}{2} \cdot (3 \cos^2 \vartheta - 1) \\
&= 1 \cdot P_0 + 0 \cdot P_1(\cos \vartheta) + \frac{1}{2} \cdot P_2(\cos \vartheta) + \sum_{n=3}^{\infty} 0 \cdot P_n(\cos \vartheta) \,,
\end{aligned}
\tag{7.11}
$$

where the first three Legendre polynomials are given in Eqs. (2.45)–(2.47). The Legendre expansion coefficients for the Rayleigh phase function are

$$
\begin{aligned}
C_{Rayl,0} &= C_0 = 1 \,, \\
C_{Rayl,1} &= 3g_{Rayl} = 0 \,, \\
C_{Rayl,2} &= \frac{1}{2} \,, \\
C_{Rayl,n} &= 0 \text{ for } n \geq 3 \,. \tag{7.12}
\end{aligned}
$$

Therefore, the first three orders of the Legendre polynomials are sufficient enough to represent the Rayleigh phase function exactly.

7.1.2
Truncation of Phase Function and Similarity Principle

The phase function associated with the scattering of VIS and NIR EM radiation by cloud particles, particularly ice crystals within cirrus clouds, usually has a pronounced peak in the forward scattering direction primarily due to diffraction.

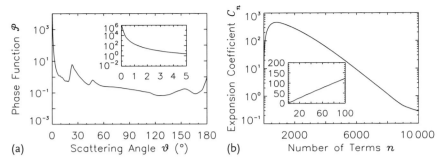

Figure 7.1 (a) The phase function of an ensemble of ice crystals with an effective particle of size of 100 μm at wavelength 0.64 μm. The strong forward peak within the first 5° scattering angle is shown with a finer resolution in the subpanel. (b) The expansion coefficients C_n, see Eq. (7.3).

Quantitatively, the forward peak can be several orders of magnitude larger than the values of the phase function at side- and backscattering angles. This feature of the phase function of cloud particles requires a significantly large number of expansion terms in Eq. (7.1) to accurately reproduce the original phase function.

As an example, Figure 7.1a shows the phase function of ice crystals with a certain habit (shape) and size distribution, computed by Baum et al. (2005), for application to the operational cloud property retrievals (Platnick et al., 2003) from the observations of the Moderate Resolution Imaging Spectroradiometer (MODIS), see King et al. (1992). For this ice crystal habit mixture scheme, it is assumed that: $D < 60$ μm is 100% droxtals; 60 μm $\leq D < 1000$ μm is 15% 3D bullet rosettes, 50% solid columns, and 35% hexagonal plates; 1000 μm $\leq D \leq 2500$ μm is 45% hollow columns, 45% solid columns, and 10% aggregates; and $D > 2500$ μm is 97% 3D bullet rosette and 3% aggregates, where D is the maximum dimension of the ice particles. The forward scattering peak is obviously more than six orders of magnitude larger than the phase function values around 120°. Figure 7.1b shows the expansion coefficients, C_n, in Eq. (7.3). From the diagram, it is evident that the expansion coefficients are not negligible even for an N value as large as several thousands.

Figure 7.2 shows the reconstruction of the phase function shown in Figure 7.1 with $N = 500$, 5000, 7000, and 10 000 terms of Legendre polynomials based on Eq. (7.1). In the cases of $N = 500$ and 5000, the reconstructed phase functions are substantially different from the original phase function. Even with $N = 7000$, some noticeable differences between the original and reconstructed phase functions are still observed near the forward angles, 120°, and the backscattering angles. An accurate representation of the original phase function requires approximately 10 000 terms of Legendre polynomials and is not practical for numerical computation.

Because the forward peak is only pronounced within a very narrow region around the forward direction (see Figure 7.1a), we can approximate the original phase func-

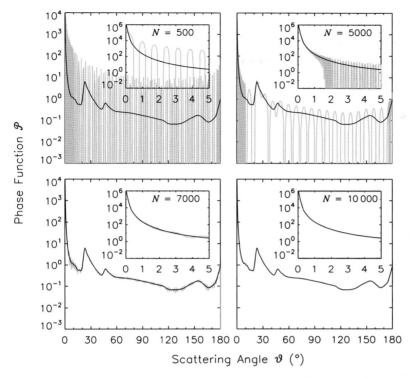

Figure 7.2 Comparison of the original phase function (black solid line) shown in Figure 7.1 with the reconstructed phase function with N terms of Legendre polynomials (gray solid line). For illustration with a logarithmic scale, the zero and negative values of the reconstructed phase functions are assigned as 10^{-19} in the plots.

tion in the form of

$$\mathcal{P}(\cos \vartheta) \approx f_{\text{fws}} \cdot 2\delta(1 - \cos \vartheta) + (1 - f_{\text{fws}}) \cdot \mathcal{P}'(\cos \vartheta) , \tag{7.13}$$

where $\mathcal{P}'(\cos \vartheta)$ is referred to as the truncated phase function. It is straightforward to prove that $\mathcal{P}'(\cos \vartheta)$ is normalized, provided the original phase function is normalized. A fraction of scattered energy, indicated in terms of the quantity f_{fws}, is assumed to be in the direction of the incident radiation and can be regarded as "unscattered." Thus, if we use $\mathcal{P}'(\cos \vartheta)$ to simulate the radiative transfer, the corresponding optical thickness and single-scattering albedo need to be scaled.

Let the original optical thickness and single-scattering albedo be τ and $\tilde{\omega}$, respectively. The optical thickness, τ, can be decomposed into τ_{sca} and τ_{abs}, which are the contributions from the scattering and absorption processes. The scaled optical thickness, τ', consistent with $\mathcal{P}'(\cos\vartheta)$ is given by

$$
\begin{aligned}
\tau' &= \tau'_{\text{sca}} + \tau_{\text{abs}} \\
&= \tau_{\text{sca}} - f_{\text{fws}} \cdot \tau_{\text{sca}} + \tau_{\text{abs}} \\
&= \tau - f_{\text{fws}} \cdot \tau_{\text{sca}} \\
&= \tau - f_{\text{fws}} \cdot \tau \cdot \tilde{\omega} \\
&= (1 - f_{\text{fws}} \cdot \tilde{\omega}) \cdot \tau .
\end{aligned}
\tag{7.14}
$$

Accordingly, the scaled single-scattering albedo is given by

$$
\begin{aligned}
\tilde{\omega}' &= \frac{\tau'_{\text{sca}}}{\tau'} \\
&= \frac{(1 - f_{\text{fws}}) \cdot \tau_{\text{sca}}}{\tau'} \\
&= \frac{(1 - f_{\text{fws}}) \cdot \tau_{\text{sca}}}{(1 - f_{\text{fws}} \cdot \tilde{\omega}) \cdot \tau} \\
&= \frac{(1 - f_{\text{fws}}) \cdot \tilde{\omega}}{1 - f_{\text{fws}} \cdot \tilde{\omega}} .
\end{aligned}
\tag{7.15}
$$

To the best of the authors' knowledge, the scaling scheme associated with Eqs. (7.14) and (7.15) was originally reported by Joseph et al. (1976), but is also known as the δ-scaling approximation (Zdunkowski et al., 2007), the similarity principle (Liou, 2002), and the similarity relations (Thomas and Stamnes, 1999; van de Hulst, 1980). A more rigorous proof of this principle based on the radiative transfer equation will be given in the following section.

7.1.3
Atmospheric Angular Coordinates

In radiative transfer calculation, the angular dependence of radiometric quantities is usually specified with respect to the zenith (θ and θ') and azimuthal (φ and φ') angles, where the pair of angles (θ', φ') indicates the incident direction, and the pair of angles (θ, φ) specifies the scattering direction. Accordingly, it is necessary to express the phase function as a function of θ, θ', φ, and φ' instead of the scattering angle ϑ. Using Eq. (2.41) for the relation between the four angles and the addition theorem for spherical harmonic functions, we obtain

$$
\begin{aligned}
P_n(\cos\vartheta) &= P_n(\mu, \varphi, \mu', \varphi') \\
&= P_n(\mu) \cdot P_n(\mu') + 2 \sum_{m=1}^{n} \frac{(n-m)!}{(n+m)!} \cdot P_n^{(m)}(\mu) \cdot P_n^{(m)}(\mu') \cdot \cos m(\varphi - \varphi') \\
&= \sum_{m=0}^{n} (2 - \delta_{m0}) \cdot \tilde{P}_n^{(m)}(\mu) \cdot \tilde{P}_n^{(m)}(\mu') \cdot \cos m(\varphi - \varphi') ,
\end{aligned}
\tag{7.16}
$$

where $\mu = \cos\theta$ and $\mu' = \cos\theta'$. The quantities $\tilde{P}_n^{(m)}(\mu)$ and $\tilde{P}_n^{(m)}(\mu')$ are the renormalized Legendre functions, defined by Eq. (2.62) and derived from the Legendre functions, see Eq. (2.52) in Section 2.6.2. Using Eq. (7.16), we can expand the phase function as, see Eq. (7.1),

$$\mathcal{P}(\cos\vartheta) \approx \sum_{n=0}^{N-1} C_n \cdot P_n(\cos\vartheta)$$

$$= \sum_{n=0}^{N-1} C_n \cdot \sum_{m=0}^{n} (2 - \delta_{m0}) \cdot \tilde{P}_n^{(m)}(\mu) \cdot \tilde{P}_n^{(m)}(\mu') \cdot \cos m(\varphi - \varphi')$$

$$= \sum_{n=0}^{N-1} \sum_{m=0}^{\infty} C_n \cdot (2 - \delta_{m0}) \cdot \tilde{P}_n^{(m)}(\mu) \cdot \tilde{P}_n^{(m)}(\mu') \cdot \cos m(\varphi - \varphi') . \tag{7.17}$$

The change of the second summation

$$\sum_{m=0}^{n} \to \sum_{m=0}^{\infty} ,$$

is a result of the following relation:

$$\tilde{P}_n^{(m)} = 0 ; \quad \text{if} \quad m > n . \tag{7.18}$$

After we invert the order of the summations in Eq. (7.17) and utilize Eq. (7.18) once more, we get

$$\mathcal{P}(\cos\vartheta) \approx \sum_{n=0}^{N-1} C_n \cdot P_n(\cos\vartheta)$$

$$= \sum_{m=0}^{\infty} \sum_{n=0}^{N-1} C_n \cdot (2 - \delta_{m0}) \cdot \tilde{P}_n^{(m)}(\mu) \cdot \tilde{P}_n^{(m)}(\mu') \cdot \cos m(\varphi - \varphi')$$

$$= \sum_{m=0}^{N-1} (2 - \delta_{m0}) \cdot \sum_{n=m}^{N-1} C_n \cdot \tilde{P}_n^{(m)}(\mu) \cdot \tilde{P}_n^{(m)}(\mu') \cdot \cos m(\varphi - \varphi') . \tag{7.19}$$

This yields the Fourier expansion of the phase function

$$\mathcal{P}(\cos\vartheta) \approx \sum_{m=0}^{N-1} P^{(m)}(\mu, \mu') \cdot \cos m(\varphi - \varphi') , \tag{7.20}$$

where

$$P^{(m)}(\mu, \mu') = (2 - \delta_{m0}) \cdot \sum_{n=m}^{N-1} C_n \cdot \tilde{P}_n^{(m)}(\mu) \cdot \tilde{P}_n^{(m)}(\mu') . \tag{7.21}$$

It is convenient to use the atmospheric angular coordinates to rigorously derive the similarity relations in Eqs. (7.14) and (7.15) in conjunction with the approximation to the original phase function in Eq. (7.13), which can be expressed in the form of, see Eq. (2.42),

$$\mathcal{P}(\cos\vartheta) \approx 2 f_{\text{fws}} \cdot \delta(1 - \cos\vartheta) + (1 - f_{\text{fws}}) \cdot \mathcal{P}'(\cos\vartheta)$$

$$= 4\pi f_{\text{fws}} \cdot \delta(\mu - \mu') \cdot \delta(\varphi - \varphi') + (1 - f_{\text{fws}}) \cdot \mathcal{P}'(\mu, \varphi, \mu', \varphi') . \tag{7.22}$$

Note, the equation for the transfer of total radiance (direct plus diffuse radiance) in a scattering atmosphere is given by

$$
\mu \frac{\mathrm{d} I_\lambda(\tau, \mu, \varphi)}{\mathrm{d}\tau} = I_\lambda(\tau, \mu, \varphi)
$$
$$
- \tilde{\omega}(\tau) \int_0^{2\pi} \int_{-1}^1 I_\lambda(\tau, \mu', \varphi') \cdot \frac{\mathcal{P}(\tau, \mu, \varphi, \mu', \varphi')}{4\pi} \, \mathrm{d}\mu' \, \mathrm{d}\varphi'
$$
$$
- [1 - \tilde{\omega}(\tau)] \cdot B_\lambda[T(\tau)] \,. \tag{7.23}
$$

Substituting Eq. (7.22) into the integral in Eq. (7.23) and using the properties of the Dirac δ-function gives us the scattering term in Eq. (7.23), that is,

$$
\tilde{\omega}(\tau) \int_0^{2\pi} \int_{-1}^1 I_\lambda(\tau, \mu', \varphi')
$$
$$
\times \frac{4\pi f_{\mathrm{fws}} \cdot \delta(\mu - \mu') \cdot \delta(\varphi - \varphi') + (1 - f_{\mathrm{fws}}) \cdot \mathcal{P}'(\tau, \mu, \varphi, \mu', \varphi')}{4\pi} \, \mathrm{d}\mu' \, \mathrm{d}\varphi'
$$
$$
= \tilde{\omega}(\tau) \cdot f_{\mathrm{fws}} \cdot I_\lambda(\tau, \mu, \varphi)
$$
$$
+ \tilde{\omega}(\tau) \cdot (1 - f_{\mathrm{fws}}) \int_0^{2\pi} \int_{-1}^1 I_\lambda(\tau, \mu', \varphi') \cdot \frac{\mathcal{P}'(\tau, \mu, \varphi, \mu', \varphi')}{4\pi} \, \mathrm{d}\mu' \, \mathrm{d}\varphi' \,. \tag{7.24}
$$

Thus, the radiative transfer equation can be rewritten in the form of

$$
\mu \frac{\mathrm{d} I_\lambda(\tau, \mu, \varphi)}{\mathrm{d}\tau} = I_\lambda(\tau, \mu, \varphi)
$$
$$
- \tilde{\omega}(\tau) \cdot f_{\mathrm{fws}} \cdot I_\lambda(\tau, \mu, \varphi)
$$
$$
- \tilde{\omega}(\tau) \cdot (1 - f_{\mathrm{fws}}) \int_0^{2\pi} \int_{-1}^1 I_\lambda(\tau, \mu', \varphi') \cdot \frac{\mathcal{P}'(\tau, \mu, \varphi, \mu', \varphi')}{4\pi} \, \mathrm{d}\mu' \, \mathrm{d}\varphi'
$$
$$
- [1 - \tilde{\omega}(\tau)] \cdot B_\lambda[T(\tau)] \,, \tag{7.25}
$$

or with some simplifications,

$$
\mu \frac{\mathrm{d} I_\lambda(\tau, \mu, \varphi)}{\mathrm{d}\tau} = I_\lambda(\tau, \mu, \varphi) \cdot [1 - f_{\mathrm{fws}} \cdot \tilde{\omega}(\tau)]
$$
$$
- \tilde{\omega}(\tau) \cdot (1 - f_{\mathrm{fws}}) \int_0^{2\pi} \int_{-1}^1 I_\lambda(\tau, \mu', \varphi') \cdot \frac{\mathcal{P}'(\tau, \mu, \varphi, \mu', \varphi')}{4\pi} \, \mathrm{d}\mu' \, \mathrm{d}\varphi'
$$
$$
- [1 - \tilde{\omega}(\tau)] \cdot B_\lambda[T(\tau)] \,. \tag{7.26}
$$

If we apply the definitions of the similarity relations, see Eqs. (7.14) and (7.15), in differential form,

$$[1 - f_{\text{fws}} \cdot \tilde{\omega}(\tau)] = \frac{d\tau'}{d\tau}$$

$$\tilde{\omega}(\tau) \cdot (1 - f_{\text{fws}}) = \tilde{\omega}'(\tau') \cdot [1 - f_{\text{fws}} \cdot \tilde{\omega}(\tau)] = \tilde{\omega}'(\tau') \frac{d\tau'}{d\tau}$$

$$[1 - \tilde{\omega}(\tau)] = [1 - \tilde{\omega}'(\tau')] \frac{d\tau'}{d\tau} , \tag{7.27}$$

we can rewrite Eq. (7.23) as

$$\mu \frac{d I_\lambda(\tau', \mu, \varphi)}{d\tau'} = I_\lambda(\tau', \mu, \varphi)$$

$$- \tilde{\omega}'(\tau') \int_0^{2\pi} \int_{-1}^1 I_\lambda(\tau', \mu', \varphi') \cdot \frac{\mathcal{P}'(\tau', \mu, \varphi, \mu', \varphi')}{4\pi} \, d\mu' \, d\varphi'$$

$$- [1 - \tilde{\omega}'(\tau')] \cdot B_\lambda[T(\tau')] . \tag{7.28}$$

Since Eqs. (7.23) and (7.28) have the same form, the similarity relations are valid. Furthermore, the diffuse radiative transfer equation, see Eq. (6.67), corresponding to Eq. (7.28), is in the form of

$$\mu \frac{d I_{\text{diff},\lambda}(\tau', \mu, \varphi)}{d\tau'} = I_{\text{diff},\lambda}(\tau', \mu, \varphi)$$

$$- \tilde{\omega}'(\tau') \cdot S_{\text{dir},\lambda,\text{TOA}} \cdot e^{-\tau'/\mu_0} \cdot \frac{\mathcal{P}'(\tau', \mu, \varphi, -\mu_0, \varphi_0)}{4\pi}$$

$$- \tilde{\omega}'(\tau') \int_0^{2\pi} \int_{-1}^1 I_{\text{diff},\lambda}(\tau', \mu', \varphi') \cdot \frac{\mathcal{P}'(\tau', \mu, \varphi, \mu', \varphi')}{4\pi} \, d\mu' \, d\varphi'$$

$$- [1 - \tilde{\omega}'(\tau')] \cdot B_\lambda[T(\tau')] . \tag{7.29}$$

In the preceding equation, the direct radiation, $S_{\text{dir},\lambda,\text{TOA}} \cdot e^{-\tau'/\mu_0}$, in the scaling scheme is larger than the corresponding unscaled value because the scaling decreases the equivalent optical thickness $\tau' < \tau$, see Eq. (7.14), when the truncated phase function is used, as noted by Zdunkowski et al. (2007). Physically, the attenuation of the direct radiation beam is smaller in the scaling scheme due to the assumption that a portion of scattered radiation in the directions near the forward direction is regarded as "being not scattered." However, the similarity principle approximately warrants the same amount of total downward radiation in the scaled and unscaled cases. The similarity principle is also a good approximation for the upward radiation.

7.1.4
The Delta-M Method (DMM) and Delta-Fit Methods (DFM)

Up to this point, we have not discussed how to truncate the phase function, or, specifically, how to define the quantities f_{fws} and $\mathcal{P}'$ in Eq. (7.13). The Delta-M

Method (DMM) developed by Wiscombe (1977) offers a logical way to truncate the phase function. Prior to Wiscombe's study, phase function truncation had been, at best, *ad hoc*. Hu et al. (2000) further enhanced Wiscombe's work and developed the Delta-Fit Method (DFM). In the following discussion, we will briefly recapture the two methods with some illustrative results.

If we use M terms of Legendre polynomials to specify the truncated counterpart for a general phase function that, in principle, requires an infinite number of Legendre polynomials in its expansion, Eq. (7.13), by using Eq. (7.1), can be written in the form

$$\mathcal{P}(\cos\vartheta) = \sum_{l=0}^{\infty} C_l \cdot P_l(\cos\vartheta)$$

$$\approx f_{\text{fws}} \cdot 2\delta(1 - \cos\vartheta) + (1 - f_{\text{fws}}) \cdot \sum_{l=0}^{M-1} C_l' \cdot P_l(\cos\vartheta) , \qquad (7.30)$$

with the truncated phase function

$$\mathcal{P}'(\cos\vartheta) = \sum_{l=0}^{M-1} C_l' \cdot P_l(\cos\vartheta) . \qquad (7.31)$$

Multiplying Eq. (7.30) with $P_k(\cos\vartheta)$ and integrating the resultant expression over $\mu = \cos\vartheta$ from -1 to 1, results in

$$\int_{-1}^{1} \sum_{l=0}^{\infty} C_l \cdot P_l(\cos\vartheta) \cdot P_k(\cos\vartheta)\mathrm{d}\mu \approx \int_{-1}^{1} f_{\text{fws}} \cdot 2\delta(1 - \cos\vartheta) \cdot P_k(\cos\vartheta)\,\mathrm{d}\mu$$

$$+ \int_{-1}^{1} (1 - f_{\text{fws}}) \cdot \sum_{l=0}^{M-1} C_l' \cdot P_l(\cos\vartheta) \cdot P_k(\cos\vartheta)\,\mathrm{d}\mu . \qquad (7.32)$$

By exchanging the integration and summation, applying the orthogonality relation for Legendre polynomials given by Eq. (2.44), and considering the following relation involving the Dirac δ-function, see Eq. (2.29), with $\phi(\mu) \equiv P_k(\mu)$; $x \equiv \mu$; $x_0 \equiv 1$,

$$\int_{-1}^{1} \delta(1 - \mu) \cdot P_k(\mu)\,\mathrm{d}\mu = P_k(1) = 1 \qquad (7.33)$$

yields

$$\frac{C_k}{2k + 1} \approx f_{\text{fws}} + (1 - f_{\text{fws}}) \cdot \frac{C_k'}{2k + 1} ;$$

$$\text{for } k = 0, 1, 2 \ldots, M - 2, M - 1 , \qquad (7.34)$$

$$\frac{C_k}{2k + 1} \approx f_{\text{fws}} ; \quad \text{for } k = M, M + 1, M + 2 \ldots \qquad (7.35)$$

It should be noted that Eq. (7.34) may exactly hold, provided the coefficients C'_k are appropriately specified; while, Eq. (7.35) is an approximation for a general phase function and leads to truncation errors. Furthermore, Eqs. (7.34) and (7.35) do not provide a constraint on the specification of the truncation factor f_{fws}. To overcome the indeterminateness, Wiscombe (1977) argued that it is more important to accurately represent the lower order expansion moments than the higher order ones and suggested specifying f_{fws} in the form of

$$f_{\text{fws}} = \frac{C_M}{2M+1} \cdot \tag{7.36}$$

An approach for the phase function truncation based on the determination of C'_k and f_{fws} in terms of Eqs. (7.34) and (7.36), respectively, is referred to as the DMM and was originally developed by Wiscombe (1977).

It should be noted that Eq. (7.30) is only an approximation. To be more specific, the left side of Eq. (7.30) represents the original phase function; whereas, the right side represents the summation of the truncated phase function and the Dirac δ-function employed as a surrogate to specify the distribution of truncated energy around the forward direction.

Using the following property of the Dirac δ-function (Morse and Feshbach, 1953) as a special case of Eq. (2.51),

$$\delta(1 - \cos\vartheta) = \frac{1}{2} \sum_{l=0}^{\infty} (2l + 1) \cdot P_l(\cos\vartheta) , \tag{7.37}$$

we can quantify the truncation error (Δ) associated with the DMM as, see Eq. (7.30),

$$\Delta = \sum_{l=0}^{\infty} C_l \cdot P_l(\cos\vartheta) - f_{\text{fws}} \cdot 2\delta(1 - \cos\vartheta) - (1 - f_{\text{fws}}) \cdot \sum_{l=0}^{M-1} C'_l \cdot P_l(\cos\vartheta)$$

$$= \sum_{l=0}^{\infty} C_l \cdot P_l(\cos\vartheta)$$

$$\quad - f_{\text{fws}} \cdot \sum_{l=0}^{\infty} (2l+1) \cdot P_l(\cos\vartheta) - (1 - f_{\text{fws}}) \cdot \sum_{l=0}^{M-1} C'_l \cdot P_l(\cos\vartheta)$$

$$= \sum_{l=M}^{\infty} \left[C_l - \left(\frac{2l+1}{2M+1} \right) \cdot C_M \right] \cdot P_l(\cos\vartheta) . \tag{7.38}$$

It is evident from Eqs. (7.34) and (7.36) that the truncated phase function $\mathcal{P}'(\cos\vartheta)$ can be specified in terms of M terms of Legendre polynomials with coefficients of

$$C'_l \approx \frac{[C_l - f_{\text{fws}} \cdot (2l+1)]}{(1 - f_{\text{fws}})}$$

$$= \left[C_l - \left(\frac{2l+1}{2M+1} \right) \cdot C_M \right] \Big/ \left(1 - \frac{C_M}{2M+1} \right) , \tag{7.39}$$

where C_l, $l = 0, 1, 2, 3, \ldots$ are the expansion coefficients for the original phase function. We obtain

$$
\mathcal{P}'(\cos \vartheta) = \sum_{l=0}^{M-1} C_l' \cdot P_l(\cos \vartheta)
$$

$$
= \sum_{l=0}^{M-1} \left[C_l - \left(\frac{2l+1}{2M+1} \right) \cdot C_M \right] \bigg/ \left(1 - \frac{C_M}{2M+1} \right) \cdot P_l(\cos \vartheta) .
$$

$$(7.40)$$

Hu et al. (2000) found that for a given number of Legendre polynomials, Eq. (7.40) may not be the best fitting to the original phase function and, instead, suggested the following least-square fitting method to determine the truncated phase function in terms of the coefficients $C_{\text{Hu},l}'$, namely,

$$
\frac{\partial}{\partial C_{\text{Hu},l}'} \sum_i W_i \left[\frac{\sum_{k=0}^{M-1} C_{\text{Hu},l}' \cdot P_k(\cos \vartheta_i)}{\mathcal{P}(\cos \vartheta_i)} - 1 \right]^2 = 0 ,
$$

$$(7.41)$$

where ϑ_i denotes the scattering angles at which the phase function values are calculated in the least-square fitting. Hu et al. (2000) also suggested 361 equally spaced scattering angles with a resolution of $0.5°$ for the least-square fitting. The term W_i in Eq. (7.41) indicates the weight for the phase function value at scattering angle ϑ_i. In the DFM, W_i is selected to be zero in a small angular region near the forward direction. For example, $\vartheta_i < 3°$ leads to the truncation of the forward peak. The truncation factor f_{fws} is given by $(1 - C_{\text{Hu},0}')$ and the truncated phase function, after normalization, is given by

$$
\mathcal{P}'(\cos \vartheta) = \frac{1}{C_{\text{Hu},0}'} \sum_{l=0}^{M-1} C_{\text{Hu},l}' \cdot P_l(\cos \vartheta) .
$$

$$(7.42)$$

As stated by Hu et al. (2000), one of the DFM's advantages is "better estimation of the phase function at large scattering angles with small phase function values" in comparison with the DMM. Furthermore, in the DFM, the first few moments can be conserved if necessary for some applications.

As an example, Figure 7.3 shows the truncated phase functions based on the DMM and the DFM for the original phase function shown in Figures 7.1 and 7.2. Note that the truncated phase functions shown in the diagram are not normalized, and the truncated phase function shown in Figure 7.2 is actually $(1 - f_{\text{fws}}) \cdot \mathcal{P}'(\cos \vartheta)$. Although about 10 000 terms are required to accurately represent the original phase function, see Section 7.1.2, the number of the expansion terms can be substantially reduced if the forward peak of the original phase function is truncated, evident in Figure 7.3. Furthermore, it is obvious that the DFM enhances the representation of the original phase function in terms of its truncated counterpart when compared with the DMM.

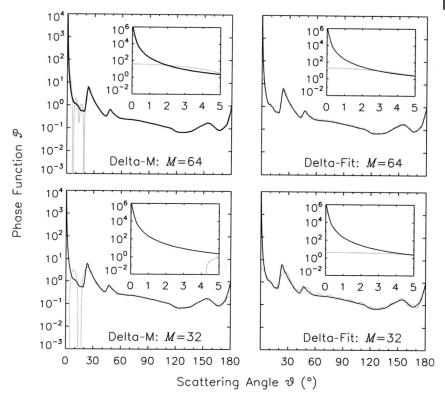

Figure 7.3 Comparison of the Delta-M Method (DMM) and the Delta-Fit Method (DFM) for phase function truncation. The original phase function (black solid line) is shown in Figure 7.1. Shown for the reconstructed phase (black dashed line) is $(1 - f_{\text{fws}}) \cdot \mathcal{P}'(\cos \vartheta)$ that is not normalized. For illustration with a logarithmic scale, the zero and negative values of the reconstructed phase functions are assigned as 10^{-19} in the plots.

7.1.5
Fourier Expansions of Diffuse Radiance and Irradiance

In accordance with the expansion of the phase function of Eq. (7.20), we expand the diffuse radiance in the form of

$$I_{\text{diff},\lambda}(\tau,\mu,\varphi) = \sum_{m=0}^{N-1} I_{\lambda}^{(m)}(\tau,\mu) \cdot \cos m\varphi \,, \tag{7.43}$$

where $I_{\lambda}^{(m)}(\tau,\mu)$ has the units of radiance and is referred to as the mth Fourier mode of the diffuse radiance. Note that in some texts, for example, Zdunkowski et al. (2007), the Fourier expansion of the diffuse radiance is expressed as

$$\sum_{m=0}^{N-1} (2 - \delta_{m0}) \cdot I_{\lambda}^{(m)}(\tau,\mu) \cdot \cos m\varphi \,, \tag{7.44}$$

giving rise to a constant factor difference in the downstream formulation.

As an example, we apply the Fourier modes of the diffuse radiance to the calculation of the corresponding diffuse irradiance. Specifically, the upward diffuse irradiance, $F_{\mathrm{diff},\lambda}^{\uparrow}(\tau)$, is given by integrating the upward radiance over the upper hemisphere as, see Eq. (3.52),

$$
\begin{aligned}
F_{\mathrm{diff},\lambda}^{\uparrow}(\tau) &= \int_0^{2\pi} \left[\int_0^1 I_{\mathrm{diff},\lambda}(\tau,\mu,\varphi) \cdot \mu \, \mathrm{d}\mu \right] \mathrm{d}\varphi \\
&= \int_0^1 \left[\int_0^{2\pi} I_{\mathrm{diff},\lambda}(\tau,\mu,\varphi)\mathrm{d}\varphi \right] \cdot \mu \, \mathrm{d}\mu \\
&= \int_0^1 \left[\int_0^{2\pi} \sum_{m=0}^{N-1} I_{\lambda}^{(m)}(\tau,\mu) \cdot \cos m\varphi \, \mathrm{d}\varphi \right] \cdot \mu \, \mathrm{d}\mu \\
&= 2\pi \int_0^1 I_{\lambda}^{(0)}(\tau,\mu) \cdot \mu \, \mathrm{d}\mu \ .
\end{aligned}
\tag{7.45}
$$

Similarly, for the downward diffuse irradiance, we get

$$
F_{\mathrm{diff},\lambda}^{\downarrow}(\tau) = 2\pi \int_0^1 I_{\lambda}^{(0)}(\tau,-\mu) \cdot \mu \, \mathrm{d}\mu \ .
\tag{7.46}
$$

In the derivation of the preceding two equations, we have used the following relation:

$$
\int_0^{2\pi} \cos m\varphi \, \mathrm{d}\varphi = 2\pi \cdot \delta_{m0} \ .
\tag{7.47}
$$

If we consider the radiance values at discrete zenith angles, for example, the $\mu = \cos\theta$ values specified at the abscissas of Gaussian quadrature, we can express, by applying Eq. (2.70), the upward irradiance in Eq. (7.45) as

$$
F_{\mathrm{diff},\lambda}^{\uparrow}(\tau) = 2\pi \sum_{i=1}^{s} I_{\lambda}^{(0)}(\tau,\mu_i) \cdot \mu_i \cdot c_i \ .
\tag{7.48}
$$

Similarly, we obtain the downward diffuse irradiance as

$$
F_{\mathrm{diff},\lambda}^{\downarrow}(\tau) = 2\pi \sum_{i=1}^{s} I_{\lambda}^{(0)}(\tau,-\mu_i) \cdot \mu_i \cdot c_i \ .
\tag{7.49}
$$

It is evident from Eqs. (7.48) and (7.49) that only the zero-order mode of the Fourier coefficients of the radiance expansion contributes to the diffuse irradiance. For this reason, in the study of the radiant energy budget in the atmosphere, we only need to consider the angular (specifically, zenith) and spatial variations of the azimuthally independent component of $I_{\mathrm{diff},\lambda}$.

7.2
Equations for Fourier Modes of Diffuse Radiance

In numerical computation, the expansion of diffuse radiance into a number of Fourier modes can substantially decrease the computational burden. To derive the governing equations for the Fourier modes of diffuse radiance, we substitute the Fourier expansions of the phase function, see Eq. (7.20), and diffuse radiance, see Eq. (7.43), into the plane-parallel RTE, see Eq. (6.67), as

$$\mu \sum_{n=0}^{N-1} \frac{d I_\lambda^{(n)}(\tau,\mu)}{d\tau} \cdot \cos n\varphi = \sum_{n=0}^{N-1} I_\lambda^{(n)}(\tau,\mu) \cdot \cos n\varphi - \frac{\tilde{\omega}(\tau)}{4\pi} \cdot S_{\text{dir},\lambda,\text{TOA}} \cdot e^{-\tau/\mu_0}$$

$$\times \sum_{n=0}^{N-1} P^{(n)}(\tau,-\mu_0,\mu) \cdot \cos n(\varphi - \varphi_0) - \frac{\tilde{\omega}(\tau)}{4\pi} \int_0^{2\pi} \int_{-1}^{1} \sum_{j=0}^{N-1} I_\lambda^{(j)}(\tau,\mu') \cdot \cos j\varphi'$$

$$\times \sum_{n=0}^{N-1} P^{(n)}(\tau,\mu,\mu') \cdot \cos n(\varphi - \varphi') \, d\mu' \, d\varphi' - [1 - \tilde{\omega}(\tau)] \cdot B_\lambda[T(\tau)] . \qquad (7.50)$$

For the integration over the azimuthal angle in the multiple-scattering term, that is, the third term on the right side of Eq. (7.50), we have

$$\int_0^{2\pi} \cos j\varphi' \cdot \cos n(\varphi - \varphi') \, d\varphi'$$

$$= \int_0^{2\pi} (\cos n\varphi \cdot \cos n\varphi' + \sin n\varphi \cdot \sin n\varphi') \cdot \cos j\varphi' \, d\varphi'$$

$$= (1 + \delta_{n0}) \cdot \delta_{nj} \cdot \pi \cdot \cos n\varphi . \qquad (7.51)$$

Substituting Eq. (7.51) into Eq. (7.50) and assuming $\varphi_0 = 0$, we obtain

$$\mu \sum_{n=0}^{N-1} \frac{d I_\lambda^{(n)}(\tau,\mu)}{d\tau} \cdot \cos n\varphi = \sum_{n=0}^{N-1} I_\lambda^{(n)}(\tau,\mu) \cdot \cos n\varphi$$

$$- \frac{\tilde{\omega}(\tau)}{4\pi} \cdot S_{\text{dir},\lambda,\text{TOA}} \cdot e^{-\tau/\mu_0} \cdot \sum_{n=0}^{N-1} P^{(n)}(\tau,-\mu_0,\mu) \cdot \cos n\varphi$$

$$- \frac{\tilde{\omega}(\tau)}{4} \sum_{n=0}^{N-1} \int_{-1}^{1} (1 + \delta_{n0}) \cdot I_\lambda^{(n)}(\tau,\mu') \cdot P^{(n)}(\tau,\mu,\mu') \, d\mu' \cdot \cos n\varphi$$

$$- [1 - \tilde{\omega}(\tau)] \cdot B_\lambda[T(\tau)] . \qquad (7.52)$$

Multiplying Eq. (7.52) with $\cos m\varphi$, integrating the resultant equation over the azimuthal angle φ, and deleting the common factor $(1 + \delta_{m0}) \cdot \pi$, gives us the governing equation for the mth Fourier mode of the diffuse radiance, that is,

$$
\begin{aligned}
\mu \frac{\mathrm{d}\, I_\lambda^{(m)}(\tau,\mu)}{\mathrm{d}\tau} = {} & I_\lambda^{(m)}(\tau,\mu) \\
& - \frac{\tilde{\omega}(\tau)}{4\pi} \cdot S_{\mathrm{dir},\lambda,\mathrm{TOA}} \cdot \mathrm{e}^{-\tau/\mu_0} \cdot P^{(m)}(\tau, -\mu_0, \mu) \\
& - \frac{\tilde{\omega}(\tau)}{4} \cdot (1 + \delta_{m0}) \int_{-1}^{1} I_\lambda^{(m)}(\tau,\mu') \cdot P^{(m)}(\tau,\mu,\mu')\, \mathrm{d}\mu' \\
& - [1 - \tilde{\omega}(\tau)] \cdot B_\lambda[T(\tau)] \,.
\end{aligned}
\tag{7.53}
$$

7.2.1
Net Radiative Flux Density in a Nonabsorbing Atmosphere

As an example of the usage of Eq. (7.53), we prove that the total (direct plus diffuse) net solar flux density is conserved for a nonabsorbing atmosphere, that is, in the case of $\tilde{\omega} = 1$. Because only flux density in the solar spectrum is concerned, the thermal emission does not need to be considered; hence, only Eq. (7.53) is required for the zero-order mode of the radiance Fourier expansion in the form of

$$
\begin{aligned}
\mu \frac{\mathrm{d}\, I_\lambda^{(0)}(\tau,\mu)}{\mathrm{d}\tau} = {} & I_\lambda^{(0)}(\tau,\mu) \\
& - \frac{1}{4\pi} \cdot S_{\mathrm{dir},\lambda,\mathrm{TOA}} \cdot \mathrm{e}^{-\tau/\mu_0} \cdot P^{(0)}(\tau, -\mu_0, \mu) \\
& - \frac{1}{2} \int_{-1}^{1} I_\lambda^{(0)}(\tau,\mu') \cdot P^{(0)}(\tau,\mu,\mu')\, \mathrm{d}\mu' \,,
\end{aligned}
\tag{7.54}
$$

where $P^{(0)}$ is the azimuthally independent phase function and can be expressed as, see Eq. (7.21),

$$
\begin{aligned}
P^{(0)}(\tau,\mu,\mu') &= \sum_{n=0}^{N-1} C_n(\tau) \cdot \tilde{P}_n^{(0)}(\mu) \cdot \tilde{P}_n^{(0)}(\mu') \\
&= \sum_{n=0}^{N-1} C_n(\tau) \cdot P_n(\mu) \cdot P_n(\mu') \,.
\end{aligned}
\tag{7.55}
$$

Substituting Eq. (7.55) into Eq. (7.54), yields

$$
\begin{aligned}
\mu \frac{\mathrm{d}\, I_\lambda^{(0)}(\tau,\mu)}{\mathrm{d}\tau} = {} & I_\lambda^{(0)}(\tau,\mu) \\
& - \frac{1}{4\pi} \cdot S_{\mathrm{dir},\lambda,\mathrm{TOA}} \cdot \mathrm{e}^{-\tau/\mu_0} \cdot \sum_{n=0}^{N-1} C_n(\tau) \cdot P_n(-\mu_0) \cdot P_n(\mu) \\
& - \frac{1}{2} \int_{-1}^{1} I_\lambda^{(0)}(\tau,\mu') \cdot \sum_{n=0}^{N-1} C_n(\tau) \cdot P_n(\mu) \cdot P_n(\mu')\, \mathrm{d}\mu' \,.
\end{aligned}
\tag{7.56}
$$

Integrating Eq. (7.56) over $\mu \in [-1, 1]$ gives

$$\int_{-1}^{1} \mu \frac{\mathrm{d} I_{\lambda}^{(0)}(\tau, \mu)}{\mathrm{d}\tau} \, \mathrm{d}\mu = \int_{-1}^{1} I_{\lambda}^{(0)}(\tau, \mu) \, \mathrm{d}\mu$$

$$- \frac{1}{4\pi} \cdot S_{\mathrm{dir},\lambda,\mathrm{TOA}} \cdot \mathrm{e}^{-\tau/\mu_0} \cdot \int_{-1}^{1} \sum_{n=0}^{N-1} C_n(\tau) \cdot P_n(-\mu_0) \cdot P_n(\mu) \, \mathrm{d}\mu$$

$$- \frac{1}{2} \int_{-1}^{1} \int_{-1}^{1} I_{\lambda}^{(0)}(\tau, \mu') \cdot \sum_{n=0}^{N-1} C_n(\tau) \cdot P_n(\mu) \cdot P_n(\mu') \, \mathrm{d}\mu \, \mathrm{d}\mu' \,. \tag{7.57}$$

Applying the orthogonality relation for Legendre polynomials, see Eq. (2.44), with $P_0 = 1$, see Eq. (2.45),

$$\int_{-1}^{1} P_n(\mu) \, \mathrm{d}\mu = 2\delta_{n0} \,, \tag{7.58}$$

we can simplify the third term on the right side of Eq. (7.57) as

$$\frac{1}{2} \int_{-1}^{1} \int_{-1}^{1} I_{\lambda}^{(0)}(\tau, \mu') \cdot \sum_{n=0}^{N-1} C_n(\tau) \cdot P_n(\mu) \cdot P_n(\mu') \, \mathrm{d}\mu \, \mathrm{d}\mu'$$

$$= \frac{1}{2} \int_{-1}^{1} I_{\lambda}^{(0)}(\tau, \mu') \cdot \sum_{n=0}^{N-1} C_n(\tau) \cdot P_n(\mu') \cdot \int_{-1}^{1} P_n(\mu) \, \mathrm{d}\mu \, \mathrm{d}\mu'$$

$$= \frac{1}{2} \int_{-1}^{1} I_{\lambda}^{(0)}(\tau, \mu') \cdot \sum_{n=0}^{N-1} C_n(\tau) \cdot P_n(\mu') \cdot 2\,\delta_{n0} \, \mathrm{d}\mu'$$

$$= \int_{-1}^{1} I_{\lambda}^{(0)}(\tau, \mu') \cdot C_0(\tau) \cdot P_0(\mu') \, \mathrm{d}\mu'$$

$$= \int_{-1}^{1} I_{\lambda}^{(0)}(\tau, \mu') \, \mathrm{d}\mu'$$

$$= \int_{-1}^{1} I_{\lambda}^{(0)}(\tau, \mu) \, \mathrm{d}\mu \,, \tag{7.59}$$

where $C_0 = P_0 = 1$, see Eqs. (7.4) and (2.45). Similarly, we can prove for the second term on the right side of Eq. (7.57) that

$$\int_{-1}^{1} \sum_{n=0}^{N-1} C_n \cdot P_n(-\mu_0) \cdot P_n(\mu) \, \mathrm{d}\mu = \sum_{n=0}^{N-1} C_n \cdot P_n(-\mu_0) \cdot 2\delta_{n0} = 2 \,. \tag{7.60}$$

Substituting Eqs. (7.60) and (7.59) into Eq. (7.57) results in

$$\int_{-1}^{1} \mu \frac{d I_\lambda^{(0)}(\tau,\mu)}{d\tau} \, d\mu = -\frac{1}{2\pi} \cdot S_{\text{dir},\lambda,\text{TOA}} \cdot e^{-\tau/\mu_0} . \tag{7.61}$$

Furthermore, we have

$$\int_{-1}^{1} \mu \frac{d I_\lambda^{(0)}(\tau,\mu)}{d\tau} \, d\mu = \int_{-1}^{0} \frac{d I_\lambda^{(0)}(\tau,\mu)}{d\tau} \cdot \mu \, d\mu + \int_{0}^{1} \frac{d I_\lambda^{(0)}(\tau,\mu)}{d\tau} \cdot \mu \, d\mu$$

$$= \frac{1}{2\pi} \frac{d}{d\tau} \left[2\pi \int_{-1}^{0} I_\lambda^{(0)}(\tau,\mu) \cdot \mu \, d\mu + 2\pi \int_{0}^{1} I_\lambda^{(0)}(\tau,\mu) \cdot \mu \, d\mu \right]$$

$$= \frac{1}{2\pi} \frac{d}{d\tau} \left[-2\pi \int_{0}^{1} I_\lambda^{(0)}(\tau,-\mu) \cdot \mu \, d\mu + 2\pi \int_{0}^{1} I_\lambda^{(0)}(\tau,\mu) \cdot \mu \, d\mu \right]$$

$$= \frac{1}{2\pi} \frac{d}{d\tau} \left[F_{\text{diff},\lambda}^{\uparrow}(\tau) - F_{\text{diff},\lambda}^{\downarrow}(\tau) \right], \tag{7.62}$$

see Eqs. (7.45) and (7.46). By combining Eqs. (7.62) and (7.61), we obtain

$$\frac{d}{d\tau} \left[F_{\text{diff},\lambda}^{\downarrow}(\tau) - F_{\text{diff},\lambda}^{\uparrow}(\tau) \right] = S_{\text{dir},\lambda,\text{TOA}} \cdot e^{-\tau/\mu_0} . \tag{7.63}$$

The preceding equation can be rewritten as

$$\frac{d}{d\tau} \left[\mu_0 \cdot S_{\text{dir},\lambda,\text{TOA}} \cdot e^{-\tau/\mu_0} + F_{\text{diff},\lambda}^{\downarrow}(\tau) - F_{\text{diff},\lambda}^{\uparrow}(\tau) \right] = 0 . \tag{7.64}$$

Therefore,

$$\mu_0 \cdot S_{\text{dir},\lambda,\text{TOA}} \cdot e^{-\tau/\mu_0} + F_{\text{diff},\lambda}^{\downarrow}(\tau) - F_{\text{diff},\lambda}^{\uparrow}(\tau) = \text{constant} , \tag{7.65}$$

where the sum of the first two terms gives the total (direct plus diffuse) downward flux density. The third term is the upward flux density. Equation (7.65) proves the difference between the total downward and the upward diffuse flux densities to be independent of the altitude in a nonabsorbing atmosphere. Using Eqs. (3.50) and (3.51), we can express Eq. (7.65) as

$$\mu_0 \cdot S_{\text{dir},\lambda,\text{TOA}} \cdot e^{-\tau/\mu_0} - \int_{0}^{2\pi} \int_{0}^{\pi} I_{\text{diff},\lambda}(\tau,\theta,\varphi) \cdot \cos\theta \cdot \sin\theta \, d\theta \, d\varphi = \text{constant} .$$
$$\tag{7.66}$$

Equation (7.66) is usually referred to as the flux integral (Liou, 2002).

7.3
Method of Successive Order of Scattering (MSOS)

To illustrate the basic principle of the Method of Successive Order of Scattering (MSOS), we only consider an isolated layer (with an optical thickness of $\Delta\tau$) illuminated by a quasicollimated radiation field from above and specified by Eq. (6.74), that is,

$$I_{\text{dir},\lambda}(\tau = 0, \mu, \varphi) = S_{\text{dir},\lambda}(\tau = 0) \cdot \delta[\mu - (-\mu_0)] \cdot \delta(\varphi - \varphi_0) . \tag{7.67}$$

Thus, there is neither downward diffuse radiation at the top of the layer nor upward diffuse radiation at the bottom of the layer. For the mth order of the diffuse radiance Fourier expansion, the boundary conditions are

$$I_\lambda^{(m)}(\tau = 0, -\mu) = 0 ; \quad \text{with} \quad \mu > 0 , \tag{7.68}$$

$$I_\lambda^{(m)}(\tau = \Delta\tau, \mu) = 0; \quad \text{with} \quad \mu > 0 . \tag{7.69}$$

In the MSOS, the radiation is decomposed into the contributions from all orders of scattering events. Specifically, we express the Fourier modes of the diffuse radiance as

$$I_\lambda^{(m)}(\tau, \mu) = \sum_{n=0}^{N-1} I_{\lambda,n}^{(m)}(\tau, \mu) , \tag{7.70}$$

where the subscript n indicates the order of scattering events. In practice, the contributions associated with $n > N$ (N is a certain integer) are negligible, particularly in an absorptive layer. Note, the MSOS does not necessarily require the use of the Fourier modes of the diffuse radiance, but the computational demand of this method will be mitigated if the Fourier expansion of diffuse radiance is utilized instead of the diffuse radiance.

To account for the contribution of the first-order scattering events, the radiative transfer equation for the Fourier modes of the diffuse radiance can be written as, see Eq. (7.53),

$$\mu \frac{d I_{\lambda,1}^{(m)}(\tau, \mu)}{d\tau} = I_{\lambda,1}^{(m)}(\tau, \mu) - \frac{1}{4\pi} \cdot S_{\text{dir},\lambda}(\tau = 0) \cdot e^{-\tau/\mu_0} \cdot P^{(m)}(-\mu_0, \mu) ;$$
$$\text{for} \quad -1 \leq \mu \leq 1 ; \text{ but } \mu \neq 0 . \tag{7.71}$$

Here, we assume that scattering takes place without absorption, that is, $\tilde{\omega} = 1$. Furthermore, we may assume the phase function is independent of τ, that is, there is no vertical variation of the phase function

$$P^{(m)}(\tau, \mu, \mu') = P^{(m)}(\mu, \mu') .$$

The solutions to Eq. (7.71), subject to the boundary conditions in Eqs. (7.68) and (7.69), for upward Fourier modes are

$$I_{\lambda,1}^{(m)}(\tau,\mu) = \frac{1}{4\pi} \cdot \left(\frac{\mu_0}{\mu + \mu_0}\right) \cdot S_{\mathrm{dir},\lambda}(\tau = 0) \cdot e^{-\tau/\mu_0}$$
$$\times \left[1 - e^{-(\Delta\tau-\tau)\cdot(1/\mu+1/\mu_0)}\right] \cdot P^{(m)}(-\mu_0,\mu) ;$$
$$\text{for} \quad \mu > 0 ,$$

(7.72)

and for the downward Fourier modes,

$$I_{\lambda,1}^{(m)}(\tau,-\mu) = \frac{1}{4\pi} \cdot \left(\frac{\mu_0}{\mu_0 - \mu}\right) \cdot S_{\mathrm{dir},\lambda}(\tau = 0) \cdot e^{-\tau/\mu_0}$$
$$\times \left[1 - e^{-\tau\cdot(1/\mu-1/\mu_0)}\right] \cdot P^{(m)}(-\mu_0,-\mu)$$
$$\text{for} \quad \mu > 0 \quad \text{and} \quad \mu \neq \mu_0$$

(7.73)

$$I_{\lambda,1}^{(m)}(\tau,-\mu) = \frac{1}{4\pi} \cdot \left(\frac{\tau}{\mu_0}\right) \cdot S_{\mathrm{dir},\lambda}(\tau = 0) \cdot e^{-\tau/\mu_0} \cdot P^{(m)}(-\mu_0,-\mu) ;$$
$$\text{for} \quad \mu > 0 \text{ and } \mu = \mu_0 .$$

(7.74)

For the contributions associated with the second-order and higher-orders of scattering events, the governing equations are given by

$$\mu \frac{d I_{\lambda,n}^{(m)}(\tau,\mu)}{d\tau} = I_{\lambda,n}^{(m)}(\tau,\mu) - J_{\lambda,n}^{(m)}(\tau,\mu) ;$$
$$\text{for} \quad -1 \leq \mu \leq 1 \text{ and } \mu \neq 0 ,$$

(7.75)

where the source function is

$$J_{\lambda,n}^{(m)}(\tau,\mu) = \frac{1}{4} \cdot (1 + \delta_{m0}) \int_{-1}^{1} I_{\lambda,n-1}^{(m)}(\tau,\mu') \cdot P^{(m)}(\mu,\mu') \, d\mu' .$$

(7.76)

The solutions to Eq. (7.75), subject to the boundary conditions in Eqs. (7.68) and (7.69), for upward Fourier modes are

$$I_{\lambda,n}^{(m)}(\tau,\mu) = \frac{1}{4} \cdot (1 + \delta_{m0}) \int_{\tau}^{\Delta\tau} e^{-(\tau'-\tau)/\mu}$$
$$\times \int_{-1}^{1} I_{\lambda,n-1}^{(m)}(\tau',\mu') \cdot P^{(m)}(\mu,\mu') \, d\mu' \, d\tau'/\mu ;$$
$$\text{for} \quad \mu > 0 ,$$

(7.77)

and for downward Fourier modes,

$$I_{\lambda,n}^{(m)}(\tau,-\mu) = \frac{1}{4} \cdot (1 + \delta_{m0}) \int_{0}^{\tau} e^{-(\tau-\tau')/\mu} \int_{-1}^{1} I_{\lambda,n-1}^{(m)}(\tau',\mu')$$
$$\times P^{(m)}(-\mu,\mu') \, d\mu' \, d\tau'/\mu ; \quad \text{for} \quad \mu > 0 .$$

(7.78)

From the preceding discussion, Eqs. (7.72), (7.73), (7.77), and (7.78) constitute a closed-form solution to the radiative transfer solution. The advantage is that the solution is expressed in an explicit, iterative form. For practical applications, the boundary conditions, absorption $\tilde{\omega} \neq 1$, and vertical variations of the medium properties of concern (in particular the phase function) can be taken into account in a straightforward manner. However, the convergence of the solution of MSOS depends on the single-scattering albedo and optical thickness. For a nonabsorptive atmosphere with a large optical thickness, for example, under optically thick water cloud conditions, the convergence of the MSOS solution requires a tremendous amount of computational effort.

7.4
Adding-Doubling Method (A-DM)

7.4.1
Simplified Example

Instead of beginning with formulating the Adding-Doubling Method (A-DM) for the computation of the Fourier coefficients of the diffuse radiance, we consider a simple example to illustrate the basic principle of this method. As shown in Figure 7.4, a quasi-collimated radiation beam (e.g., a collimated laser beam) impinges on two parallel layers of continuous media (e.g., glass) stacked over a dark surface whose reflection can be neglected. The radiant power (radiant flux in units of Watt) of the incident beam is Φ_{inc}. The reflectivity and transmissivity of the upper layer are $\mathcal{R}_1$ and $\mathcal{T}_1$, respectively; whereas, their counterparts for the lower layer are $\mathcal{R}_2$ and $\mathcal{T}_2$. Note that the effect of the multiple reflections between the upper and lower surfaces of the top (or bottom) layer in the isolated situation has been incorporated into $\mathcal{R}_1$ and $\mathcal{T}_1$ (or $\mathcal{R}_2$ and $\mathcal{T}_2$).

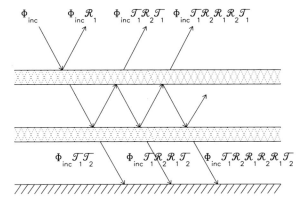

Figure 7.4 A schematic diagram showing the reflection and transmission of a quasi-collimated radiation beam through two optically homogeneous, parallel, and stacked layers over a dark (i.e., nonreflective) surface.

From the configuration in Figure 7.4, the reflected radiant power observed above the upper layer is given by

$$
\begin{aligned}
\Phi_{\mathrm{ref}} &= \Phi_{\mathrm{inc}} \cdot \mathcal{R}_1 \\
&\quad + \Phi_{\mathrm{inc}} \cdot \mathcal{T}_1 \cdot \mathcal{R}_2 \cdot \mathcal{T}_1 \\
&\quad + \Phi_{\mathrm{inc}} \cdot \mathcal{T}_1 \cdot \mathcal{R}_2 \cdot \mathcal{R}_1 \cdot \mathcal{R}_2 \cdot \mathcal{T}_1 \\
&\quad + \Phi_{\mathrm{inc}} \cdot \mathcal{T}_1 \cdot \mathcal{R}_2 \cdot \mathcal{R}_1 \cdot \mathcal{R}_2 \cdot \mathcal{R}_1 \cdot \mathcal{R}_2 \cdot \mathcal{T}_1 \\
&\quad + \dots \\
&= \Phi_{\mathrm{inc}} \cdot \left[\mathcal{R}_1 + \mathcal{T}_1 \cdot \mathcal{R}_2 \cdot \mathcal{T}_1 \cdot \left(1 + \mathcal{R}_1 \cdot \mathcal{R}_2 + \mathcal{R}_1^2 \cdot \mathcal{R}_2^2 + \dots \right) \right].
\end{aligned} \tag{7.79}
$$

We use the equality of

$$
\sum_{i=0}^{\infty} x^i = \frac{1}{1-x} ; \quad \text{for} \quad |x| < 1 , \tag{7.80}
$$

where x is an arbitrary scalar. Hence, applying Eq. (7.80) with $x = \mathcal{R}_1 \cdot \mathcal{R}_2$ to Eq. (7.79) obtains

$$
\Phi_{\mathrm{ref}} = \Phi_{\mathrm{inc}} \cdot \left(\mathcal{R}_1 + \frac{\mathcal{T}_1 \cdot \mathcal{R}_2 \cdot \mathcal{T}_1}{1 - \mathcal{R}_1 \cdot \mathcal{R}_2} \right). \tag{7.81}
$$

Thus, the total reflectivity associated with the combination of the two layers is

$$
\mathcal{R} = \frac{\Phi_{\mathrm{ref}}}{\Phi_{\mathrm{inc}}} = \mathcal{R}_1 + \frac{\mathcal{T}_1 \cdot \mathcal{R}_2 \cdot \mathcal{T}_1}{1 - \mathcal{R}_1 \cdot \mathcal{R}_2} . \tag{7.82}
$$

Similarly, the transmitted radiant power observed at the underlying surface is

$$
\begin{aligned}
\Phi_{\mathrm{tra}} &= \Phi_{\mathrm{inc}} \cdot \mathcal{T}_1 \cdot \mathcal{T}_2 \\
&\quad + \Phi_{\mathrm{inc}} \cdot \mathcal{T}_1 \cdot \mathcal{R}_2 \cdot \mathcal{R}_1 \cdot \mathcal{T}_2 \\
&\quad + \Phi_{\mathrm{inc}} \cdot \mathcal{T}_1 \cdot \mathcal{R}_2 \cdot \mathcal{R}_1 \cdot \mathcal{R}_2 \cdot \mathcal{R}_1 \cdot \mathcal{T}_2 \\
&\quad + \Phi_{\mathrm{inc}} \cdot \mathcal{T}_1 \cdot \mathcal{R}_2 \cdot \mathcal{R}_1 \cdot \mathcal{R}_2 \cdot \mathcal{R}_1 \cdot \mathcal{R}_2 \cdot \mathcal{R}_1 \cdot \mathcal{T}_2 \\
&\quad + \dots \\
&= \Phi_{\mathrm{inc}} \cdot \mathcal{T}_1 \cdot \mathcal{T}_2 \cdot \left(1 + \mathcal{R}_2 \cdot \mathcal{R}_1 + \mathcal{R}_2^2 \cdot \mathcal{R}_1^2 + \dots \right).
\end{aligned} \tag{7.83}
$$

Applying Eqs. (7.80) to (7.83) result in

$$
\Phi_{\mathrm{tra}} = \Phi_{\mathrm{inc}} \cdot \left(\frac{\mathcal{T}_1 \cdot \mathcal{T}_2}{1 - \mathcal{R}_2 \cdot \mathcal{R}_1} \right). \tag{7.84}
$$

The total transmissivity of the two layers without accounting for the effect of the underlying surface is

$$
\mathcal{T} = \frac{\Phi_{\mathrm{tra}}}{\Phi_{\mathrm{inc}}} = \frac{\mathcal{T}_1 \cdot \mathcal{T}_2}{1 - \mathcal{R}_2 \cdot \mathcal{R}_1} . \tag{7.85}
$$

Furthermore, it can be shown from Figure 7.4 that the radiant power associated with downward radiant energy between the two layers is

$$
\begin{aligned}
\Phi_{\text{tra},0} &= \Phi_{\text{inc}} \cdot \mathcal{T}_1 \\
&\quad + \Phi_{\text{inc}} \cdot \mathcal{T}_1 \cdot \mathcal{R}_2 \cdot \mathcal{R}_1 \\
&\quad + \Phi_{\text{inc}} \cdot \mathcal{T}_1 \cdot \mathcal{R}_2 \cdot \mathcal{R}_1 \cdot \mathcal{R}_2 \cdot \mathcal{R}_1 \\
&\quad + \Phi_{\text{inc}} \cdot \mathcal{T}_1 \cdot \mathcal{R}_2 \cdot \mathcal{R}_1 \cdot \mathcal{R}_2 \cdot \mathcal{R}_1 \cdot \mathcal{R}_2 \cdot \mathcal{R}_1 \\
&\quad + \ldots \\
&= \Phi_{\text{inc}} \cdot \mathcal{T}_1 \cdot \left(1 + \mathcal{R}_2 \cdot \mathcal{R}_1 + \mathcal{R}_2^2 \cdot \mathcal{R}_1^2 + \ldots \right) \\
&= \Phi_{\text{inc}} \cdot \left(\frac{\mathcal{T}_1}{1 - \mathcal{R}_2 \cdot \mathcal{R}_1} \right),
\end{aligned}
\tag{7.86}
$$

and we define

$$
D = \frac{\mathcal{T}_1}{1 - \mathcal{R}_2 \cdot \mathcal{R}_1} .
\tag{7.87}
$$

Similarly, the power associated with upward radiant energy between the two layers is

$$
\begin{aligned}
\Phi_{\text{ref},0} &= \Phi_{\text{inc}} \cdot \mathcal{T}_1 \cdot \mathcal{R}_2 \\
&\quad + \Phi_{\text{inc}} \cdot \mathcal{T}_1 \cdot \mathcal{R}_2 \cdot \mathcal{R}_1 \cdot \mathcal{R}_2 \\
&\quad + \Phi_{\text{inc}} \cdot \mathcal{T}_1 \cdot \mathcal{R}_2 \cdot \mathcal{R}_1 \cdot \mathcal{R}_2 \cdot \mathcal{R}_1 \cdot \mathcal{R}_2 \\
&\quad + \ldots \\
&= \Phi_{\text{inc}} \cdot \mathcal{T}_1 \cdot \mathcal{R}_2 \cdot \left(1 + \mathcal{R}_1 \cdot \mathcal{R}_2 + \mathcal{R}_1^2 \cdot \mathcal{R}_2^2 + \ldots \right) \\
&= \Phi_{\text{inc}} \cdot \left(\frac{\mathcal{T}_1 \cdot \mathcal{R}_2}{1 - \mathcal{R}_1 \cdot \mathcal{R}_2} \right).
\end{aligned}
\tag{7.88}
$$

Furthermore, we define

$$
U = \frac{\mathcal{T}_1 \cdot \mathcal{R}_2}{1 - \mathcal{R}_1 \cdot \mathcal{R}_2} ,
\tag{7.89}
$$

where U should not be confused with the third component of the Stokes vector, see Eq. (4.16). Comparing Eq. (7.82) with Eq. (7.89), and Eq.(7.85) with Eq. (7.87), respectively, it follows for the combined reflectivity and transmissivity of the two layers that

$$
\mathcal{R} = \mathcal{R}_1 + \mathcal{T}_1 \cdot U ,
\tag{7.90}
$$

$$
\mathcal{T} = \mathcal{T}_2 \cdot D .
\tag{7.91}
$$

Equations (7.87), (7.89), (7.90), and (7.91) constitute the basic formalism of the principle of the A-DM whose originality can be traced to the work of Stokes (1862). Specifically, if the reflectivities and transmissivities of two stacked layers parallel to each other are known, the total reflectivity and transmissivity of the composite system can be calculated. If the two layers have the same reflection and transmission properties, the method is referred to as the doubling method; otherwise, it is referred to as the adding method.

7.4.2
Generalization for Radiances

The A-DM is an efficient approach to account for the multiple-scattering process in a vertically inhomogeneous atmosphere. The standard formalism of this method, in the form of matrix operators or the star products involving angular integration, can be found in numerous existing texts, for example, Liou (2002), Mishchenko et al. (2006), and Thomas and Stamnes (1999). Hansen and Travis (1974) presented a comprehensive tutorial review of the A-DM for the solution of the transfer of polarized radiation. Here, we formulate the A-DM in a discrete form following Yang et al. (2001). Specifically, we select a certain quadrature scheme in a form similar to the Gauss–Lobatto quadrature, see Eq. (2.70), that is,

$$\int_0^1 \phi(\mu)\, d\mu = \sum_{j=1}^{s} c_j \cdot \phi(\mu_j) , \tag{7.92}$$

where μ_j and c_j are the abscissas and weights of the quadrature scheme, and ϕ an arbitrary scalar function. For the mth Fourier mode of the reflected spectral radiance, we abbreviate

$$I_{\lambda,\text{ref}}^{(m)}(\tau,\mu_i) = I_{\text{ref},i}^{(m)} . \tag{7.93}$$

For the mth Fourier mode of diffuse spectral transmitted radiance, we write

$$I_{\lambda,\text{tra}}^{(m)}(\tau,\mu_i) = I_{\text{tra},i}^{(m)} , \tag{7.94}$$

and for the mth Fourier mode of incident spectral radiance, we get

$$I_{\lambda,\text{inc}}^{(m)}(\tau,\mu_i) = I_{\text{inc},i}^{(m)} , \tag{7.95}$$

where the subscript λ is partially omitted. We define the discrete star multiplication, denoted in terms of the symbol $\star$, as

$$I_{\text{ref},i}^{(m)} = I_{\text{inc},j}^{(m)} \star R_{ji}^{(m)} = (1 + \delta_{m0}) \cdot \sum_{j=1}^{s} c_j \cdot \mu_j \cdot R_{ji}^{(m)} \cdot I_{\text{inc},j}^{(m)} , \tag{7.96}$$

$$I_{\text{tra},i}^{(m)} = I_{\text{inc},j}^{(m)} \star T_{ji}^{(m)} = (1 + \delta_{m0}) \cdot \sum_{j=1}^{s} c_j \cdot \mu_j \cdot T_{ji}^{(m)} \cdot I_{\text{inc},j}^{(m)} , \tag{7.97}$$

where the subscripts "inc," "ref," and "tra," indicate the incident, reflected and transmitted radiation associated with a layer of a scattering medium whose optical thickness is assumed to be τ. Because the discrete star multiplication is defined, all the radiometric quantities must be specified with respect to the discrete values of μ in accordance with the selected quadrature scheme. Accordingly, the quantity $I_{\text{ref},i}^{(m)}$ indicates $I_{\text{ref}}^{(m)}(\tau,\mu_i)$, and $I_{\text{tra},i}^{(m)}$ replaces $I_{\text{tra}}^{(m)}(\tau,\mu_i)$. The conventional bidirectional reflection and transmission functions are given, applying Eq. (7.20), by the Fourier series of the form

$$\mathcal{R}(\mu_i, \varphi, \mu_j, \varphi') \approx \sum_{m=0}^{N-1} R_{ji}^{(m)} \cdot \cos m(\varphi - \varphi') , \tag{7.98}$$

$$\mathcal{T}(\mu_i, \varphi, \mu_j, \varphi') \approx \sum_{m=0}^{N-1} T_{ji}^{(m)} \cdot \cos m(\varphi - \varphi') , \tag{7.99}$$

where $R_{ji}^{(m)} = R^{(m)}(\mu_j, \mu_i)$ and $T_{ji}^{(m)} = T^{(m)}(\mu_j, \mu_i)$. From Eqs. (7.98) and (7.99), the subscripts j and i of $R_{ji}^{(m)}$ or $T_{ji}^{(m)}$ indicate that the direction of the radiation beam is changed from μ_j to μ_i via the multiple-scattering events.

In Eqs. (7.96) and (7.97), the incident radiation field is assumed to have a continuous distribution, for example, a diffuse radiation field characterized as a continuous function of μ and φ. If the incident radiation field is a collimated radiation field, see Eq. (6.47), in the form of

$$\tilde{I}_{\text{inc}}(\mu, \varphi) = S_{\text{dir}} \cdot \delta(\mu - \mu_k) \cdot \delta(\varphi) , \tag{7.100}$$

then the corresponding Fourier modes of the preceding radiance in the discrete form, consistent with the star multiplication, is given by

$$\tilde{I}_{\text{inc},j}^{(m)} = \frac{1}{(1 + \delta_{m0}) \cdot \pi} \cdot \frac{1}{c_j} \cdot S_{\text{dir}} \cdot \delta_{jk} , \tag{7.101}$$

where δ_{jk} is the Kronecker symbol, see Eq. (2.14). In the derivation of Eq. (7.101) from Eq. (7.100), we use δ_{jk}/c_j to discretize the Dirac δ-function $\delta(\mu_j - \mu_k)$, see Section 2.4, to assure the discretization is consistent with the star multiplication. Substituting Eq. (7.101) into Eq. (7.96), yields

$$I_{\text{ref},i}^{(m)} = (1 + \delta_{m0}) \cdot \sum_{j=1}^{s} c_j \cdot \mu_j \cdot R_{ji}^{(m)} \cdot I_{\text{inc},j}^{(m)}$$

$$= (1 + \delta_{m0}) \cdot \sum_{j=1}^{s} c_j \cdot \mu_j \cdot R_{ji}^{(m)} \cdot \frac{1}{(1 + \delta_{m0}) \cdot \pi} \cdot \frac{1}{c_j} \cdot S_{\text{dir}} \cdot \delta_{jk}$$

$$= \sum_{j=1}^{s} \mu_j \cdot R_{ji}^{(m)} \cdot \frac{1}{\pi} \cdot S_{\text{dir}} \cdot \delta_{jk}$$

$$= \mu_k \cdot R_{ki}^{(m)} \cdot \frac{1}{\pi} \cdot S_{\text{dir}} . \tag{7.102}$$

In a similar way, by substituting Eq. (7.101) into Eq. (7.97), we have

$$I_{\text{tra},i}^{(m)} = (1 + \delta_{m0}) \cdot \sum_{j=1}^{s} c_j \cdot \mu_j \cdot T_{ji}^{(m)} \cdot I_{\text{inc},j}^{(m)}$$

$$= \mu_k \cdot T_{ki}^{(m)} \cdot \frac{1}{\pi} \cdot S_{\text{dir}} . \tag{7.103}$$

From Eqs. (7.102) and (7.103), we get the Fourier modes of the discrete values of the bidirectional reflection and transmission functions

$$R_{ki}^{(m)} = \frac{\pi \cdot I_{\mathrm{ref},i}^{(m)}}{\mu_k \cdot S_{\mathrm{dir}}} , \qquad (7.104)$$

$$T_{ki}^{(m)} = \frac{\pi \cdot I_{\mathrm{tra},i}^{(m)}}{\mu_k \cdot S_{\mathrm{dir}}} . \qquad (7.105)$$

For the direct (first-order scattering) transmissivity, the corresponding Fourier modes $t_{ji}^{(m)}$ can be expressed in the form

$$t_{ji}^{(m)} = \frac{1}{(1 + \delta_{m0})} \cdot \frac{1}{c_j \cdot \mu_j} \cdot e^{-\tau/\mu_j} \cdot \delta_{ij} . \qquad (7.106)$$

It can be proven that the Fourier modes, $t_{ji}^{(m)}$, of the direct transmissivity satisfy the following relations in terms of the star product defined between any mth Fourier mode, see also Eqs. (7.96) and (7.97), that is,

$$I_{\mathrm{inc},j}^{(m)} \star t_{ji}^{(m)} = e^{-\tau/\mu_i} \cdot I_{\mathrm{inc},i}^{(m)} , \qquad (7.107)$$

$$t_{ji}^{(m)} \star T_{jk}^{(m)} = e^{-\tau/\mu_i} \cdot T_{ik}^{(m)} , \qquad (7.108)$$

$$T_{ij}^{(m)} \star t_{jk}^{(m)} = e^{-\tau/\mu_k} \cdot T_{ik}^{(m)} . \qquad (7.109)$$

We show the validity of Eqs. (7.107)–(7.109). As an example of Eq. (7.107), following the definition of the star product, see Eq. (7.96), we obtain

$$I_{\mathrm{inc},j}^{(m)} \star t_{ji}^{(m)} = (1 + \delta_{m0}) \cdot \sum_{j=1}^{s} c_j \cdot \mu_j \cdot t_{ji}^{(m)} \cdot I_{\mathrm{inc},j}^{(m)} . \qquad (7.110)$$

We substitute Eq. (7.106) into the right side of Eq. (7.110) to give

$$I_{\mathrm{inc},j}^{(m)} \star t_{ji}^{(m)} = (1 + \delta_{m0}) \cdot \sum_{j=1}^{s} c_j \cdot \mu_j \cdot \left[\cdot \frac{1}{(1 + \delta_{m0})} \cdot \frac{1}{c_j \cdot \mu_j} \cdot e^{-\tau/\mu_j} \cdot \delta_{ij} \right] \cdot I_{\mathrm{inc},j}^{(m)}$$

$$= \sum_{j=1}^{s} e^{-\tau/\mu_j} \cdot \delta_{ij} \cdot I_{\mathrm{inc},j}^{(m)}$$

$$= e^{-\tau/\mu_i} \cdot I_{\mathrm{inc},i}^{(m)} ,$$

$$(7.111)$$

which corresponds to Eq. (7.107).

The mth Fourier modes of the diffuse and the first-order scattering transmissivities $T_{ji}^{(m)}$ and $t_{ji}^{(m)}$ are distinct in terms of their relevant physical processes. Specifically, $T_{ji}^{(m)}$ is associated with diffuse radiation contributed by multiple-scattering events; whereas, $t_{ji}^{(m)}$ is directly associated with the attenuation of the radiation

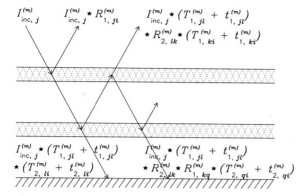

Figure 7.5 Schematic diagram illustrating the adding procedure for calculating the bidirectional reflection and transmission functions of a system consisting of two layers over a dark surface, that is, the surface reflection is negligible, provided that the reflection and transmission characteristics of the two layers are given.

beam by single scattering or absorption without changing the propagation direction of the radiation beam.

Referring to Figure 7.5, let us again consider two stacked, parallel scattering medium layers over a dark surface, that is, the surface reflection can be neglected. The reflection and transmission properties of the upper layer are characterized in terms of $R_{1,ji}^{(m)}$ (bidirectional reflection function), $T_{1,ji}^{(m)}$ (bidirectional transmission function), and $t_{1,ji}^{(m)}$ (bidirectional transmissivity for single scattering or absorption). However, their counterparts for the lower layer are $R_{2,ji}^{(m)}$, $T_{2,ji}^{(m)}$, and $t_{2,ji}^{(m)}$. From the schematic illustration in Figure 7.5 and analogous with Eq. (7.84), the Fourier mode of the transmitted radiance is given by

$$
\begin{aligned}
I_{\mathrm{tra},i}^{(m)} &= I_{\mathrm{inc},j}^{(m)} \star \left[T_{1,jl}^{(m)} + t_{1,jl}^{(m)} \right] \star \left[T_{2,li}^{(m)} + t_{2,li}^{(m)} \right] \\
&\quad + I_{\mathrm{inc},j}^{(m)} \star \left[T_{1,jl}^{(m)} + t_{1,jl}^{(m)} \right] \star R_{2,lk}^{(m)} \star R_{1,kq}^{(m)} \star \left[T_{2,qi}^{(m)} + t_{2,qi}^{(m)} \right] \\
&\quad + I_{\mathrm{inc},j}^{(m)} \star \left[T_{1,jl}^{(m)} + t_{1,jl}^{(m)} \right] \star R_{2,lk}^{(m)} \star R_{1,kq}^{(m)} \star R_{2,qr}^{(m)} \star R_{1,rs}^{(m)} \star \left[T_{2,si}^{(m)} + t_{2,si}^{(m)} \right] \\
&\quad + I_{\mathrm{inc},j}^{(m)} \star \left[T_{1,jl}^{(m)} + t_{1,jl}^{(m)} \right] \star R_{2,lk}^{(m)} \star R_{1,kq}^{(m)} \star R_{2,qr}^{(m)} \star R_{1,rs}^{(m)} \star R_{2,st}^{(m)} \star R_{1,tu}^{(m)} \\
&\quad \star \left[T_{2,ui}^{(m)} + t_{2,ui}^{(m)} \right] \\
&\quad + \ldots \\
&= I_{\mathrm{inc},j}^{(m)} \star \left\{ \left[T_{1,jl}^{(m)} + t_{1,jl}^{(m)} \right] \star \left[T_{2,li}^{(m)} + t_{2,li}^{(m)} \right] + \left[T_{1,jl}^{(m)} + t_{1,jl}^{(m)} \right] \star D_{lk}^{(m)} \right. \\
&\quad \left. \star \left[T_{2,ki}^{(m)} + t_{2,ki}^{(m)} \right] \right\} \\
&= I_{\mathrm{inc},j}^{(m)} \star \left\{ T_{1,jl}^{(m)} \star T_{2,li}^{(m)} + \mathrm{e}^{-\tau_2/\mu_i} \cdot T_{1,ji}^{(m)} + \mathrm{e}^{-\tau_1/\mu_j} \cdot T_{2,ji}^{(m)} + t_{1,jl}^{(m)} \star t_{2,li}^{(m)} \right. \\
&\quad + T_{1,jl}^{(m)} \star D_{lk}^{(m)} \star T_{2,ki}^{(m)} + \mathrm{e}^{-\tau_1/\mu_j} \cdot D_{jk}^{(m)} \star T_{2,ki}^{(m)} \\
&\quad \left. + \mathrm{e}^{-\tau_2/\mu_i} \cdot T_{1,jl}^{(m)} \star D_{li}^{(m)} + \mathrm{e}^{-\tau_1/\mu_j} \cdot D_{ji}^{(m)} \cdot \mathrm{e}^{-\tau_2/\mu_i} \right\},
\end{aligned}
\qquad (7.112)
$$

where

$$D_{lk}^{(m)} = \sum_{n=1}^{\infty} D_{lk,n}^{(m)} , \qquad (7.113)$$

$$D_{lk,1}^{(m)} = R_{2,lq}^{(m)} \star R_{1,qk}^{(m)} , \qquad (7.114)$$

$$D_{lk,n}^{(m)} = D_{lq,n}^{(m)} \star D_{qk,1}^{(m)} . \qquad (7.115)$$

In Eq. (7.112), the term $t_{1,jl}^{(m)} \star t_{2,li}^{(m)}$ is associated with the direct transmissivity, that is, the attenuation of the incident radiation beam, and should not be included in calculating the diffuse transmission function of the combined layers. In a numerical computation, the higher order ($n > 10$) terms in the summation in Eq. (7.113) can usually be approximated in terms of a geometric series. This simplification leads to a significant decrease in computational effort.

Similarly, in analogy with Eq. (7.81), the Fourier modes of the reflected radiance are given by

$$
\begin{aligned}
I_{\text{ref},i}^{(m)} &= I_{\text{inc},j}^{(m)} \star R_{1,ji}^{(m)} \\
&+ I_{\text{inc},j}^{(m)} \star \left[T_{1,jl}^{(m)} + t_{1,jl}^{(m)} \right] \star R_{2,lk}^{(m)} \star \left[T_{1,ki}^{(m)} + t_{1,ki}^{(m)} \right] \\
&+ I_{\text{inc},j}^{(m)} \star \left[T_{1,jl}^{(m)} + t_{1,jl}^{(m)} \right] \star R_{2,lk}^{(m)} \star R_{1,kq}^{(m)} \star R_{2,qu}^{(m)} \star \left[T_{1,ui}^{(m)} + t_{1,ui}^{(m)} \right] \\
&+ I_{\text{inc},j}^{(m)} \star \left[T_{1,jl}^{(m)} + t_{1,jl}^{(m)} \right] \star R_{2,lk}^{(m)} \star R_{1,kq}^{(m)} \star R_{2,qr}^{(m)} \star R_{1,rs}^{(m)} \star R_{2,st}^{(m)} \\
&\quad \star \left[T_{1,ti}^{(m)} + t_{1,ti}^{(m)} \right] \\
&+ \dots \\
&= I_{\text{inc},j}^{(m)} \star \left\{ R_{1,ji}^{(m)} + \left[T_{1,jl}^{(m)} + t_{1,jl}^{(m)} \right] \star U_{lk}^{(m)} \star \left[T_{1,ki}^{(m)} + t_{1,ki}^{(m)} \right] \right\} \\
&= I_{\text{inc},j}^{(m)} \star \left\{ R_{1,ji}^{(m)} + T_{1,jl}^{(m)} \star U_{lk}^{(m)} \star T_{1,ki}^{(m)} + e^{-\tau_1/\mu_j} \cdot U_{lk}^{(m)} \star T_{1,ki}^{(m)} \right. \\
&\quad \left. + e^{-\tau_1/\mu_i} \cdot T_{1,jl}^{(m)} \star U_{li}^{(m)} + e^{-\tau_1/\mu_j} \cdot U_{ji}^{(m)} \cdot e^{-\tau_1/\mu_i} \right\} , \qquad (7.116)
\end{aligned}
$$

where

$$U_{lk}^{(m)} = R_{2,lk}^{(m)} + D_{lq}^{(m)} \star R_{2,qk}^{(m)} . \qquad (7.117)$$

From Eqs. (7.116) and (7.112), the Fourier modes of the bidirectional reflection and diffuse transmission functions of the two layer system, $R_{ji}^{(m)}$ and $T_{jl}^{(m)}$, are given by the expressions in brackets $\{\dots\}$, that is,

$$
\begin{aligned}
R_{ji}^{(m)} &= R_{1,ji}^{(m)} + T_{1,jl}^{(m)} \star U_{lk}^{(m)} \star T_{1,ki}^{(m)} + e^{-\tau_1/\mu_j} \cdot U_{lk}^{(m)} \star T_{1,ki}^{(m)} \\
&\quad + e^{-\tau_1/\mu_i} \cdot T_{1,jl}^{(m)} \star U_{li}^{(m)} + e^{-\tau_1/\mu_j} \cdot U_{ji}^{(m)} \cdot e^{-\tau_1/\mu_i} , \qquad (7.118)
\end{aligned}
$$

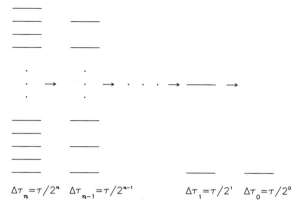

$$\Delta\tau_n = \tau/2^n \quad \Delta\tau_{n-1} = \tau/2^{n-1} \qquad \Delta\tau_1 = \tau/2^1 \quad \Delta\tau_0 = \tau/2^0$$

Figure 7.6 Schematic diagram illustrating the doubling procedure for computing the reflection and transmission of an optically homogeneous layer.

and

$$
\begin{aligned}
T_{jl}^{(m)} =\ & T_{1,jl}^{(m)} \star T_{2,li}^{(m)} + e^{-\tau_2/\mu_i} \cdot T_{1,ji}^{(m)} + e^{-\tau_1/\mu_j} \cdot T_{2,ji}^{(m)} \\
& + T_{1,jl}^{(m)} \star D_{lk}^{(m)} \star T_{2,ki}^{(m)} + e^{-\tau_1/\mu_j} \cdot D_{jk}^{(m)} \star T_{2,ki}^{(m)} \\
& + e^{-\tau_2/\mu_i} \cdot T_{1,jl}^{(m)} \star D_{li}^{(m)} + e^{-\tau_1/\mu_j} \cdot D_{ji}^{(m)} \cdot e^{-\tau_2/\mu_i} .
\end{aligned}
\tag{7.119}
$$

As previously explained, the direct transmissivity term $t_{1,jl}^{(m)} \star t_{2,li}^{(m)}$ is omitted in Eq. (7.119).

Equations (7.113)–(7.115) combined with Eqs. (7.117)–(7.119) formulate a basic algorithm of the A-DM for computing Fourier modes of the bidirectional reflection and diffuse transmission functions.

For an optically thin layer with an optical thickness of $\Delta\tau$, we can neglect the contribution of multiple-scattering events to the reflectivity and diffuse transmissivity. Under this approximation, we can use Eqs. (7.75) and (7.76) (including absorption: $1/4 \longrightarrow \tilde{\omega}/4$) to obtain

$$
R_{ji}^{(m)} = \frac{\tilde{\omega}}{4} \cdot \frac{\Delta\tau}{\mu_i \cdot \mu_j} \cdot P^{(m)}(\mu_i, -\mu_j) ,
\tag{7.120}
$$

$$
T_{ji}^{(m)} = \frac{\tilde{\omega}}{4} \cdot \frac{\Delta\tau}{\mu_i \cdot \mu_j} \cdot P^{(m)}(-\mu_i, -\mu_j) .
\tag{7.121}
$$

For an optically homogeneous thick layer with an optical thickness of τ, we can divide the layer into 2^n layers where n is of the order 10. Thus, by using Eqs. (7.120) and (7.121), we can obtain the reflection and diffuse transmission functions of a sublayer with an optical thickness of $\tau/2^n$. After conducting the doubling calculation, we can get the reflection and transmission characteristics of a sublayer with an optical thickness of $\tau/2^{n-1}$. We repeat the doubling calculation until the optical thickness of a sublayer equals that of the original layer. The process of the doubling procedure is schematically illustrated in Figure 7.6. We can use the zero-order

Fourier modes of the bidirectional reflection and diffuse transmission functions to calculate the planetary (or local) albedo and the planetary transmissivity according to Eqs. (7.98) and (7.99), that is,

$$\mathcal{R}_p(\mu_i) = 2 \sum_{j=1}^{s} c_j \cdot \mu_j \cdot R_{ij}^{(0)} , \tag{7.122}$$

$$\mathcal{T}_p(\mu_i) = 2 \sum_{j=1}^{s} c_j \cdot \mu_j \cdot T_{ij}^{(0)} , \tag{7.123}$$

where μ_i indicates the cosine of the incident zenith angle. At the surface, the planetary reflectivity corresponds to the planetary albedo, see Eq. (6.78). Furthermore, the spherical (or global) albedo, see Eq. (6.79), and the spherical diffuse transmissivity are defined as

$$\overline{\mathcal{R}} = 2 \sum_{i=1}^{s} c_i \cdot \mu_i \cdot \mathcal{R}_p(\mu_i) , \tag{7.124}$$

$$\overline{\mathcal{T}} = 2 \sum_{i=1}^{s} c_i \cdot \mu_i \cdot \mathcal{T}_p(\mu_i) . \tag{7.125}$$

7.4.3
Application to Flux Densities

As an example of applying the planetary and spherical albedos and transmissivities to radiative transfer simulations, we will show how to approximately calculate the vertical variation of flux density in the atmosphere in a computationally efficient way. Referring to Figure 7.7, a scattering atmosphere is divided into $(N - 1)$ layers.

The transmitted and reflected diffuse flux densities associated with the multiple-scattering events within layer l are given by

$$\tilde{F}_{2l+1} = \mu_k \cdot \mathcal{T}_p(\mu_k) \cdot S_{dir} \cdot \exp\left(-\frac{1}{\mu_k} \cdot \sum_{i=1}^{l-1} \tau_i\right) , \tag{7.126}$$

$$\tilde{F}_{2l} = \mu_k \cdot \mathcal{R}_p(\mu_k) \cdot S_{dir} \cdot \exp\left(-\frac{1}{\mu_k} \cdot \sum_{i=1}^{l-1} \tau_i\right) , \tag{7.127}$$

where $\mu_k = \cos \theta_k$ and θ_k is the solar zenith angle. In Figure 7.7, the upward and downward diffuse flux densities between layers $(l - 1)$ and l are denoted as F_{2l-1} and F_{2l} and include multiple-scattering. From the configuration in Figure 7.7, we

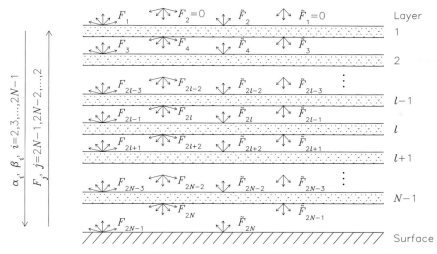

Figure 7.7 Schematic diagram illustrating the calculation of the vertical distribution of diffuse flux density.

have the following relations:

$$F_1 = \tilde{F}_2 + \left(\overline{\mathcal{T}}_1 + e^{-\tau_1/\overline{\mu}}\right) \cdot F_3 ,$$ (7.128)

$$F_2 = 0 ,$$ (7.129)

$$F_{2l-1} = \tilde{F}_{2l} + \left(\overline{\mathcal{T}}_l + e^{-\tau_l/\overline{\mu}}\right) \cdot F_{2l+1} + \overline{\mathcal{R}}_l \cdot F_{2l} ,$$
$$\text{for} \quad l = 2, 3, 4, \ldots, N-1 ,$$ (7.130)

$$F_{2l} = \tilde{F}_{2l-1} + \left(\overline{\mathcal{T}}_{l-1} + e^{-\tau_{l-1}/\overline{\mu}}\right) \cdot F_{2l-2} + \overline{\mathcal{R}}_{l-1} \cdot F_{2l-1} ,$$
$$\text{for} \quad l = 2, 3, 4, \ldots, N-1 ,$$ (7.131)

$$F_{2N-1} = \tilde{F}_{2N} + \gamma_g \cdot F_{2N} ,$$ (7.132)

$$F_{2N} = \tilde{F}_{2N-1} + \overline{\mathcal{R}}_{N-1} \cdot F_{2N-1} + \left(\overline{\mathcal{T}}_{N-1} + e^{-\tau_{N-1}/\overline{\mu}}\right) \cdot F_{2N-2} ,$$ (7.133)

where $e^{-\tau_j/\overline{\mu}}$, $j = 1, 2, 3, \ldots, N-1$, is the effective direct transmissivity. The quantity $1/\overline{\mu}$ is referred to as the diffusivity factor (Liou, 2002) and equals two in the case of a hemispherically isotropic radiation field (Thomas and Stamnes, 1999). In Eq. (7.132), γ_g indicates the surface albedo, also called the spherical or global albedo, see Eq. (6.79). To solve for the downward and upward diffuse flux densities, we let

$$F_{j-1} = \alpha_{j-1} \cdot F_j + \beta_{j-1} .$$ (7.134)

Thus, it follows that

$$F_{2l-1} = \alpha_{2l-1} \cdot F_{2l} + \beta_{2l-1} .$$ (7.135)

From Eqs. (7.130) and (7.135), we obtain

$$\tilde{F}_{2l} + \left(\overline{\mathcal{T}}_l + e^{-\tau_l/\overline{\mu}}\right) \cdot F_{2l+1} + \overline{\mathcal{R}}_l \cdot F_{2l} = \alpha_{2l-1} \cdot F_{2l} + \beta_{2l-1} . \tag{7.136}$$

The preceding equation can be rewritten as

$$F_{2l} = \frac{\overline{\mathcal{T}}_l + e^{-\tau_l/\overline{\mu}}}{\alpha_{2l-1} - \overline{\mathcal{R}}_l} \cdot F_{2l+1} + \frac{\tilde{F}_{2l} - \beta_{2l-1}}{\alpha_{2l-1} - \overline{\mathcal{R}}_l} . \tag{7.137}$$

Similarly, from Eq. (7.134), we obtain

$$F_{2l-2} = \alpha_{2l-2} \cdot F_{2l-1} + \beta_{2l-2} . \tag{7.138}$$

Substituting Eq. (7.138) into Eq. (7.131), we can rewrite the resultant expression in the form of

$$F_{2l-1} = \frac{1}{\left(\overline{\mathcal{T}}_{l-1} + e^{-\tau_{l-1}/\overline{\mu}}\right) \cdot \alpha_{2l-2} + \overline{\mathcal{R}}_{l-1}} \cdot F_{2l}$$
$$- \frac{\tilde{F}_{2l-1} + \left(\overline{\mathcal{T}}_{l-1} + e^{-\tau_{l-1}/\overline{\mu}}\right) \cdot \beta_{2l-2}}{\left(\overline{\mathcal{T}}_{l-1} + e^{-\tau_{l-1}/\overline{\mu}}\right) \cdot \alpha_{2l-2} + \overline{\mathcal{R}}_{l-1}} . \tag{7.139}$$

From Eqs. (7.139) and (7.137), in comparison with Eq. (7.134), we get

$$\alpha_{2l-1} = \frac{1}{\left(\overline{\mathcal{T}}_{l-1} + e^{-\tau_{l-1}/\overline{\mu}}\right) \cdot \alpha_{2l-2} + \overline{\mathcal{R}}_{l-1}} , \tag{7.140}$$

$$\beta_{2l-1} = -\frac{\tilde{F}_{2l-1} + \left(\overline{\mathcal{T}}_{l-1} + e^{-\tau_{l-1}/\overline{\mu}}\right) \cdot \beta_{2l-2}}{\left(\overline{\mathcal{T}}_{l-1} + e^{-\tau_{l-1}/\overline{\mu}}\right) \cdot \alpha_{2l-2} + \overline{\mathcal{R}}_{l-1}} , \tag{7.141}$$

$$\alpha_{2l} = \frac{\overline{\mathcal{T}}_l + e^{-\tau_l/\overline{\mu}}}{\alpha_{2l-1} - \overline{\mathcal{R}}_l} , \tag{7.142}$$

$$\beta_{2l} = \frac{\tilde{F}_{2l} - \beta_{2l-1}}{\alpha_{2l-1} - \overline{\mathcal{R}}_l} . \tag{7.143}$$

Furthermore, from Eq. (7.129), we have the following conditions:

$$\alpha_2 = 0 , \tag{7.144}$$

$$\beta_2 = 0 . \tag{7.145}$$

Using the conditions in Eqs. (7.144) and (7.145), and the recurrence relations in Eqs. (7.140) and (7.143), we can calculate α_j and β_j for $j = 3, 4, 5, \ldots, N-1$.

To calculate the flux densities, let us consider the following expression based on Eq. (7.134):

$$F_{2N-1} = \alpha_{2N-1} \cdot F_{2N} + \beta_{2N-1} . \tag{7.146}$$

From Eqs. (7.146) and (7.132), we have

$$F_{2N} = \frac{\beta_{2N-1} - \tilde{F}_{2N}}{\gamma_g - \alpha_{2N-1}} . \tag{7.147}$$

After F_{2N} is calculated, we can apply the recurrence relation Eq. (7.134) to calculate the flux densities at all the layers, provided that all coefficients α_j and β_j for $j = 3, 4, 5, \ldots, N - 1$ are also calculated. Note that $\tilde{F}_j$ for $j = 3, 4, 5, \ldots, N - 1$ can be calculated using the explicit expressions in Eqs. (7.126) and (7.127). The arrows on the left of Figure 7.7 indicate the sequences for calculating the coefficients and flux densities. Specifically, the calculation of the coefficients is from the TOA to the surface. On the contrary, the calculation of the flux densities is from the surface to the TOA.

After the diffuse flux densities are calculated from the preceding procedure, the total (direct plus diffuse) downward flux density between layers $(l-1)$ and l is given by

$$
\mu_k \cdot S_{\text{dir}} \cdot \exp\left(-\frac{1}{\mu_k} \cdot \sum_{j=1}^{l-1} \tau_i \right) + F_{2l} ; \tag{7.148}
$$

whereas, the corresponding upward flux density is F_{2l-1}.

7.5
Discrete Ordinate Method (DOM)

The Discrete Ordinate Method (DOM) was originally developed by Chandrasekhar (1950), who employed a technique introduced by Wick (1943) to discretize the equation of radiative transfer in integrodifferential form into a set of ordinary differential equations. For this reason, this method is also known as the Wick–Chandrasekhar DOM. Rybicki (1996) briefly explained this method in a concise review of the mathematical elegance and beauty which Chandrasekhar brought to the discipline of radiative transfer.

Numerically, Liou (1973) used the DOM to simulate irradiance in scattering atmospheres. Stamnes et al. (1988) developed a comprehensive radiative transfer model based on this method, commonly known as DISORT, to compute the angular variation of radiance. Weng (1992a,b) extended the numerical implementation of the DOM to a polarized radiation field.

To illustrate the basic principle of the DOM, we first consider the azimuth-independent component of the radiance, that is, the $m = 0$ component of the Fourier expansion of the diffuse radiance, $I_\lambda^{(0)}$, in the diffuse region where the direct radiation can be neglected and emission omitted. This case can be valid, for example, in the region below a thick water cloud layer with a large optical thickness that essentially attenuates the direct radiation. Furthermore, for simplicity, the phase function is assumed to be isotropic, that is, $\mathcal{P}(\cos \vartheta) = 1$. With these simplifications, the equation of the radiative transfer in the form of Eq. (7.53) is given by

$$
\mu \frac{\mathrm{d} I_\lambda^{(0)}(\tau, \mu)}{\mathrm{d}\tau} = I_\lambda^{(0)}(\tau, \mu) - \frac{\tilde{\omega}(\tau)}{2} \int_{-1}^{1} I_\lambda^{(0)}(\tau, \mu') \, \mathrm{d}\mu' , \tag{7.149}
$$

where $P^{(m)}(\mu, \mu') = 1$ for $\mathcal{P}(\cos \vartheta) = 1$. In the DOM, the Gauss–Lobatto quadrature is used to discretize the integral in Eq. (7.149), see Eq. (2.70). To do so, the angular dependence of the quantity is specified at Gauss quadrature abscissa as

$$\int_{-1}^{1} I_\lambda^{(0)}(\tau, \mu') \, \mathrm{d}\mu' = \sum_{j=-s}^{s} c_j \cdot I_\lambda^{(0)}(\tau, \mu_j), \tag{7.150}$$

where μ_j and c_j are the Gauss quadrature abscissas and weights of the quadrature scheme, respectively. Thus, Eq. (7.149) can be approximated in the form of

$$\mu_i \frac{\mathrm{d} I_\lambda^{(0)}(\tau, \mu_i)}{\mathrm{d}\tau} = I_\lambda^{(0)}(\tau, \mu_i) - \frac{\tilde{\omega}(\tau)}{2} \sum_{j=-s}^{s} c_j \cdot I_\lambda^{(0)}(\tau, \mu_j). \tag{7.151}$$

For the zero-order Fourier mode of the diffuse radiance, we introduce

$$I_i(\tau) = I_\lambda^{(0)}(\tau, \mu_i), \tag{7.152}$$

and obtain

$$\mu_i \frac{\mathrm{d} I_i(\tau)}{\mathrm{d}\tau} = I_i(\tau) - \frac{\tilde{\omega}(\tau)}{2} \sum_{j=-s}^{s} c_j \cdot I_j(\tau). \tag{7.153}$$

In Eq. (7.153), the summation excludes the case of $j = 0$. Furthermore, a negative value of j corresponds to a downward stream; whereas, a positive j denotes an upward stream (flux density). The above equation gives the $2s$-stream approximation whose special cases with $s = 1$ (two-stream) and $s = 2$ (four-stream) have been widely used in practical applications.

If we denote $I_i(\tau)$ and $I_{-i}(\tau)$ in terms of $I_j^\uparrow(\tau)$ and $I_j^\downarrow(\tau)$ for upward and downward radiation, respectively, we can rewrite Eq. (7.153) separately for upward and downward radiation in the form of

$$\mu_i \frac{\mathrm{d} I_i^\uparrow(\tau)}{\mathrm{d}\tau} = I_i^\uparrow(\tau) - \frac{\tilde{\omega}(\tau)}{2} \sum_{j=1}^{s} c_j \cdot \left[I_j^\uparrow(\tau) + I_j^\downarrow(\tau) \right], \tag{7.154}$$

$$-\mu_i \frac{\mathrm{d} I_i^\downarrow(\tau)}{\mathrm{d}\tau} = I_i^\downarrow(\tau) - \frac{\tilde{\omega}(\tau)}{2} \sum_{j=1}^{s} c_j \cdot \left[I_j^\uparrow(\tau) + I_j^\downarrow(\tau) \right], \tag{7.155}$$

where $c_i = c_{-i}$ and $\mu_i > 0$ are implied. The above two equations are usually consolidated into a concise matrix form as (Liou, 2002; Thomas and Stamnes, 1999; Zdunkowski et al., 2007)

$$\frac{\mathrm{d}}{\mathrm{d}\tau} \begin{pmatrix} \vec{\mathbf{I}}^\uparrow(\tau) \\ \vec{\mathbf{I}}^\downarrow(\tau) \end{pmatrix} = \begin{pmatrix} \mathsf{IM}^{\uparrow\uparrow} & \mathsf{IM}^{\uparrow\downarrow} \\ \mathsf{IM}^{\downarrow\uparrow} & \mathsf{IM}^{\downarrow\downarrow} \end{pmatrix} \cdot \begin{pmatrix} \vec{\mathbf{I}}^\uparrow(\tau) \\ \vec{\mathbf{I}}^\downarrow(\tau) \end{pmatrix}, \tag{7.156}$$

where

$$
\vec{I}^{\uparrow}(\tau) = \begin{pmatrix} I_1^{\uparrow}(\tau) \\ I_2^{\uparrow}(\tau) \\ \vdots \\ I_{s-1}^{\uparrow}(\tau) \\ I_s^{\uparrow}(\tau) \end{pmatrix}, \quad \text{and} \quad \vec{I}^{\downarrow}(\tau) = \begin{pmatrix} I_1^{\downarrow}(\tau) \\ I_2^{\downarrow}(\tau) \\ \vdots \\ I_{s-1}^{\downarrow}(\tau) \\ I_s^{\downarrow}(\tau) \end{pmatrix}. \tag{7.157}
$$

In Eq. (7.156), $IM^{\uparrow\uparrow}, IM^{\uparrow\downarrow}, IM^{\downarrow\uparrow}$, and $IM^{\downarrow\downarrow}$ are $s \times s$ matrices and satisfy the following relations:

$$
IM^{\downarrow\downarrow} = -IM^{\uparrow\uparrow}, \quad \text{and} \quad IM^{\downarrow\uparrow} = -IM^{\uparrow\downarrow}. \tag{7.158}
$$

The elements of the matrices are given by

$$
M_{ij}^{\uparrow\uparrow} = \frac{\delta_{i,j} - c_j \cdot \tilde{\omega}(\tau)/2}{\mu_i}, \tag{7.159}
$$

$$
M_{ij}^{\uparrow\downarrow} = \frac{-c_j \cdot \tilde{\omega}(\tau)}{2\mu_i}. \tag{7.160}
$$

To solve Eq. (7.156), we assume $\vec{I}^{\uparrow}(\tau)$ and $\vec{I}^{\downarrow}(\tau)$ are given in terms of eigenvectors $\vec{I}_0^{\uparrow}$ and $\vec{I}_0^{\downarrow}$ and eigenvalue k as

$$
\begin{pmatrix} \vec{I}^{\uparrow}(\tau) \\ \vec{I}^{\downarrow}(\tau) \end{pmatrix} = \begin{pmatrix} \vec{I}_0^{\uparrow} \\ \vec{I}_0^{\downarrow} \end{pmatrix} \cdot e^{-k \cdot \tau}. \tag{7.161}
$$

Substituting Eq. (7.161) into Eq. (7.156), we obtain

$$
-k \cdot \begin{pmatrix} \vec{I}_0^{\uparrow} \\ \vec{I}_0^{\downarrow} \end{pmatrix} = \begin{pmatrix} IM^{\uparrow\uparrow} & IM^{\uparrow\downarrow} \\ -IM^{\uparrow\downarrow} & -IM^{\uparrow\uparrow} \end{pmatrix} \cdot \begin{pmatrix} \vec{I}_0^{\uparrow} \\ \vec{I}_0^{\downarrow} \end{pmatrix}. \tag{7.162}
$$

Up to this point, solving the equation of radiative transfer reduces to the problem of determining $\vec{I}_0^{\uparrow}$ and $\vec{I}_0^{\downarrow}$ and eigenvalue k, a standard eigenvalue-eigenvector problem. Note that in the preceding discussion, we did not consider the contribution of the first-order scattering of the direct radiation and the resultant equations are homogeneous. If we include the term associated with first-order scattering of the direct radiation, the resultant ordinary differential equations are inhomogeneous. In this case, the solution can be obtained by using the standard technique to solve inhomogeneous ordinary differential equations, that is, the solution is the superposition of the general solution to the homogeneous equations and a specific solution to the inhomogeneous equations. The mathematical details are not presented here.

It has been found that the double Gauss quadrature scheme improves the accuracy of the DOM (Kourganoff, 1963; Sykes, 1951; Thomas and Stamnes, 1999, and references cited therein). To illustrate the double Gauss quadrature scheme, let us consider the integral in Eq. (7.149), which can be decomposed into two parts as

$$\int_{-1}^{1} I_\lambda^{(0)}(\tau,\mu')\,d\mu' = \int_0^1 I_\lambda^{(0)}(\tau,\mu')\,d\mu' + \int_0^1 I_\lambda^{(0)}(\tau,-\mu')\,d\mu'$$

$$= \frac{1}{2}\int_{-1}^{1}\left[I_\lambda^{(0)}\left(\tau,\frac{\mu''+1}{2}\right) + I_\lambda^{(0)}\left(\tau,-\frac{\mu''+1}{2}\right)\right]d\mu''.$$

$$(7.163)$$

Applying the conventional Gauss–Lobatto quadrature scheme with s terms to the above equation, we have

$$\int_{-1}^{1} I_\lambda^{(0)}(\tau,\mu')\,d\mu' = \frac{1}{2}\sum_{j=1}^{s} c_j \cdot \left[I_\lambda^{(0)}\left(\tau,\frac{\mu_j+1}{2}\right) + I_\lambda^{(0)}\left(\tau,-\frac{\mu_j+1}{2}\right)\right]$$

$$= \sum_{j=1}^{s} c_{d,j} \cdot \left[I_\lambda^{(0)}\left(\tau,\mu_{d,j}\right) + I_\lambda^{(0)}\left(\tau,-\mu_{d,j}\right)\right]$$

$$= \sum_{j=1}^{s} c_{d,j} \cdot \left[I_{d,j}^{\uparrow}(\tau) + I_{d,j}^{\downarrow}(\tau)\right],$$

$$(7.164)$$

where the subscript d indicates that the discretization is based on the double Gauss quadrature scheme. It is evident from Eq. (7.164) that the following relations hold:

$$c_{d,j} = \frac{1}{2}\cdot c_j\,,$$

$$(7.165)$$

$$\mu_{d,j} = \pm\frac{1}{2}\cdot(\mu_j+1)\,.$$

$$(7.166)$$

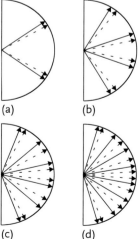

(a)   (b)

(c)   (d)

Figure 7.8 Discrete streams based on the conventional (black solid arrows) and double (black dashed arrows) Gauss quadrature scheme for the (a) 2-, (b) 4-, (c) 6-, and (d) 8-stream approximations.

Table 7.1 The quadrature points based on the conventional and the double Gauss quadrature scheme.

	Conventional Gauss quadrature μ_i	Double Gauss quadrature $\mu_{d,i}$
Two streams $s = 1$		
$i = \pm 1$	$\mu_{\pm 1} = \pm 0.577\,350$	$\mu_{d,\pm 1} = \pm 0.50\,000$
Four streams $s = 2$		
$i = \pm 1$	$\mu_{\pm 1} = \pm 0.339\,981$	$\mu_{d,\pm 1} = \pm 0.211\,325$
$i = \pm 2$	$\mu_{\pm 2} = \pm 0.861\,136$	$\mu_{d,\pm 2} = \pm 0.788\,675$
Six streams $s = 3$		
$i = \pm 1$	$\mu_{\pm 1} = \pm 0.238\,619$	$\mu_{d,\pm 1} = \pm 0.112\,702$
$i = \pm 2$	$\mu_{\pm 2} = \pm 0.661\,209$	$\mu_{d,\pm 2} = \pm 0.500\,000$
$i = \pm 3$	$\mu_{\pm 3} = \pm 0.932\,470$	$\mu_{d,\pm 3} = \pm 0.887\,298$
Eight streams $s = 4$		
$i = \pm 1$	$\mu_{\pm 1} = \pm 0.183\,435$	$\mu_{d,\pm 1} = \pm 0.069\,432$
$i = \pm 2$	$\mu_{\pm 2} = \pm 0.525\,532$	$\mu_{d,\pm 2} = \pm 0.330\,009$
$i = \pm 3$	$\mu_{\pm 3} = \pm 0.796\,666$	$\mu_{d,\pm 3} = \pm 0.669\,991$
$i = \pm 4$	$\mu_{\pm 4} = \pm 0.960\,290$	$\mu_{d,\pm 4} = \pm 0.930\,568$

Table 7.1 lists the quadrature points, see Table 2.4, based on the conventional Gauss quadrature scheme and the double Gauss quadrature scheme for the two-, four-, six-, and eight-stream approximations. Figure 7.8 schematically illustrates the discrete streams based on the quadrature points listed in Table 7.1. Evidently, the radiation streams based on the double Gauss quadrature scheme are closer to $\theta = 90°$ than their conventional Gauss quadrature counterparts. Thus, the double Gauss quadrature scheme is more accurate to account for the variation of radiance from upper hemisphere to lower hemisphere, which can be important for some applications.

7.6
Spherical Harmonics Method (SHM)

In addition to the DOM, the Spherical Harmonic Method (SHM) (Kourganoff, 1963; Liou, 2002; Zdunkowski et al., 2007) offers a numerically rigorous approach to the solution of the radiation transfer in a scattering atmosphere. By representing the radiance field in terms of spherical harmonics and discrete ordinates, user-friendly and computationally efficient software packages have been developed by Evans (1998, 2007) in 1D and 3D cases.

To illustrate the basic principles of the SHM, we consider a simple case given by Eq. (7.149). To account for the angular dependence of $I_\lambda^{(0)}(\tau, \mu)$, we expand $I_\lambda^{(0)}(\tau, \mu)$ in terms of Legendre polynomials, see Eq. (7.1), in the form of

$$I_\lambda^{(0)}(\tau, \mu) = \sum_{j=0}^{\infty} C_j(\tau) \cdot P_j(\mu) \approx \sum_{j=0}^{N-1} C_j(\tau) \cdot P_j(\mu) . \tag{7.167}$$

A special case with $N = 2$ for the preceding approximation,

$$I_\lambda^{(0)}(\tau, \mu) = C_0(\tau) + C_1(\tau) \cdot \mu , \tag{7.168}$$

represents the well-known Eddington approximation (Eddington, 1916). Note, the Legendre coefficients $C_j(\tau)$ in Eq. (7.167) are sometimes formulated as $C_j(\tau) \cdot (2j + 1)/2$ in literature reports, for example, Zdunkowski et al. (2007), for convenience in normalizing the Legendre polynomials.

Substituting Eq. (7.167) into Eq. (7.149), yields

$$\mu \sum_{j=0}^{N-1} \frac{dC_j(\tau)}{d\tau} P_j(\mu) = \sum_{j=0}^{N-1} C_j(\tau) \cdot P_j(\mu)$$

$$- \frac{\tilde{\omega}(\tau)}{2} \sum_{j=0}^{N-1} C_j(\tau) \cdot \int_{-1}^{1} P_j(\mu') d\mu' . \tag{7.169}$$

Using the orthogonality of Legendre polynomials, see Eq. (2.44), the integral in Eq. (7.169) reduces to $2\delta_{j0}$. Thus, Eq. (7.169) is simplified as

$$\mu \sum_{j=0}^{N-1} \frac{dC_j(\tau)}{d\tau} P_j(\mu) = \sum_{j=0}^{N-1} C_j(\tau) \cdot P_j(\mu) - \tilde{\omega}(\tau) \cdot C_0(\tau) . \tag{7.170}$$

Furthermore, using the recurrence relation given by Eq. (2.61) in the form of

$$P_j(\mu) = \frac{j \cdot P_{j-1}(\mu) + (j + 1) \cdot P_{j+1}(\mu)}{(2j + 1) \cdot \mu} , \tag{7.171}$$

we can rewrite the left side of Eq. (7.170) as

$$\sum_{j=0}^{N-1} \frac{dC_j(\tau)}{d\tau} \left[\left(\frac{j}{2j + 1} \right) \cdot P_{j-1}(\mu) + \left(\frac{j+1}{2j + 1} \right) \cdot P_{j+1}(\mu) \right]$$

$$= \sum_{j=0}^{N-1} C_j(\tau) \cdot P_j(\mu) - \tilde{\omega}(\tau) \cdot C_0(\tau) . \tag{7.172}$$

After multiplying Eq. (7.172) with $P_i(\mu)$ and integrating the resultant expression from -1 to 1 over μ, we will apply the orthogonality of Legendre polynomials, see

Eq. (2.53), in the following forms:

$$\int_{-1}^{1} P_i(\mu) \cdot P_{j-1}(\mu) d\mu = \delta_{i,j-1} \cdot \left(\frac{2}{2i+1}\right); \quad \text{for} \quad \mu \in [-1,1], \qquad (7.173)$$

$$\int_{-1}^{1} P_i(\mu) \cdot P_{j+1}(\mu) d\mu = \delta_{i,j+1} \cdot \left(\frac{2}{2i+1}\right); \quad \text{for} \quad \mu \in [-1,1], \qquad (7.174)$$

$$\int_{-1}^{1} P_i(\mu) \cdot P_j(\mu) d\mu = \delta_{ij} \cdot \left(\frac{2}{2i+1}\right); \quad \text{for} \quad \mu \in [-1,1], \qquad (7.175)$$

thus obtaining

$$\sum_{j=0}^{N-1} \frac{dC_j(\tau)}{d\tau} \left[\left(\frac{j}{2j+1}\right) \cdot \delta_{i,j-1} \cdot \left(\frac{2}{2i+1}\right) \right.$$
$$\left. + \left(\frac{j+1}{2j+1}\right) \cdot \delta_{i,j+1} \cdot \left(\frac{2}{2i+1}\right) \right]$$
$$= \sum_{j=0}^{N-1} C_j(\tau) \cdot \delta_{ij} \cdot \left(\frac{2}{2i+1}\right) - \tilde{\omega}(\tau) \cdot \delta_{0i} \cdot \left(\frac{2}{2i+1}\right) \cdot C_0(\tau). \qquad (7.176)$$

Dividing by the factor $2/(2i+1)$ and applying the definition of the Kronecker symbol, from Eq. (7.176) we obtain

$$\frac{dC_{i+1}(\tau)}{d\tau} \left[\frac{(i+1)}{2(i+1)+1} \right] + \frac{dC_{i-1}(\tau)}{d\tau} \left[\frac{(i-1)+1}{2(i-1)+1} \right]$$
$$= C_i(\tau) - \tilde{\omega}(\tau) \cdot \delta_{0i} \cdot C_0(\tau), \qquad (7.177)$$

which yields

$$\left(\frac{i+1}{2i+3}\right) \frac{dC_{i+1}(\tau)}{d\tau} + \left(\frac{i}{2i-1}\right) \frac{dC_{i-1}(\tau)}{d\tau}$$
$$= C_i(\tau) - \tilde{\omega}(\tau) \cdot \delta_{0i} \cdot C_0(\tau). \qquad (7.178)$$

Furthermore, let

$$\vec{C}(\tau) = \begin{pmatrix} C_0(\tau) \\ C_1(\tau) \\ \vdots \\ C_{N-1}(\tau) \\ C_N(\tau) \end{pmatrix}, \qquad (7.179)$$

then Eq. (7.178) can be written in a concise matrix form as

$$\frac{d\vec{C}(\tau)}{d\tau} = IL \cdot \vec{C}(\tau), \qquad (7.180)$$

with

$$IL = IA^{-1}IB(\tau) ,$$ (7.181)

where

$$IA = \begin{pmatrix} 0 & \frac{1}{3} & 0 & 0 & 0 & \cdots & 0 & 0 \\ 1 & 0 & \frac{2}{5} & 0 & 0 & \cdots & 0 & 0 \\ 0 & \frac{2}{3} & 0 & \frac{3}{7} & 0 & \cdots & 0 & 0 \\ 0 & 0 & \frac{3}{5} & 0 & \frac{4}{9} & \cdots & 0 & 0 \\ 0 & 0 & 0 & \frac{4}{7} & 0 & \cdots & 0 & 0 \\ \vdots & \vdots & \vdots & \vdots & \vdots & \ddots & \vdots & \vdots \\ 0 & 0 & 0 & 0 & 0 & \cdots & 0 & N/(2N+1) \\ 0 & 0 & 0 & 0 & 0 & \cdots & N/(2N-1) & 0 \end{pmatrix} ,$$ (7.182)

and

$$IB(\tau) = \begin{pmatrix} [1 - \tilde{\omega}(\tau)] & 0 & 0 & 0 & 0 & \cdots & 0 & 0 \\ 0 & 1 & 0 & 0 & 0 & \cdots & 0 & 0 \\ 0 & 0 & 1 & 0 & 0 & \cdots & 0 & 0 \\ 0 & 0 & 0 & 1 & 0 & \cdots & 0 & 0 \\ 0 & 0 & 0 & 0 & 1 & \cdots & 0 & 0 \\ \vdots & \vdots & \vdots & \vdots & \vdots & \ddots & \vdots & \vdots \\ 0 & 0 & 0 & 0 & 0 & \cdots & 1 & 0 \\ 0 & 0 & 0 & 0 & 0 & \cdots & 0 & 1 \end{pmatrix} .$$ (7.183)

Let the formal solution to Eq. (7.180) be

$$\vec{C}(\tau) = \vec{C}_0 \cdot e^{-k \cdot \tau} .$$ (7.184)

Substituting Eq. (7.184) into Eq. (7.180) yields

$$-k \cdot \vec{C}_0 = IL \cdot \vec{C}_0 .$$ (7.185)

The preceding equation represents a standard eigenvalue and eigenvector problem that can be solved using some ordinary numerical techniques which are described by Press et al. (1992) and not recaptured here.

7.7
Monte Carlo Method (MCM)

The Monte Carlo Method (MCM) is a general statistical method that can be used to solve a wide range of problems, for example, being applied to the transfer of radiation for the calculation of radiance. A large number of photons are traced from the incident radiation source to receivers which are produced by sources located at given altitudes and aimed in specific viewing directions. Note, the concept of the "photon" is different here from its counterpart in quantum mechanics. The "photons" used in classic radiative transfer theory are essentially "rays" or "beams" of radiation. The number of photons required is determined by the statistical significance of the photons arriving at the receivers. For optically thick atmospheres, the number of incident photons may be quite large in order to warrant a statistically significant number of photons to reach the receivers. As photons propagate through the atmosphere, each single interaction event between the photons and air molecules, aerosol particles, cloud droplets, ice crystals, or precipitation particulates is modeled. The interactions between the photons and the particles are described by the phase function, the single-scattering albedo, and the optical thickness. To trace photons propagating in a reverse direction from the receiver to the source, we use the inverse (or backward) Monte Carlo approach.

7.7.1
Basic Principle

To illustrate the procedure with some mathematical detail, let us consider the calculation of the reflectivity and transmissivity of an optically homogeneous layer with the optical depth, single-scattering albedo, and phase function specified by τ^*, $\tilde{\omega}$, and $\mathcal{P}(\theta)$, respectively. Figure 7.9 illustrates the propagation of a reflected photon. The directions of the photon at the various stages of propagation are denoted in terms of unit vectors $\hat{Z}_i$, $(i = 0, 1, 2 \ldots)$.

In conjunction with the incident direction $\hat{Z}_0$, we define two unit vectors, $\hat{X}_0$ and $\hat{Y}_0$, in Figure 7.9. The three unit vectors, $\hat{X}_0$, $\hat{Y}_0$, and $\hat{Z}_0$, constitute a right-handed coordinate system. Additionally, we define a laboratory coordinate system indicated by the three unit vectors $\hat{X}_{lab}$, $\hat{Y}_{lab}$, and $\hat{Z}_{lab}$, where $\hat{Z}_{lab}$ points in the direction of the zenith.

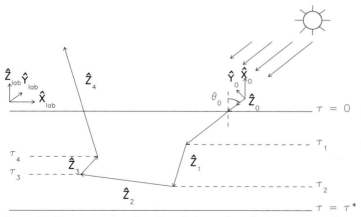

Figure 7.9 The propagation of a reflected photon through a homogeneous layer.

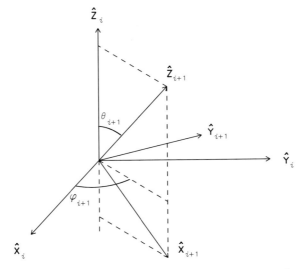

Figure 7.10 The relative orientations of two consecutive coordinate systems with $\hat{\mathbf{Z}}_i$-directions along the propagation of a photon (see Figure 7.9).

Following Eq. (2.1), a vector $\vec{a}$ can be decomposed into three Cartesian components within the laboratory or incident coordinate systems as

$$\vec{a} = a_{1,\text{lab}} \cdot \hat{\mathbf{X}}_{\text{lab}} + a_{2,\text{lab}} \cdot \hat{\mathbf{Y}}_{\text{lab}} + a_{3,\text{lab}} \cdot \hat{\mathbf{Z}}_{\text{lab}}$$
$$= a_{1,0} \cdot \hat{\mathbf{X}}_0 + a_{2,0} \cdot \hat{\mathbf{Y}}_0 + a_{3,0} \cdot \hat{\mathbf{Z}}_0 \ . \tag{7.186}$$

From the geometric configuration shown in Figure 7.9, it follows that

$$\begin{pmatrix} a_{1,\text{lab}} \\ a_{2,\text{lab}} \\ a_{3,\text{lab}} \end{pmatrix} = \mathsf{IL}_0 \cdot \begin{pmatrix} a_{1,0} \\ a_{2,0} \\ a_{3,0} \end{pmatrix} , \tag{7.187}$$

with

$$\mathsf{IL}_0 = \begin{pmatrix} -\cos\theta_0 & 0 & -\sin\theta_0 \\ 0 & 1 & 0 \\ \sin\theta_0 & 0 & -\cos\theta_0 \end{pmatrix} . \tag{7.188}$$

We also define a number of coordinate systems whose directions are specified in terms of unit vectors $\hat{\mathbf{X}}_i$, $\hat{\mathbf{Y}}_i$, and $\hat{\mathbf{Z}}_i$. Note that $\hat{\mathbf{Z}}_1$, $\hat{\mathbf{Z}}_2$, $\hat{\mathbf{Z}}_3$, and $\hat{\mathbf{Z}}_4$ are indicated in Figure 7.9 where $\hat{\mathbf{X}}_i$ and $\hat{\mathbf{Y}}_i$ are not shown.

Figure 7.10 shows the relative positions of two consecutive coordinate systems, $(\hat{\mathbf{X}}_i, \hat{\mathbf{Y}}_i, \hat{\mathbf{Z}}_i)$ and $(\hat{\mathbf{X}}_{i+1}, \hat{\mathbf{Y}}_{i+1}, \hat{\mathbf{Z}}_{i+1})$. With respect to these two coordinates, a vector $\vec{a}$ can be decomposed, that is,

$$\vec{a} = a_{1,i} \cdot \hat{\mathbf{X}}_i + a_{2,i} \cdot \hat{\mathbf{Y}}_i + a_{3,i} \cdot \hat{\mathbf{Z}}_i$$
$$= a_{1,i+1} \cdot \hat{\mathbf{X}}_{i+1} + a_{2,i+1} \cdot \hat{\mathbf{Y}}_{i+1} + a_{3,i+1} \cdot \hat{\mathbf{Z}}_{i+1} , \tag{7.189}$$

where

$$\begin{pmatrix} a_{1,i} \\ a_{2,i} \\ a_{3,i} \end{pmatrix} = \mathsf{IL}_{i+1} \cdot \begin{pmatrix} a_{1,i+1} \\ a_{2,i+1} \\ a_{3,i+1} \end{pmatrix} , \tag{7.190}$$

and

$$\mathsf{IL}_{i+1} = \begin{pmatrix} \cos\varphi_{i+1} & -\sin\varphi_{i+1} & 0 \\ \sin\varphi_{i+1} & \cos\varphi_{i+1} & 0 \\ 0 & 0 & 1 \end{pmatrix} \cdot \begin{pmatrix} \cos\theta_{i+1} & 0 & \sin\theta_{i+1} \\ 0 & 1 & 0 \\ -\sin\theta_{i+1} & 0 & \cos\theta_{i+1} \end{pmatrix} . \tag{7.191}$$

From Eqs. (7.187) and (7.190), we obtain

$$\begin{pmatrix} a_{1,\text{lab}} \\ a_{2,\text{lab}} \\ a_{3,\text{lab}} \end{pmatrix} = \mathsf{IG}_i \cdot \begin{pmatrix} a_{1,i} \\ a_{2,i} \\ a_{3,i} \end{pmatrix} , \tag{7.192}$$

where

$$\mathsf{IG}_0 = \mathsf{IL}_0 , \tag{7.193}$$

$$\mathsf{IG}_i = \mathsf{IG}_{i-1} \cdot \mathsf{IL}_i . \tag{7.194}$$

The forward Monte Carlo algorithm begins with the determination of whether a photon transmits through the layer and whether the extinction is due to scattering or absorption. To numerically perform the decision, we use a random number generator to generate a series of pseudorandom numbers (the term "pseudo" is used because the numbers are produced with a computer program and are not truly random numbers) $\varsigma_1, \varsigma_2, \varsigma_3, \dots$, which are uniformly distributed in (0, 1). Then, we use the following equations to determine the parameter τ_1 in Figure 7.9:

$$\varsigma_1 = e^{-\tau_1/\mu_0} , \tag{7.195}$$

or

$$\tau_1 = -\mu_0 \cdot \ln\varsigma_1 , \tag{7.196}$$

where $\mu_0 = \cos\theta_0$. If $\tau_1 > \tau^*$, the photon is directly transmitted through the layer and is counted for calculating the transmissivity. If $\tau_1 < \tau$, the photon is either scattered or absorbed, and if $\varsigma_2 < \tilde{\omega}$, a photon scattering event occurs. In the latter case, the new direction for the propagation of the photon with respect to the coordinate system $(\hat{\mathbf{X}}_0, \hat{\mathbf{Y}}_0, \hat{\mathbf{Z}}_0)$ is determined by the two angles φ_1 and θ_1, see Figure 7.10, and can be calculated by

$$\varphi_1 = 2\pi \cdot \varsigma_3 , \tag{7.197}$$

$$\varsigma_4 = \frac{1}{2} \int_0^{\theta_1} \mathcal{P}(\theta) \cdot \sin\theta \, d\theta . \tag{7.198}$$

In practice, we can use precalculated look-up tables to solve for θ_1 in Eq. (7.198). After the two angles are determined, we can calculate the matrix IG_i in Eq. (7.192) with $i = 1$. If the (3,3) element of IG_1 (denoted as $\mathsf{IG}_1(3,3)$) is positive, then the scattered photon propagates upward, and τ_2 (see Figure 7.9) is determined as

$$
\begin{aligned}
\tau_2 &= \tau_1 + |\mathsf{IG}_1(3,3)| \cdot \ln \varsigma_5 \\
&= \tau_1 + \mathsf{IG}_1(3,3) \cdot \ln \varsigma_5 \, .
\end{aligned}
\tag{7.199}
$$

If $\tau_2 < 0$ for the upward photon, then the photon is scattered out of the layer and counted as a reflected photon. If $\mathsf{IG}_1(3,3)$ is negative (the case shown in Figure 7.9), the photon is scattered downward, and we have

$$
\begin{aligned}
\tau_2 &= \tau_1 - |\mathsf{IG}_1(3,3)| \cdot \ln \varsigma_5 \\
&= \tau_1 + \mathsf{IG}_1(3,3) \cdot \ln \varsigma_5 \, .
\end{aligned}
\tag{7.200}
$$

If $\tau_2 > \tau^*$, the photon travels through the layer and is counted as a transmitted photon.

If the photon is not scattered out of the layer, that is, $\tau_2 > 0$ in the case of an upward photon $\mathsf{IG}_1(3,3) > 0$, or $\tau_2 < \tau^*$ in the case of a downward photon $\mathsf{IG}_1(3,3) < 0$, we need to determine whether the photon is scattered or absorbed and to determine the new direction of the photon by using a procedure similar to Eqs. (7.197) and (7.198).

Repeating the preceding procedure for a large number of photons allows us to obtain statistics of the reflection, transmission, and absorption characteristics of the medium. As an example, let us calculate the reflected radiance at the top of a scattering medium illuminated by a quasicollimated radiation field with irradiance S_{dir} that impinges on the medium at a zenith distance of $\mu_0 = \cos \theta_0$ with θ_0 the zenith angle. At the top of the medium, the radiant flux crossing a horizontal area element ΔA is $\mu_0 \cdot S_{\mathrm{dir}} \cdot \Delta A$. We assign this amount of flux as the weight for an incident photon. After we trace the paths of all incident photons with a total number of N_{tot}, let us assume that $\Delta N(\mu, \varphi)$ photons emerge within a small solid angle element $\Delta \Omega(\mu, \varphi)$ in the direction of (μ, φ). According to the definition of radiance, we obtain

$$
\begin{aligned}
I_{\mathrm{ref}}(\tau = 0, \mu, \varphi) &= \frac{\Delta N(\mu, \varphi) \cdot \mu_0 \cdot S_{\mathrm{dir}} \cdot \Delta A}{\mu \cdot \Delta A \cdot \Delta \Omega(\mu, \varphi)} \\
&= \frac{\Delta N(\mu, \varphi)}{\Delta \Omega(\mu, \varphi) \cdot N_{\mathrm{tot}}} \cdot \left(\frac{\mu_0}{\mu} \right) \cdot S_{\mathrm{dir}} \, .
\end{aligned}
\tag{7.201}
$$

7.7.2
Backward (Inverse) Monte Carlo Method (BMCM)

The preceding formalism outlines the basic principles of the MCM for application to radiative transfer simulation. In practice, various techniques have been developed to enhance the computational efficiency and accuracy. As an example, we describe the backward Monte Carlo technique in conjunction with the successive order scattering formulation. For simplicity, we consider the reflection of solar radiation by a plane-parallel atmosphere over a dark surface (i.e., the reflection of the surface is neglected). The Schwarzschild–Emden form of the RTE, see Eq. (6.90), for the upward radiance is given by

$$\mu \frac{d I_{\text{diff},\lambda}^{\uparrow}(\tau,\mu,\varphi)}{d\tau} = I_{\text{diff},\lambda}^{\uparrow}(\tau,\mu,\varphi) - J_{\lambda}(\tau,\mu,\varphi); \quad \text{for} \quad \mu > 0, \qquad (7.202)$$

where $J_{\lambda}(\tau,\mu,\varphi)$ denotes the source function

$$J_{\lambda}(\tau,\mu,\varphi) = J_{\text{dir},\lambda}(\tau,\mu,\varphi) + J_{\text{diff},\lambda}(\tau,\mu,\varphi), \qquad (7.203)$$

and we have omitted emission ($J_{\text{emi},\lambda} = 0$). Multiplying Eq. (7.202) with $e^{-\tau/\mu}$, rearranging the terms, and considering the chain rule of differentiation, we obtain

$$d\left[I_{\text{diff},\lambda}^{\uparrow}(\tau,\mu,\varphi) \cdot e^{-\tau/\mu} \right] = -e^{-\tau/\mu} \cdot J_{\lambda}(\tau,\mu,\varphi) \frac{d\tau}{\mu}; \quad \text{for} \quad \mu > 0. \quad (7.204)$$

If we change τ to τ' in Eq. (7.204) and integrate the resultant expression from τ to τ^* where τ^* is the optical thickness value at the surface, we have

$$I_{\text{diff},\lambda}^{\uparrow}(\tau^*,\mu,\varphi) \cdot e^{-\tau^*/\mu} - I_{\text{diff},\lambda}^{\uparrow}(\tau,\mu,\varphi) \cdot e^{-\tau/\mu}$$

$$= -\int_{\tau}^{\tau^*} e^{-\tau'/\mu} \cdot J_{\lambda}(\tau',\mu,\varphi) \frac{d\tau'}{\mu}, \quad \text{for} \quad \mu > 0. \qquad (7.205)$$

Because the surface reflection is neglected, the lower boundary condition for the upward radiation is

$$I_{\text{diff},\lambda}^{\uparrow}(\tau^*,\mu,\varphi) = 0. \qquad (7.206)$$

Thus, for the upward diffuse radiance we write

$$I_{\text{diff},\lambda}^{\uparrow}(\tau,\mu,\varphi) = \int_{\tau}^{\tau^*} e^{-(\tau'-\tau)/\mu} \cdot J_{\lambda}(\tau',\mu,\varphi) \frac{d\tau'}{\mu}; \quad \text{for} \quad \mu > 0. \qquad (7.207)$$

The governing equation and boundary condition for the downward radiation, see Eq. (6.80), is given by

$$-\mu \frac{d I_{\text{diff},\lambda}^{\downarrow}(\tau,-\mu,\varphi)}{d\tau} = I_{\text{diff},\lambda}^{\downarrow}(\tau,-\mu,\varphi) - J_{\lambda}(\tau,-\mu,\varphi);$$

$$\text{for} \quad \mu > 0, \qquad (7.208)$$

where

$$I_{\text{diff},\lambda}^{\downarrow}(0,-\mu,\varphi) = 0. \qquad (7.209)$$

Using the same technique as in the upward case, we get

$$
I_{\text{diff},\lambda}^{\downarrow}(\tau, -\mu, \varphi) = \int_0^{\tau} e^{-(\tau-\tau')/\mu} \cdot J_{\lambda}(\tau', -\mu, \varphi) \frac{d\tau'}{\mu} \; ; \quad \text{for} \quad \mu > 0 \; . \quad (7.210)
$$

To unify the expressions in Eqs. (7.207) and (7.210), we define a boundary optical thickness value, τ_b, that depends on the propagation direction of the radiation beam in the form of

$$
\tau_b(\mu) = \frac{1}{2} \tau^* \cdot \left(1 + \frac{\mu}{|\mu|} \right) ; \quad \text{for} \quad -1 \leq \mu \leq 1 , \quad \text{and} \quad \mu \neq 0 , \quad (7.211)
$$

where $|\mu|$ indicates the absolute value of μ. From Eq. (7.211), τ_b is τ^* for upward radiation; whereas, τ_b is zero for downward radiation. With τ_b, Eqs. (7.207) and (7.210) can be written in a unified form as

$$
I_{\text{diff},\lambda}(\tau, \mu, \varphi) = \int_{\tau}^{\tau_b(\mu)} e^{-(\tau'-\tau)/\mu} \cdot J_{\lambda}(\tau', \mu, \varphi) \frac{d\tau'}{\mu} \; ;
$$

$$
\text{for} \quad -1 \leq \mu \leq 1 , \quad \mu \neq 0 . \quad (7.212)
$$

Note, the domain of μ in Eq. (7.212) is different from that of Eqs. (7.207) and (7.210).

In the same essence of the MSOS, we can decompose the diffuse radiance in Eq. (7.212) into the contributions of various orders of scattering events, that is,

$$
I_{\text{diff},\lambda}(\tau, \mu, \varphi) = \sum_{n=0}^{\infty} I_n(\tau, \mu, \varphi) . \quad (7.213)
$$

I_n is given by

$$
I_n(\tau, \mu, \varphi) = \int_{\tau}^{\tau_b(\mu)} e^{-(\tau'-\tau)/\mu} \cdot J_n(\tau', \mu, \varphi) \frac{d\tau'}{\mu} \; ;
$$

$$
\text{for} \quad -1 \leq \mu \leq 1 , \quad \mu \neq 0 , \quad (7.214)
$$

where the subscript n indicates the order of scattering events. Similarly, the source function is decomposed as

$$
J_{\lambda}(\tau', \mu, \varphi) = \sum_{n=1}^{\infty} J_n(\tau', \mu, \varphi) = J_1(\tau', \mu, \varphi) + \sum_{n=2}^{\infty} J_n(\tau', \mu, \varphi) . \quad (7.215)
$$

J_1 corresponds to the term representing the first-order scattering of direct solar radiation, $J_{\text{dir},\lambda}$, given in Eq. (6.69),

$$
J_1(\tau', \mu, \varphi) = J_{\text{dir},\lambda}(\tau', \mu, \varphi)
$$

$$
= \tilde{\omega}(\tau') \cdot S_{\text{dir},\lambda,\text{TOA}} \cdot e^{-\tau'/\mu_0} \cdot \frac{\mathcal{P}(\tau', -\mu_0, \varphi_0, \mu, \varphi)}{4\pi} . \quad (7.216)
$$

The expression for J_n corresponds to the diffuse source term (multiple-scattering) $J_{\text{diff},\lambda}$, given in Eq. (6.70),

$$
\begin{aligned}
J_n(\tau',\mu,\varphi) &= J_{n,\text{dir},\lambda}(\tau',\mu,\varphi) \\
&= \tilde{\omega}(\tau') \int_0^{2\pi} \int_{-1}^1 I_{n-1}(\tau',\mu',\varphi') \cdot \frac{\mathcal{P}(\tau',\mu',\varphi',\mu,\varphi)}{4\pi}\, d\mu'\, d\varphi' \; ;
\end{aligned}
$$

$$
\text{for} \quad n = 2,3,4,\dots \tag{7.217}
$$

Equation (7.216) corresponds to Eq. (7.217) by applying the term for the direct solar radiation, see Eq. (6.74), in the form of

$$
I_0(\tau',-\mu,\varphi) = S_{\text{dir},\lambda,\text{TOA}} \cdot e^{-\tau'/\mu_0} \cdot \delta[\mu-(-\mu_0)] \cdot \delta(\varphi-\varphi_0) \; .
$$

Thus, the diffuse radiance component from the first-order scattering events is

$$
\begin{aligned}
I_1(0,\mu,\varphi) &= \int_0^{\tau_b(\mu)} e^{-\tau'/\mu} \cdot J_1(\tau',\mu,\varphi)\frac{d\tau'}{\mu} \\
&= \int_0^{\tau^*} e^{-\tau'/\mu} \cdot \tilde{\omega}(\tau') \cdot S_{\text{dir},\lambda,\text{TOA}} \cdot e^{-\tau'/\mu_0} \cdot \frac{\mathcal{P}(\tau',-\mu_0,\varphi_0,\mu,\varphi)}{4\pi}\frac{d\tau'}{\mu} \; ;
\end{aligned}
$$

$$
\text{for} \quad -1 \le \mu \le 1, \quad \mu \ne 0. \tag{7.218}
$$

Furthermore, we consider the following integral,

$$
\int_0^{\tau^*} e^{-\tau'/\mu}\frac{d\tau'}{\mu} = 1 - e^{-\tau^*/\mu} \; . \tag{7.219}
$$

With the result in Eq. (7.219), we will rewrite Eq. (7.218) as

$$
\begin{aligned}
I_1(0,\mu,\varphi) &= \left(1 - e^{-\tau^*/\mu}\right) \cdot S_{\text{dir},\lambda,\text{TOA}} \int_0^{\tau^*} f(\tau') \cdot \tilde{\omega}(\tau') \cdot e^{-\tau'/\mu_0} \\
&\quad \times \frac{\mathcal{P}(\tau',-\mu_0,\varphi_0,\mu,\varphi)}{4\pi}\frac{d\tau'}{\mu} \; ;
\end{aligned}
$$

$$
\text{for} \quad -1 \le \mu \le 1, \quad \mu \ne 0, \tag{7.220}
$$

where

$$
f(\tau') = \frac{e^{-\tau'/\mu}}{1 - e^{-\tau^*/\mu}} \; . \tag{7.221}
$$

The function $f(\tau')$ can be regarded as the normalized probability distribution of τ'. Thus, we can use $f(\tau')$ to sample τ' as

$$
\begin{aligned}
\xi_i &= \int_0^{\tau'} f(\tau'') \frac{d\tau''}{\mu} \\
&= \frac{1}{1 - e^{-\tau^*/\mu}} \cdot (1 - e^{-\tau'/\mu}) ,
\end{aligned}
\tag{7.222}
$$

where ξ_i indicates random numbers uniformly distributed in the region of $[0, 1]$. From Eq. (7.222), we obtain the randomly sampled optical thickness as

$$
\tau_i' = -\mu \cdot \ln[1 - \xi_i \cdot (1 - e^{-\tau^*/\mu})] .
\tag{7.223}
$$

With the optical thickness determined by Eq. (7.223), the radiance in Eq. (7.220) can be calculated, that is,

$$
\begin{aligned}
I_1(0, \mu, \varphi) &= (1 - e^{-\tau^*/\mu}) \cdot S_{\text{dir},\lambda,\text{TOA}} \cdot \frac{1}{N_1} \cdot \sum_{i=1}^{N_1} \tilde{\omega}(\tau_i') \cdot e^{-\tau_i'/\mu_0} \\
&\quad \times \frac{\mathcal{P}(\tau_i', -\mu_0, \varphi_0, \mu, \varphi)}{4\pi} \frac{d\tau'}{\mu} ; \\
&\quad \text{for} \quad -1 \leq \mu \leq 1, \quad \mu \neq 0 .
\end{aligned}
\tag{7.224}
$$

Note that in Eq. (7.222), μ instead of μ_0 is involved, that is, the sampling starts from the observation direction rather than the incident direction. Because the sampling sequence is from the observer (or sensor) to the source, this method is called the Backward Monte Carlo Method (BMCM). To make this point clearer, we consider the second-order scattering contribution in the following discussion.

From Eq. (7.217) for $n = 2$, Eq. (7.214) for $n = 1$, and Eq. (7.216), we have

$$
\begin{aligned}
J_2(\tau', \mu, \varphi) &= \tilde{\omega}(\tau') \int_0^{2\pi} \int_{-1}^{1} I_1(\tau', \mu', \varphi') \cdot \frac{\mathcal{P}(\tau', \mu', \varphi', \mu, \varphi)}{4\pi} \, d\mu' \, d\varphi' \\
&= \tilde{\omega}(\tau') \int_0^{2\pi} \int_{-1}^{1} \left[\int_{\tau'}^{\tau_b(\mu')} e^{-(\tau''-\tau')/\mu'} \cdot J_1(\tau'', \mu, \varphi) \frac{d\tau''}{\mu'} \right] \\
&\quad \times \frac{\mathcal{P}(\tau', \mu', \varphi', \mu, \varphi)}{4\pi} \, d\mu' \, d\varphi' \\
&= \tilde{\omega}(\tau') \int_0^{2\pi} \int_{-1}^{1} \left[\int_{\tau'}^{\tau_b(\mu')} e^{-(\tau''-\tau')/\mu'} \cdot \left\{ \tilde{\omega}(\tau'') \cdot S_{\text{dir},\lambda,\text{TOA}} \cdot e^{-\tau''/\mu_0} \right. \right. \\
&\quad \left. \left. \times \frac{\mathcal{P}(\tau'', -\mu_0, \varphi_0, \mu', \varphi')}{4\pi} \right\} \frac{d\tau''}{\mu'} \right] \cdot \frac{\mathcal{P}(\tau', \mu', \varphi', \mu, \varphi)}{4\pi} \, d\mu' \, d\varphi' .
\end{aligned}
\tag{7.225}
$$

From Eq. (7.214), with $n = 2$, we are able to solve

$$
I_2(0, \mu, \varphi) = \int_0^{\tau^*} e^{-\tau'/\mu} \cdot J_2(\tau', \mu, \varphi) \frac{d\tau'}{\mu}
$$

$$
= \int_0^{\tau^*} e^{-\tau'/\mu} \cdot \tilde{\omega}(\tau')
$$

$$
\times \int_0^{2\pi} \int_{-1}^{1} \left[\int_{\tau'}^{\tau_b(\mu')} e^{-(\tau''-\tau')/\mu'} \cdot \left\{ \tilde{\omega}(\tau'') \cdot S_{\mathrm{dir},\lambda,\mathrm{TOA}} \cdot e^{-\tau''/\mu_0} \right. \right.
$$

$$
\times \left. \frac{P(\tau'', -\mu_0, \varphi_0, \mu', \varphi')}{4\pi} \right\} \frac{d\tau''}{\mu'} \left. \right] \cdot \frac{P(\tau', \mu', \varphi', \mu, \varphi)}{4\pi} \, d\mu' \, d\varphi' \, \frac{d\tau'}{\mu} .
$$

$$
(7.226)
$$

After slightly rearranging Eq. (7.226), we can write

$$
I_2(0, \mu, \varphi) = \left(1 - e^{-\tau^*/\mu} \right) \cdot S_{\mathrm{dir},\lambda,\mathrm{TOA}}
$$

$$
\times \int_0^{\tau^*} f(\tau') \cdot \tilde{\omega}(\tau') \int_0^{2\pi} \int_{-1}^{1} \frac{P(\tau', \mu', \varphi', \mu, \varphi)}{4\pi}
$$

$$
\times \left\{ 1 - e^{-[\tau_b(\mu')-\tau']/\mu'} \right\} \cdot \int_{\tau'}^{\tau_b(\mu')} \tilde{\omega}(\tau'') \cdot e^{-\tau''/\mu_0}
$$

$$
\times \frac{P(\tau'', -\mu_0, \varphi_0, \mu', \varphi')}{4\pi} \cdot f(\tau'' - \tau') \, d\tau'' \, d\mu' \, d\varphi' \, d\tau' . \quad (7.227)
$$

According to the statistical meanings of $f(\tau')$, $P(\tau', \mu', \varphi', \mu, \varphi)$, and $f(\tau'' - \tau')$, we can sequentially sample τ' based on $f(\tau')$, (μ', φ') based on $P(\tau', \mu', \varphi', \mu, \varphi)$, and τ'' based on $f(\tau'' - \tau')$. $I_2(0, \mu, \varphi)$ can then be found, see also Eq. (7.224),

$$
I_2(0, \mu, \varphi) = \left(1 - e^{-\tau^*/\mu} \right) \cdot S_{\mathrm{dir},\lambda,\mathrm{TOA}}
$$

$$
\times \frac{1}{N_2} \cdot \sum_{i=1}^{N_2} \tilde{\omega}(\tau_i') \cdot \left\{ 1 - e^{-[\tau_b(\mu_i')-\tau_i']/\mu_i'} \right\}
$$

$$
\times \tilde{\omega}(\tau_i'') \cdot e^{-\tau_i''/\mu_0} \cdot \frac{P(\tau_i'', -\mu_0, \varphi_0, \mu, \varphi)}{4\pi} . \quad (7.228)
$$

Again, it is evident from the aforementioned sampling procedure that the sampling sequence from the observer to the source, that is, the BMCM, is applied. Note, the sampling of μ_i' and φ_i' is similar to that described by Eqs. (7.197) and (7.198). The only difference is that in the backward sampling, the direction opposite to (μ, φ)

is treated as the "incident direction"; whereas, the direction opposite to (μ', φ') is treated as the "scattering direction" in the procedure for sampling μ' and φ'.

The contributions from higher orders of scattering events can be solved in the same way as above. The reader is referred to the works of Collins et al. (1972); Marshak and Davis (2005); Plass and Kattawar (1968), and references therein for an indepth explanation of the Monte Carlo techniques for radiative transfer simulation.

The strength of the MCM is that any possible cloud geometry can be assumed, which makes this method very well suited for 3D radiative transfer. Complicated phase functions, for example, those for ice crystals, can be treated. The MCM is suitable for calculating spectral radiances and irradiances but is less appropriate for broadband considerations. Other useful quantities, such as, the photon path length or the penetration depth of the photons, can be easily derived. Additionally, a computer program for this method can be written in a straightforward manner; however, long computing time may be encountered due to a large number of photons traced and counted by the receiver before reaching a statistical significance in optically thick atmospheres such as cloudy atmospheres. The statistical nature of the method involves uncertainties and the accuracy depends on the number of photons; different runs may give rise to slightly different results.

7.8
Two-Stream Approximation (TSA)

As computational burden is a major concern for radiative transfer simulations for many applications, rigorous schemes such as the DOM implemented with multiple streams are not practical, particularly for applications to general circulation models and numerical weather prediction models. The two-stream approximation (TSA) provides an efficient way to simulate the radiative transfer in scattering atmospheres. Various two-stream schemes have been developed in the literature and can be expressed in a unified form but with different coefficients for the two-stream radiative transfer equations, which has been demonstrated by Meador and Weaver (1980). The history of the development of the TSA is quite rich, and the origin of this method can be traced to the work of Schuster (1905) who presented the two-stream equations based on phenomenological reasoning regarding the transfer of radiation in a foggy atmosphere. From historical and pedagogical perspectives, we briefly recapture Schuster's formalism in modern terminology.

7.8.1
Classical Approach

As shown in Figure 7.11, consider upward radiation impinging on a thin layer of a foggy atmosphere whose thickness is indicated by dz. After passing the layer dz, the upward irradiance, $F^\uparrow$, is reduced in an amount of $(b_{abs} \cdot F^\uparrow dz)$ due to absorption and b_{abs} is the volume absorption coefficient. Furthermore, $F^\uparrow$ is also reduced

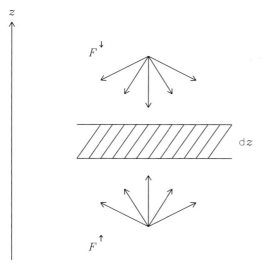

Figure 7.11 Schematic configuration for the two-stream approximation derived by Schuster (1905). A diffuse radiation field was implied in Schuster's derivation.

by scattering. Schuster (1905) did not explicitly use the term "phase function" to account for the angular distribution of scattered energy, but he assumed symmetric scattering in the forward and backward hemispheres. Thus, the decrease in $F^\uparrow$ due to scattering is $(0.5 b_{\text{sca}} \cdot F^\uparrow \mathrm{d}z)$, and b_{sca} is the volume scattering coefficient. Moreover, $F^\uparrow$ is enhanced in an amount of $(b_{\text{abs}} \cdot F_{\text{emi}} \mathrm{d}z)$ by the thermal emission and in an amount of $(0.5 b_{\text{sca}} \cdot F^\downarrow \mathrm{d}z)$ by the upward scattering of downward radiation, and F_{emi} indicates the emitted irradiance. Because emission is assumed to be isotropic, F_{emi} is given by $\pi \cdot \overline{B}_\lambda$ and $\overline{B}_\lambda$ is the mean Planck function averaged over the spectral interval of interest. Thus, the total variation of the upward irradiance is given by

$$\mathrm{d}F^\uparrow = -b_{\text{abs}} \cdot F^\uparrow \, \mathrm{d}z - \frac{1}{2} b_{\text{sca}} \cdot F^\uparrow \, \mathrm{d}z + b_{\text{abs}} \cdot F_{\text{emi}} \, \mathrm{d}z$$
$$+ \frac{1}{2} b_{\text{sca}} \cdot F^\downarrow \, \mathrm{d}z \ . \tag{7.229}$$

Similarly, we have

$$\mathrm{d}F^\downarrow = -b_{\text{abs}} \cdot F^\downarrow(-\mathrm{d}z) - \frac{1}{2} b_{\text{sca}} \cdot F^\downarrow(-\mathrm{d}z) + b_{\text{abs}} \cdot F_{\text{emi}}(-\mathrm{d}z)$$
$$+ \frac{1}{2} b_{\text{sca}} \cdot F^\uparrow(-\mathrm{d}z) \ . \tag{7.230}$$

The rearrangement of the terms in Eqs. (7.229) and (7.230) leads to the following form presented by Schuster (1905):

$$\frac{dF^{\uparrow}}{dz} = b_{abs} \cdot \left(F_{emi} - F^{\uparrow}\right) + \frac{1}{2} b_{sca} \cdot \left(F^{\downarrow} - F^{\uparrow}\right), \tag{7.231}$$

$$\frac{dF^{\downarrow}}{dz} = b_{abs} \cdot \left(F^{\downarrow} - F_{emi}\right) + \frac{1}{2} b_{sca} \cdot \left(F^{\downarrow} - F^{\uparrow}\right). \tag{7.232}$$

Although Eqs. (7.231) and (7.232) were derived more than 100 years ago with substantial simplification, the equations reflect the major physical processes involved in the transfer of radiation within a scattering and emitting medium. Major weaknesses of Eqs. (7.231) and (7.232) are the negligence of the asymmetric scattering of radiation and a lack of separation of the direct and diffuse radiation, which may lead to substantial errors in numerical calculation. If we omit the emission terms in Eqs. (7.231) and (7.232), and express the resultant equations in terms of optical thickness, see Eq. (6.8),

$$d\tau = -(b_{abs} + b_{sca}) \, dz \, ,$$

and the single-scattering albedo

$$\tilde{\omega} = \frac{b_{sca}}{b_{abs} + b_{sca}} \, ,$$

we obtain

$$\frac{dF^{\uparrow}}{d\tau} = F^{\uparrow} - \frac{\tilde{\omega}}{2} \cdot (F^{\downarrow} + F^{\uparrow}) \, , \tag{7.233}$$

$$\frac{dF^{\downarrow}}{d\tau} = -F^{\downarrow} + \frac{\tilde{\omega}}{2} \cdot (F^{\downarrow} + F^{\uparrow}) \, . \tag{7.234}$$

The preceding equations can be written in a more concise matrix form as

$$\frac{d}{d\tau} \left(F^{\uparrow} \ F^{\downarrow}\right) = \begin{pmatrix} (1 - \tilde{\omega}/2) & -\tilde{\omega}/2 \\ \tilde{\omega}/2 & -(1 - \tilde{\omega}/2) \end{pmatrix} \cdot \begin{pmatrix} F^{\uparrow} \\ F^{\downarrow} \end{pmatrix}. \tag{7.235}$$

To solve Eq. (7.235), we let the formal solution be in the form of

$$\begin{pmatrix} F^{\uparrow}(\tau) \\ F^{\downarrow}(\tau) \end{pmatrix} = \begin{pmatrix} F_0^{\uparrow} \\ F_0^{\downarrow} \end{pmatrix} \cdot e^{\Lambda \cdot \tau} \, , \tag{7.236}$$

where

$$\begin{pmatrix} F_0^{\uparrow} \\ F_0^{\downarrow} \end{pmatrix}$$

represents the eigenvector and Λ is the eigenvalue of the solution. Substituting Eq. (7.236) into Eq. (7.235), we aquire the determinant equation governing the eigenvalue as

$$\begin{vmatrix} \Lambda - (1 - \tilde{\omega}/2) & -\tilde{\omega}/2 \\ \tilde{\omega}/2 & \Lambda + (1 - \tilde{\omega}/2) \end{vmatrix} = 0 \, . \tag{7.237}$$

If we expand the determinant in Eq. (7.237), we have

$$\varLambda^2 - \left(1 - \frac{\tilde{\omega}}{2}\right)^2 + \frac{\tilde{\omega}^2}{4} = 0 .$$ (7.238)

The solution to the preceding equation is

$$\varLambda = \pm\varLambda_+ = \pm\sqrt{1 - \tilde{\omega}} .$$ (7.239)

Thus, following Eq. (7.236), the upward and downward irradiances can be expressed in the form of

$$F^\uparrow = F^\uparrow_{0+} \cdot e^{\varLambda_+ \cdot \tau} + F^\uparrow_{0-} \cdot e^{-\varLambda_+ \cdot \tau} ,$$ (7.240)

$$F^\downarrow = F^\downarrow_{0+} \cdot e^{\varLambda_+ \cdot \tau} + F^\downarrow_{0-} \cdot e^{-\varLambda_+ \cdot \tau} .$$ (7.241)

In the preceding equations, coefficients $F^\uparrow_{0+}$ and $F^\downarrow_{0+}$ are not independent and neither are $F^\uparrow_{0-}$ and $F^\downarrow_{0-}$. To determine their relations, we substitute Eqs. (7.240) and (7.241) into Eq. (7.233), rearrange the terms, thus obtaining

$$\left[\left(\varLambda - 1 + \frac{\tilde{\omega}}{2}\right) \cdot F^\uparrow_{0+} + \left(\frac{\tilde{\omega}}{2}\right) \cdot F^\downarrow_{0+}\right] \cdot e^{\varLambda_+ \cdot \tau}$$
$$= \left[\left(\varLambda + 1 - \frac{\tilde{\omega}}{2}\right) \cdot F^\uparrow_{0-} - \left(\frac{\tilde{\omega}}{2}\right) \cdot F^\downarrow_{0-}\right] \cdot e^{-\varLambda_+ \cdot \tau} .$$ (7.242)

In Eq. (7.242), $e^{\varLambda_+ \cdot \tau}$ and $e^{-\varLambda_+ \cdot \tau}$ are linearly independent. In order for Eq. (7.242) to hold for an arbitrary optical depth τ, the following conditions must be true:

$$\left(\varLambda - 1 + \frac{\tilde{\omega}}{2}\right) \cdot F^\uparrow_{0+} + \left(\frac{\tilde{\omega}}{2}\right) \cdot F^\downarrow_{0+} = 0 ,$$ (7.243)

$$\left(\varLambda + 1 - \frac{\tilde{\omega}}{2}\right) \cdot F^\uparrow_{0-} - \left(\frac{\tilde{\omega}}{2}\right) \cdot F^\downarrow_{0-} = 0 .$$ (7.244)

From Eq. (7.243), we obtain

$$F^\downarrow_{0+} = -\left(\frac{\varLambda - 1 + \tilde{\omega}/2}{\tilde{\omega}/2}\right) \cdot F^\uparrow_{0+}$$
$$= \left(\frac{2 - \sqrt{1 - \tilde{\omega}} - \tilde{\omega}}{\tilde{\omega}}\right) \cdot F^\uparrow_{0+}$$
$$= \mathcal{R}_\infty \cdot F^\uparrow_{0+} ,$$ (7.245)

with:

$$\mathcal{R}_\infty = \frac{2 - \sqrt{1 - \tilde{\omega}} - \tilde{\omega}}{\tilde{\omega}} .$$ (7.246)

It is obvious that $\mathcal{R}_\infty = 1$ if $\tilde{\omega} = 1$ and no absorption is involved. If $\tilde{\omega} \to 0$, that is, there is little scattering, $\mathcal{R}_\infty$ can be calculated by using the Taylor expansion technique as

$$\mathcal{R}_\infty \approx \frac{2 - 2[1 - 1/(2\tilde{\omega}) - 1/(8\tilde{\omega}^2)] - \tilde{\omega}}{\tilde{\omega}} = \frac{\tilde{\omega}}{4} . \tag{7.247}$$

Evidently, the $\mathcal{R}_\infty$ expression in Eq. (7.246) does not lead to a singularity for a pure absorbing medium. From Eq. (7.244), we obtain

$$F_{0-}^\uparrow = \left(\frac{\tilde{\omega}/2}{\Lambda + 1 - \tilde{\omega}/2} \right) \cdot F_{0-}^\downarrow = \left(\frac{2 - \sqrt{1 - \tilde{\omega}} - \tilde{\omega}}{\tilde{\omega}} \right) \cdot F_{0-}^\downarrow$$

$$= \mathcal{R}_\infty \cdot F_{0-}^\downarrow . \tag{7.248}$$

Therefore, from Eqs. (7.240) and (7.241), we obtain

$$F^\uparrow = F_{0+}^\uparrow \cdot e^{\Lambda + \cdot \tau} + \mathcal{R}_\infty \cdot F_{0-}^\downarrow \cdot e^{-\Lambda + \cdot \tau} , \tag{7.249}$$

$$F^\downarrow = \mathcal{R}_\infty \cdot F_{0+}^\uparrow \cdot e^{\Lambda + \cdot \tau} + F_{0-}^\downarrow \cdot e^{-\Lambda + \cdot \tau} . \tag{7.250}$$

If we apply Eqs. (7.249) and (7.250) to a scattering medium over a surface which is illuminated by diffuse radiation from above, the boundary conditions are

$$F^\downarrow(\tau = 0) = F^\downarrow(0) , \tag{7.251}$$

$$F^\uparrow(\tau^*) = \mathcal{R}_s \cdot F^\downarrow(\tau^*) , \tag{7.252}$$

where τ^* is the optical thickness of the medium and $\mathcal{R}_s$ is the reflectivity of the surface. Applying the boundary conditions to Eqs. (7.249) and (7.250) gives

$$F^\uparrow(\tau) = \frac{F^\downarrow(0)}{1 + \left(\frac{\mathcal{R}_s - \mathcal{R}_\infty}{1 - \mathcal{R}_s \cdot \mathcal{R}_\infty} \right) \cdot e^{-2\Lambda + \cdot \tau^*}}$$
$$\times \left[\left(\frac{\mathcal{R}_s - \mathcal{R}_\infty}{1 - \mathcal{R}_s \cdot \mathcal{R}_\infty} \right) \cdot e^{-2\Lambda + \cdot (\tau^* - \tau/2)} + \mathcal{R}_\infty \cdot e^{-\Lambda + \cdot \tau} \right], \tag{7.253}$$

$$F^\downarrow(\tau) = \frac{F^\downarrow(0)}{1 + \left(\frac{\mathcal{R}_s - \mathcal{R}_\infty}{1 - \mathcal{R}_s \cdot \mathcal{R}_\infty} \right) \cdot e^{-2\Lambda + \cdot \tau^*}}$$
$$\times \left[\mathcal{R}_\infty \cdot \left(\frac{\mathcal{R}_s - \mathcal{R}_\infty}{1 - \mathcal{R}_s \cdot \mathcal{R}_\infty} \right) \cdot e^{-2\Lambda + \cdot (\tau^* - \tau/2)} + e^{-\Lambda + \cdot \tau} \right]. \tag{7.254}$$

If the scattering medium is optically very thick, that is, $\tau^* \to \infty$, the reflectivity at the top of the medium $\mathcal{R}_{top}$ is given by

$$\mathcal{R}_{top} = \lim_{\tau^* \to \infty} \frac{F^\uparrow(\tau = 0)}{F^\downarrow(\tau = 0)} = \mathcal{R}_\infty . \tag{7.255}$$

Thus, $\mathcal{R}_\infty$ is the reflectivity of a semi-infinitely thick layer of a scattering medium. Note, $\mathcal{R}_\infty$ is independent of $\mathcal{R}_s$. This is understandable because the effect of the surface reflection cannot be observed at the top of a semi-infinite layer above the surface regardless of the single-scattering albedo of the layer.

7.8.2
TSA Based on RTE

To derive a more rigorous two-stream scheme that accounts for the asymmetry of scattering, we can rewrite the equation of radiative transfer, see Eq. (7.53), as follows, neglecting emission and setting $m = 0$ for the upward direction,

$$
\mu \frac{\mathrm{d} I_\lambda^{(0)}(\tau, \mu)}{\mathrm{d}\tau} = I_\lambda^{(0)}(\tau, \mu)
$$

$$
- \frac{\tilde{\omega}(\tau)}{4\pi} \cdot S_{\mathrm{dir},\lambda,\mathrm{TOA}} \cdot \mathrm{e}^{-\tau/\mu_0} \cdot P^{(0)}(\tau, -\mu_0, \mu)
$$

$$
- \frac{\tilde{\omega}(\tau)}{2} \int_0^1 I_\lambda^{(0)}(\tau, \mu') \cdot P^{(0)}(\tau, \mu, \mu') \, \mathrm{d}\mu'
$$

$$
- \frac{\tilde{\omega}(\tau)}{2} \int_0^1 I_\lambda^{(0)}(\tau, -\mu') \cdot P^{(0)}(\tau, \mu, -\mu') \, \mathrm{d}\mu' , \qquad (7.256)
$$

and for the downward direction,

$$
-\mu \frac{\mathrm{d} I_\lambda^{(0)}(\tau, -\mu)}{\mathrm{d}\tau} = I_\lambda^{(0)}(\tau, -\mu)
$$

$$
- \frac{\tilde{\omega}(\tau)}{4\pi} \cdot S_{\mathrm{dir},\lambda,\mathrm{TOA}} \cdot \mathrm{e}^{-\tau/\mu_0} \cdot P^{(0)}(\tau, -\mu_0, -\mu)
$$

$$
- \frac{\tilde{\omega}(\tau)}{2} \int_0^1 I_\lambda^{(0)}(\tau, \mu') \cdot P^{(0)}(\tau, -\mu, \mu') \, \mathrm{d}\mu'
$$

$$
- \frac{\tilde{\omega}(\tau)}{2} \int_0^1 I_\lambda^{(0)}(\tau, -\mu') \cdot P^{(0)}(\tau, -\mu, -\mu') \, \mathrm{d}\mu' ,
$$
$$(7.257)$$

where $0 < \mu \leq 1$. Note, the upward and downward diffuse irradiances can be expressed, see Eqs. (7.45) and (7.46), in terms of

$$
F_{\mathrm{diff},\lambda}^{\uparrow}(\tau) = 2\pi \int_0^1 I_\lambda^{(0)}(\tau, \mu) \cdot \mu \, \mathrm{d}\mu , \qquad (7.258)
$$

$$
F_{\mathrm{diff},\lambda}^{\downarrow}(\tau) = 2\pi \int_0^1 I_\lambda^{(0)}(\tau, -\mu) \cdot \mu \, \mathrm{d}\mu . \qquad (7.259)
$$

The zeroth Fourier coefficients $I_\lambda^{(0)}(\tau, \mu)$ correspond to the azimuthally averaged diffuse radiance, thus

$$
I_\lambda^{(0)}(\tau, \mu) = \overline{I}_{\mathrm{diff},\lambda}^{\varphi}(\tau, \mu) . \qquad (7.260)
$$

The same interpretation holds for the phase function and the zeroth Fourier coefficients $P^{(0)}(\tau, \mu, \mu')$ correspond to the azimuthally averaged phase function

$$P^{(0)}(\tau, \mu, \mu') = \overline{\mathcal{P}}^{\varphi}(\tau, \mu, \mu') . \tag{7.261}$$

Integrating Eqs. (7.257) and (7.256) over the interval $\mu \in (0, 1]$, for the upward direction, we obtain

$$
\frac{1}{2\pi} \frac{\mathrm{d} F^{\uparrow}_{\mathrm{diff},\lambda}(\tau)}{\mathrm{d}\tau} = \int_0^1 \overline{I}^{\varphi}_{\mathrm{diff},\lambda}(\tau, \mu) \, \mathrm{d}\mu
$$

$$
- \frac{\tilde{\omega}(\tau)}{4\pi} \cdot S_{\mathrm{dir},\lambda,\mathrm{TOA}} \cdot \mathrm{e}^{-\tau/\mu_0} \int_0^1 \overline{\mathcal{P}}^{\varphi}(\tau, -\mu_0, \mu) \, \mathrm{d}\mu
$$

$$
- \frac{\tilde{\omega}(\tau)}{2} \int_0^1 \int_0^1 \overline{I}^{\varphi}_{\mathrm{diff},\lambda}(\tau, \mu') \cdot \overline{\mathcal{P}}^{\varphi}(\tau, \mu, \mu') \, \mathrm{d}\mu' \, \mathrm{d}\mu
$$

$$
- \frac{\tilde{\omega}(\tau)}{2} \int_0^1 \int_0^1 \overline{I}^{\varphi}_{\mathrm{diff},\lambda}(\tau, -\mu') \cdot \overline{\mathcal{P}}^{\varphi}(\tau, \mu, -\mu') \, \mathrm{d}\mu' \, \mathrm{d}\mu , \tag{7.262}
$$

and for the downward direction,

$$
-\frac{1}{2\pi} \frac{\mathrm{d} F^{\downarrow}_{\mathrm{diff},\lambda}(\tau)}{\mathrm{d}\tau} = \int_0^1 \overline{I}^{\varphi}_{\mathrm{diff},\lambda}(\tau, -\mu) \, \mathrm{d}\mu
$$

$$
- \frac{\tilde{\omega}(\tau)}{4\pi} \cdot S_{\mathrm{dir},\lambda,\mathrm{TOA}} \cdot \mathrm{e}^{-\tau/\mu_0} \int_0^1 \overline{\mathcal{P}}^{\varphi}(\tau, -\mu_0, -\mu) \, \mathrm{d}\mu
$$

$$
- \frac{\tilde{\omega}(\tau)}{2} \int_0^1 \int_0^1 \overline{I}^{\varphi}_{\mathrm{diff},\lambda}(\tau, \mu') \cdot \overline{\mathcal{P}}^{\varphi}(\tau, -\mu, \mu') \, \mathrm{d}\mu' \, \mathrm{d}\mu
$$

$$
- \frac{\tilde{\omega}(\tau)}{2} \int_0^1 \int_0^1 \overline{I}^{\varphi}_{\mathrm{diff},\lambda}(\tau, -\mu') \cdot \overline{\mathcal{P}}^{\varphi}(\tau, -\mu, -\mu') \, \mathrm{d}\mu' \, \mathrm{d}\mu .
$$

$$\tag{7.263}$$

To simplify Eqs. (7.262) and (7.263), we introduce the hemispheric constant approximation, assuming both the upward and downward radiances are independent of direction (isotropic radiances). Thus, we have the following relations:

$$
\overline{I}^{\varphi}_{\mathrm{diff},\lambda}(\tau, \mu) = \frac{F^{\uparrow}_{\mathrm{diff},\lambda}(\tau)}{\pi} ; \quad \text{for} \quad \mu > 0 , \tag{7.264}
$$

and

$$
\overline{I}^{\varphi}_{\mathrm{diff},\lambda}(\tau, -\mu) = \frac{F^{\downarrow}_{\mathrm{diff},\lambda}(\tau)}{\pi} ; \quad \text{for} \quad \mu > 0 . \tag{7.265}
$$

With the preceding relations, we can simplify the equation of radiative transfer. For the upward direction, we obtain

$$\frac{1}{2\pi} \frac{d F_{\text{diff},\lambda}^{\uparrow}(\tau)}{d\tau} = \frac{F_{\text{diff},\lambda}^{\uparrow}(\tau)}{\pi}$$

$$- \frac{\tilde{\omega}(\tau)}{4\pi} \cdot S_{\text{dir},\lambda,\text{TOA}} \cdot e^{-\tau/\mu_0} \int_0^1 \overline{\mathcal{P}}^{\varphi}(\tau, -\mu_0, \mu)\, d\mu$$

$$- \frac{\tilde{\omega}(\tau)}{2} \cdot \frac{F_{\text{diff},\lambda}^{\uparrow}(\tau)}{\pi} \int_0^1 \int_0^1 \overline{\mathcal{P}}^{\varphi}(\tau, \mu, \mu')\, d\mu'\, d\mu$$

$$- \frac{\tilde{\omega}(\tau)}{2} \cdot \frac{F_{\text{diff},\lambda}^{\downarrow}(\tau)}{\pi} \int_0^1 \int_0^1 \overline{\mathcal{P}}^{\varphi}(\tau, \mu, -\mu')\, d\mu'\, d\mu \ . \tag{7.266}$$

For the downward direction, we have

$$-\frac{1}{2\pi} \frac{d F_{\text{diff},\lambda}^{\downarrow}(\tau)}{d\tau} = \frac{F_{\text{diff},\lambda}^{\downarrow}(\tau)}{\pi}$$

$$- \frac{\tilde{\omega}(\tau)}{4\pi} \cdot S_{\text{dir},\lambda,\text{TOA}} \cdot e^{-\tau/\mu_0} \int_0^1 \overline{\mathcal{P}}^{\varphi}(\tau, -\mu_0, -\mu)\, d\mu$$

$$- \frac{\tilde{\omega}(\tau)}{2} \cdot \frac{F_{\text{diff},\lambda}^{\uparrow}(\tau)}{\pi} \int_0^1 \int_0^1 \overline{\mathcal{P}}^{\varphi}(\tau, -\mu, \mu')\, d\mu'\, d\mu$$

$$- \frac{\tilde{\omega}(\tau)}{2} \cdot \frac{F_{\text{diff},\lambda}^{\downarrow}(\tau)}{\pi} \int_0^1 \int_0^1 \overline{\mathcal{P}}^{\varphi}(\tau, -\mu, -\mu')\, d\mu'\, d\mu \ . \tag{7.267}$$

We define the backscattering fraction as

$$b(\mu) = \frac{1}{2} \int_0^1 \overline{\mathcal{P}}^{\varphi}(\tau, -\mu, \mu')\, d\mu' \ . \tag{7.268}$$

Furthermore, from the normalization of the phase function and the properties of Legendre polynomials involved in the phase function expansion, we can show

$$1 - b(\mu) = \frac{1}{2} \int_0^1 \overline{\mathcal{P}}^{\varphi}(\tau, -\mu, -\mu')\, d\mu' \ , \tag{7.269}$$

and

$$b(\mu) = \frac{1}{2} \int_0^1 \overline{\mathcal{P}}^{\varphi}(\tau, \mu, -\mu')\, d\mu' \ . \tag{7.270}$$

We define the mean backscattering fraction as

$$\bar{b} = \int_0^1 b(\mu')\, \mathrm{d}\mu' \,. \tag{7.271}$$

Thus, we obtain the two-stream equations

$$\frac{1}{2}\frac{\mathrm{d}F_{\mathrm{diff},\lambda}^{\uparrow}(\tau)}{\mathrm{d}\tau} = F_{\mathrm{diff},\lambda}^{\uparrow}(\tau) - \frac{\tilde{\omega}(\tau)}{2} \cdot b(\mu_0) \cdot S_{\mathrm{dir},\lambda,\mathrm{TOA}} \cdot \mathrm{e}^{-\tau/\mu_0}$$
$$- \tilde{\omega}(\tau) \cdot (1 - \bar{b}) \cdot F_{\mathrm{diff},\lambda}^{\uparrow}(\tau) - \tilde{\omega}(\tau) \cdot \bar{b} \cdot F_{\mathrm{diff},\lambda}^{\downarrow}(\tau)\,, \tag{7.272}$$

$$-\frac{1}{2}\frac{\mathrm{d}F_{\mathrm{diff},\lambda}^{\downarrow}(\tau)}{\mathrm{d}\tau} = F_{\mathrm{diff},\lambda}^{\downarrow}(\tau) - \frac{\tilde{\omega}(\tau)}{2} \cdot [1 - b(\mu_0)] \cdot S_{\mathrm{dir},\lambda,\mathrm{TOA}} \cdot \mathrm{e}^{-\tau/\mu_0}$$
$$- \tilde{\omega}(\tau) \cdot \bar{b} \cdot F_{\mathrm{diff},\lambda}^{\uparrow}(\tau) - \tilde{\omega}(\tau) \cdot (1 - \bar{b}) \cdot F_{\mathrm{diff},\lambda}^{\downarrow}(\tau)\,. \tag{7.273}$$

The solution to the preceding equations is the summation of the general solution to the corresponding homogeneous equation and the special solution to the inhomogeneous equation; a standard technique for solving ordinary differential equations but not recaptured here.

Problems

Problem 7.1 Truncated Phase Function

Please show that the truncated phase function $\mathcal{P}'(\cos\vartheta)$, see Eq. (7.13), is normalized, provided the normalization of the original phase function.

Problem 7.2 Transformation to Atmospheric Angular Coordinates

Please provide evidence of Eq. (7.16) using the addition theorem for spherical harmonic functions.

Problem 7.3 Monte Carlo Extinction

a) A photon travels in a medium that has a volumetric extinction coefficient b_{ext}. What is the probability that it will be subject to an extinction event (a scattering or absorption process) on its way?

b) In a Monte Carlo model, many photons transfer through the same medium. Using the result from (a), how can we use a random number generator (which produces random values uniformly distributed between 0 and 1) to decide whether a photon should be subjected to an extinction process or not?

c) Once the Monte Carlo program has decided to start the extinction process, by virtue of (b), how would we decide whether a photon is scattered or absorbed?

d) If the fate of the photon is to be scattered, how can we determine the new propagation direction using random numbers?

Problem 7.4 Monte Carlo Noise

Assume a Monte Carlo model is run with N_0 photons. "Detectors" are located within the model domain to "measure" (i.e., count) all the photons that arrive at the location (x, y) of the detectors. Whenever a photon hits the detector location along its modeled path, the "measurement" variable of that detector is increased accordingly; the final value of that variable is the "measured" value. Not all photons that enter the model atmosphere arrive at the detectors; some are absorbed, and many are scattered out of the model domain (e.g., back into space). Assume that a fraction f_{det} of the N_0 photons is detected and contributes to the final result. Due to the statistical nature of this model, not all detectors receive the same number of photons and the result is noisy.

What kind of statistical distribution governs the radiation "measurement?" Give an estimate of the result's noise (in terms of relative standard deviation)!

Problem 7.5 Diffuse Fraction of Radiation

The total solar radiation reaching the Earth's surface is commonly divided into two components, the direct $F_{dir}^{\downarrow}$ and diffuse radiation $F_{diff}^{\downarrow}$:

$$F^{\downarrow} = F_{dir}^{\downarrow} + F_{diff}^{\downarrow}, \tag{7.274}$$

$$F_{dir}^{\downarrow} = f_{dir} \cdot F^{\downarrow}, \tag{7.275}$$

$$F_{diff}^{\downarrow} = (1 - f_{dir}) \cdot F^{\downarrow}, \tag{7.276}$$

with f_{dir} the fractional part of the direct radiation and $f_{diff} = 1 - f_{dir}$ the fractional part of the diffuse radiation.

a) Find a simple relationship for f_{diff} measured below a cloud of optical thickness τ, assuming the cloud is nonabsorbing and illuminated by a collimated radiation beam of a zenith angle $\mu = 0.5$. In this case, the two-stream approximation gives the following solutions for the total (direct plus diffuse) and direct transmissivity:

$$\mathcal{T} = \frac{F^{\downarrow}}{F_0^{\downarrow}} = \frac{1}{1 + (1 - g) \cdot \tau}, \tag{7.277}$$

$$\mathcal{T}_{dir} = \frac{F_{dir}^{\downarrow}}{F_0^{\downarrow}} = \exp\left(-\frac{\tau}{\mu}\right). \tag{7.278}$$

Calculate f_{diff} as a function of g, the asymmetry factor of the cloud, and τ, the optical thickness of the cloud.

b) Discuss qualitatively whether f_{diff} increases or decreases when the cloud optical thickness increases.

c) f_{diff} has been measured in the presence of two clouds, each of optical thickness $\tau = 1$. For cloud A, $f_{diff,A} = 0.845$, and for cloud B, $f_{diff,B} = 0.831$ were observed. Decide (by calculating the asymmetry factor) which cloud was a liquid cloud and which was an ice cloud containing nonspherical ice crystals.

Problem 7.6 Cloud Transmissivity

Start from Eq. (7.64) and derive a TSA formula for the transmissivity of a stratiform, homogeneous cloud layer of optical thickness τ_C in a nonabsorbing atmosphere. Assume hemispherical isotropy. Let the cloud be illuminated from above by purely diffuse irradiance F_0, and ignore photons that are scattered back from the atmosphere below the cloud.

8
Absorption and Emission by Atmospheric Gases

8.1
Interactions of Photons and Gas Molecules

8.1.1
Types of Molecular Energy E_{mol}

The energy of a molecule, E_{mol}, is given by the sum of several types of energy, that is,

$$E_{mol} = E_{trl} + E_{n_{rot}} + E_{n_{vib}} + E_{n_{orb}}$$
$$n_{rot}, n_{orb} = 1, 2, 3, \ldots ; \quad n_{vib} = 0, 1, 2, \ldots , \tag{8.1}$$

where the integer indices, n_{rot}, n_{vib}, and n_{orb}, indicate that the associated energy types are quantized and only certain levels are allowed. On the contrary, the term E_{trl}, the transitional energy, a type of kinetic energy of the translational movement of the molecule, is not quantized. The different types of molecular energy are illustrated in Figure 8.1.

In Eq. (8.1), $E_{n_{rot}}$ is the molecular rotational energy; $E_{n_{vib}}$ is the energy associated with the vibration of the components of the molecule around their equilibrium positions; $E_{n_{orb}}$ is the electron orbital energy, also known as electronic or potential energy, associated with the orbital states of the electrons circling the atomic nucleus. $E_{n_{orb}}$ is dependent on the arrangement of the electrons of the atom or molecule; E_{trl} is the translational kinetic energy exchanged during collisions of atoms and molecules. For a gas molecule, it generally holds that

$$E_{n_{rot}} < E_{trl} < E_{n_{vib}} < |E_{n_{orb}}| ; n_{rot}, n_{orb} = 1, 2, 3, \ldots ; \quad n_{vib} = 0, 1, 2, \ldots , \tag{8.2}$$

where $E_{n_{orb}}$ is normally defined as negative with respect to a reference energy level, see Eq. (8.28). Furthermore, the transitions between electronic energy levels are responsible for the UV, VIS, and NIR spectrum. Vibrational energy transitions are related to the NIR to FIR spectrum, and pure rotational energy transitions occur in the FIR and MW region (Goody and Yung, 1989).

Theory of Atmospheric Radiative Transfer, First Edition. Manfred Wendisch and Ping Yang
© 2012 WILEY-VCH Verlag GmbH & Co. KGaA. Published 2012 by WILEY-VCH Verlag GmbH & Co. KGaA.

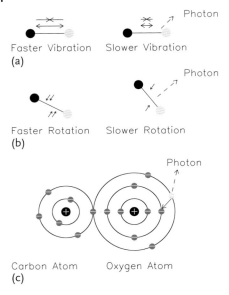

Carbon Atom   Oxygen Atom

(c)

Figure 8.1 Schematic illustration of the different mechanisms for photon emission: Decrease of (a) vibrational, (b) rotational and (c) electronic energies. Example associated with the diatomic molecule CO.

8.1.2
Photon Absorption and Emission

If a photon strikes a molecule, the photon can be absorbed by the molecule only if its energy, E_{phot}, corresponds to an allowed transition between two energy states of the molecule and if the photon energy increases the molecular energy to one of the molecule's possible energy levels. Similarly, a photon emitted by a molecule corresponds to one of the molecule's allowed transition energy levels. Therefore, molecules absorb and emit at specific, discontinuous wavelengths or frequencies, called absorption and emission lines, corresponding to the transitions between quantized energy levels. The absorption and emission lines may overlap, be very close together, and require a high spectral resolution to resolve the lines.

Several types of bands comprised of lines are formed when absorption and emission lines overlap. Combined vibration-rotational bands are created by photon absorption and emission with the energy corresponding to the allowed transition levels of vibrational and rotational energies ($E_{n_{\text{vib}}}$, $E_{n_{\text{rot}}}$). A vibrational energy transition is often jointly associated with a rotational transition. A change in translational energy, E_{trl}, from kinetic collisions, which is usually much larger than rotational energy, $E_{n_{\text{rot}}}$, but smaller than vibrational energy, $E_{n_{\text{vib}}}$, has a strong influence on the rotational lines, a weaker influence on the vibrational lines, and negligible influence on the electronic lines (Goody and Yung, 1989).

8.1.3
Allowed Quantized Energies and Frequencies (Wavelengths)

The local thermodynamic equilibrium assumption guarantees that, at any temperature above 0 K, a sufficient number of molecules in any given state of energy $E_{mol,n} = E_{mol,0}, E_{mol,1}, \ldots, E_{mol,\infty}$ is available. Allowed quantized transition energies are given by

$$\Delta E_{mol,n,m} = E_{mol,m} - E_{mol,n} \ ; \quad n = 0, 1, 2, \ldots ; \quad m = 1, 2, 3 \ldots \quad (8.3)$$

If a photon is absorbed or emitted, the following equality must be true:

$$E_{phot,n,m} = \Delta E_{mol,n,m} \ ; \quad n = 0, 1, 2, \ldots ; \quad m = 1, 2, 3, \ldots \quad (8.4)$$

Each of the possible transition energies corresponds to a respective wavelength $\lambda_{n,m}$ or frequency, $\nu_{n,m}$, equaling an allowed transition between molecular energy levels, that is,

$$\lambda_{n,m} = \frac{c}{\nu_{n,m}} \ ; \quad n = 0, 1, 2, \ldots ; \quad m = 1, 2, 3 \ldots \quad (8.5)$$

Photon energy is a function of either wavelength or frequency. Thus, there are distinct wavelengths $\lambda_{n,m}$, or spectral lines, at which photon absorption or emission may occur by specific molecules, and at all other wavelengths, the molecule neither absorbs nor emits. The spectral positions of the absorption or emission lines are defined by the allowable energy differences. However, line intensity is determined by the fraction of molecules in the particular initial state required for the transition and by the intrinsic likelihood of a photon having the correct energy and encountering a molecule in the required energy state to perform the transition. Furthermore, more than one transition can contribute to a single absorption line if the energy changes are equal for these transitions, that is,

$$\Delta E_{mol,n,m} = \Delta E_{mol,k,l} \ . \quad (8.6)$$

8.1.4
Energy Level Probability in Thermal Equilibrium

The probability density function, N_{mol}, describes the appearance of the energy levels of gaseous molecules in thermodynamic equilibrium. The function quantifies the number distribution of the allowable energy states for respective energy levels E_{mol}. According to statistical mechanics, the probability density function for E_{tra} is given for a large number of atoms or molecules in thermal equilibrium by the Maxwell–Boltzmann distribution (Bohren and Clothiaux, 2006)

$$N_{mol}(T, E_{tra}) = \frac{2\sqrt{E_{tra}}}{\sqrt{\pi} \cdot (k_B \cdot T)^{3/2}} \cdot \exp\left[\frac{-E_{tra}}{(k_B \cdot T)}\right], \quad (8.7)$$

where k_B represents the Boltzmann constant, and T is the absolute temperature in Kelvin. The units of N_{mol} are J^{-1}; the units of E_{mol} are joules or eV. N_{mol} quantifies the abundance of certain energy levels. The Maxwell–Boltzmann distribution is

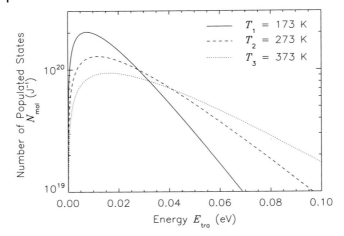

Figure 8.2 Maxwell–Boltzmann distribution showing the relation between the number of populated states and the energy levels for three temperatures $T_1 < T_2 < T_3$.

normalized such that

$$\int_0^\infty N_{\mathrm{mol}} \, \mathrm{d}E_{\mathrm{tra}} = 1 \ . \tag{8.8}$$

The most probable kinetic energy (maximum of the Maxwell–Boltzmann distribution, $N_{\mathrm{mol}} \longrightarrow$ maximum) is given by

$$E_{\mathrm{tra,max}}(T) = \frac{k_{\mathrm{B}} \cdot T}{2} \ . \tag{8.9}$$

As an example in the case of $T = 300$ K, we have

$$E_{\mathrm{tra,max}}(300 \ \mathrm{K}) \approx 1.25 \times 10^{-2} \ \mathrm{eV} \ . \tag{8.10}$$

From the Maxwell–Boltzmann distribution, the higher the temperature, the fewer the number of the maximum energy levels are allowed. To illustrate Eq. (8.7), Figure 8.2 shows N_{mol} as a function of E_{tra} for three different temperature values. Beyond the energy corresponding to the occurence of higher energy is less frequent than that of lower energy. We apply Eq. (8.7) to the allowed, quantized energy levels $E_{\mathrm{mol},n}$

$$N_{\mathrm{mol},n} = \frac{f_n \cdot \exp[-E_{\mathrm{mol},n}/(k_{\mathrm{B}} \cdot T)]}{\sum_{n=0}^{\infty} f_n \cdot \exp[-E_{\mathrm{mol},n}/(k_{\mathrm{B}} \cdot T)]} \ ; \qquad n = 0, 1, 2, \ldots , \tag{8.11}$$

where f_n is the density of degeneracy and $N_{\mathrm{mol},n}$ is dimensionless (Bohren and Clothiaux, 2006). From Eq. (8.11), we can derive an expression for the ratio of the probability density functions of two adjacent, allowable quantized energy states $E_{\mathrm{mol},n}$ and $E_{\mathrm{mol},m}$:

$$\frac{N_{\mathrm{mol},m}}{N_{\mathrm{mol},n}} \approx \exp\left[\frac{-\Delta E_{\mathrm{mol},n,m}}{(k_{\mathrm{B}} \cdot T)}\right] \ ; \qquad n = 0, 1, 2, \ldots ; \quad m = 1, 2, 3, \ldots , \tag{8.12}$$

where $\Delta E_{mol,n,m} = E_{mol,m} - E_{mol,n}$. From Eq. (8.12), we conclude that, if the following condition holds

$$\Delta E_{mol,n,m} \ll (k_B \cdot T) , \tag{8.13}$$

the probability density function of the allowed quantized energy levels is approximately constant, that is,

$$N_{mol,n} \approx N_{mol,m} ; \quad n = 0, 1, 2, \ldots ; \quad m = 1, 2, 3, \ldots \tag{8.14}$$

8.2
Examples of Energy Transitions

8.2.1
Structure of Gas Molecules

The molecular structure plays an important role in determining the various energy levels of photons absorbed or emitted by gas molecules. Many types of molecular structures are common in the atmosphere. N_2 and O_2 are examples of diatomic molecules and have a symmetric structure. The structure of the triatomic molecules CO_2 and N_2O is linear. Nonlinear molecular structures are present for symmetric-top molecules, such as, NH_3, CH_3Cl, and $CFCl_3$; symmetric-spherical-top molecules, for example, CH_4; and, asymmetric-top molecules, for example, H_2O and O_3. Figure 8.3 illustrates the structures of several molecules.

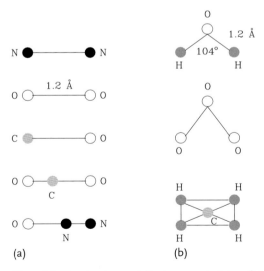

(a)   (b)

Figure 8.3 The structure and dipole moment status of atmospheric molecules. (a) Molecules with linear structures; (b) molecules with asymmetric-top structures (H_2O and O_3) and symmetric-spherical-top structure (CH_4).

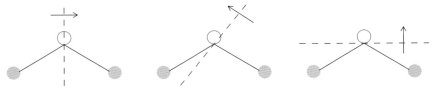

Figure 8.4 Three types of rotations for a triatomic molecule.

8.2.2
Molecular Rotational Energy $E_{n_{rot}}$

Pure rotational transitions require that molecules exhibit permanent electric or magnetic dipole moments, see Figure 8.3. Linear symmetric molecules, for example, N_2 and O_2, do not have permanent dipole moments. Thus, for these gases, no rotational energy transitions are possible in the MW to the FIR spectral region. Spherical top molecules, such as, CH_4, also do not have permanent dipole moments and, thus, allow no purely rotational energy transitions. However, molecular vibrations within these types of molecules create oscillating dipole moments and lead to vibrational-rotational spectral bands. Nonlinear asymmetric molecules, such as, H_2O and O_3, exhibit permanent dipole moments and have purely rotational spectra. As an example, three types of rotation in a H_2O molecule are illustrated in Figure 8.4.

8.2.3
Molecular Vibrational Energy $E_{n_{vib}}$

Vibrational photon energy transitions require a changing or an oscillating dipole moment, see Figure 8.5 for examples. Symmetric linear molecules, such as, N_2 and O_2, are symmetric in structure and have no vibrational energy transitions. Thus, we can conclude that these gases possess no photon energy transitions in the IR to NIR spectral regions. Nevertheless, these molecules are very important in the UV and VIS spectrum due to electron orbital energy transitions. CO has a single vibrational mode only, and CO_2 has a symmetric stretching mode (a) which

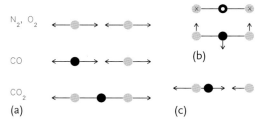

Figure 8.5 Vibrational modes of various diatomic and triatomic atmospheric molecules. The crosses and the white dot (upper right molecule) indicate vibration perpendicular to the paper sheet, the cross means vibration into the sheet, and the dot out of the sheet. (a) Symmetric stretching mode, (b) two bending modes and (c) asymmetric stretching mode.

is inactive from a radiative point of view. The molecular structures shown on the right side of Figure 8.5 have two bending modes (b) and one asymmetric stretching mode (c).

We define the number of atoms in a molecule by n_{ato}. The total number of independent vibrational frequencies, or the normal modes, of a molecule with $n_{ato} > 2$ atoms is $(3 \times n_{ato} - 6)$ for nonlinear molecules and $(3 \times n_{ato} - 5)$ for linear molecules (Goody and Yung, 1989; Thomas and Stamnes, 1999). Thus, the nonlinear molecules of H_2O and O_3 each have three important normal modes, and the nonlinear molecules of CH_4 have nine normal modes, but only two are active in the IR spectral region.

8.3
Line Spectra for Single-Atomic Gases

8.3.1
Molecular Electron Orbital Energy $E_{n_{orb}}$

In the case of a single atom, absorption or emission of EM radiation can occur only by a change in orbital energy of the electron ($E_{n_{orb}}$; $n_{orb} = 1, 2, 3, \ldots$), that is, vibrational and rotational energy transitions for single-atomic gases are not possible. Furthermore, we may neglect translational energy and the only allowable energy transition levels for single-atomic gases are given by $E_{n_{orb}}$, where the integer subscript n_{orb} indicates the allowed orbital level of the respective atom. When an atom undergoes a transition from one energy state with energy $E_{m_{orb}}$ to another state with lower or higher energy $E_{n_{orb}}$ allowed by the selection rules in quantum mechanics, we obtain the following energy:

$$\Delta E_{n_{orb}, m_{orb}} = E_{m_{orb}} - E_{n_{orb}}$$
$$= h \cdot (\nu_{m_{orb}} - \nu_{n_{orb}}) = h \cdot \nu_{n_{orb}, m_{orb}} . \tag{8.15}$$

For absorption, the atom gains energy and we have $\Delta E_{n_{orb}, m_{orb}} < 0$, and for emission, the atom loses energy and we have $\Delta E_{n_{orb}, m_{orb}} > 0$. Figure 8.6 is an illustration of the photon absorption and emission processes. Upon the absorption of a photon, an electron moves to an outer orbit and increases the orbital energy of the atom.

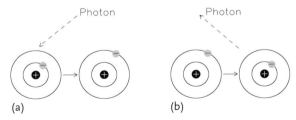

Figure 8.6 The absorption process (a) and the emission process (b) associated with the hydrogen atom.

Upon emission of a photon, an electron transits from a higher to a lower orbital energy state and decreases the orbital energy.

8.3.2
Line Spectrum of the Hydrogen Atom

Here, we briefly recapture the well-known Bohr model (Bohr, 1913) to explain the hydrogen spectrum. The text by Jain (2007) is recommended for an introduction to this subject.

Hydrogen has the simplest atomic structure of all the elements and, thus, the most simple absorption or emission line spectrum, which is partially due to the fact that the hydrogen atom possesses a single electron orbiting the nucleus. Molecules with more protons and electrons interact in much more complicated ways with photons. The electron is forced to move around the proton nucleus which has an elementary charge of $+e$. The allowed orbital level of a hydrogen atom is indicated by the integer subscript $n_{orb,H}$. The electron orbiting the nucleus has an orbiting velocity of v_{ele}, an orbital radius of $r_{n_{orb,H}}$, an elementary charge of $-e$, and a mass of m_{ele}. The Coulomb force $F_{Col,n_{orb,H}}$, and the centrifugal (radial) force, $F_{cen,n_{orb,H}}$, are given by

$$F_{Col,n_{orb,H}} = \frac{(+e) \cdot (-e)}{4\pi \cdot \epsilon_0 \cdot r^2_{n_{orb,H}}} = -\frac{e^2}{4\pi \cdot \epsilon_0 \cdot r^2_{n_{orb,H}}} , \tag{8.16}$$

$$F_{cen,n_{orb,H}} = \frac{m_{ele} \cdot v^2_{ele}}{r_{n_{orb,H}}} ;$$

$$n_{orb,H} = 1, 2, 3, \ldots , \tag{8.17}$$

where $e = 1.602 \times 10^{-19}$ C is the elementary charge, ϵ_0 represents the dielectric constant (electric permittivity of a vacuum), and $m_{ele} = 9.108 \times 10^{-31}$ kg indicates the static mass of an electron. The balance of the two forces leads to

$$-F_{Col,n_{orb,H}} = F_{cen,n_{orb,H}} ;$$

$$\frac{e^2}{4\pi \cdot \epsilon_0 \cdot r^2_{n_{orb,H}}} = \frac{m_{ele} \cdot v^2_{ele}}{r_{n_{orb,H}}} ;$$

$$n_{orb,H} = 1, 2, 3, \ldots \tag{8.18}$$

Bohr's first postulate, also called the Bohr–Sommerfeld quantum restriction, states that only those electron circular orbits are permitted for which the electron's angular momentum $\mathcal{L}_{n_{orb,H}}$ is a multiple of $\hbar = h/(2\pi)$. Thus, for the hydrogen atom (subscript H) electron, we get the allowed electron's angular momentum,

$$\mathcal{L}_{n_{orb,H}} = n_{orb,H} \cdot \hbar = n_{orb,H} \cdot \left(\frac{h}{2\pi}\right); \quad n_{orb,H} = 1, 2, 3, \ldots , \tag{8.19}$$

where $\hbar = 1.054\,59 \times 10^{-34}$ J s and $n_{orb,H}$ is the orbital quantum number of the hydrogen atom. The angular momentum of the hydrogen atom electron is given by

$$\mathcal{L}_{n_{\mathrm{orb,H}}} = m_{\mathrm{ele}} \cdot v_{\mathrm{ele}} \cdot r_{n_{\mathrm{orb,H}}} \;; \quad n_{\mathrm{orb,H}} = 1, 2, 3, \ldots \tag{8.20}$$

Multiplying Eq. (8.18) by m_{ele} gives

$$\frac{m_{\mathrm{ele}}^2 \cdot v_{\mathrm{ele}}^2}{r_{n_{\mathrm{orb,H}}}} = \frac{m_{\mathrm{ele}} \cdot e^2}{4\pi \cdot \epsilon_0 \cdot r_{n_{\mathrm{orb,H}}}^2} \;; \quad n_{\mathrm{orb,H}} = 1, 2, 3, \ldots \tag{8.21}$$

Canceling $r_{n_{\mathrm{orb,H}}}$ in the preceding expression yields

$$m_{\mathrm{ele}}^2 \cdot v_{\mathrm{ele}}^2 = \frac{m_{\mathrm{ele}} \cdot e^2}{4\pi \cdot \epsilon_0 \cdot r_{n_{\mathrm{orb,H}}}} \;; \quad n_{\mathrm{orb,H}} = 1, 2, 3, \ldots \tag{8.22}$$

Dividing Eq. (8.20) with $r_{n_{\mathrm{orb,H}}}$ and squaring the resultant equation leads to

$$\frac{\mathcal{L}_{n_{\mathrm{orb,H}}}^2}{r_{n_{\mathrm{orb,H}}}^2} = m_{\mathrm{ele}}^2 \cdot v_{\mathrm{ele}}^2 = \frac{m_{\mathrm{ele}} \cdot e^2}{4\pi \cdot \epsilon_0 \cdot r_{n_{\mathrm{orb,H}}}} \;; \quad n_{\mathrm{orb,H}} = 1, 2, 3, \ldots \tag{8.23}$$

The application of Bohr's first postulate to the left side of Eq. (8.23) produces

$$\frac{[n_{\mathrm{orb,H}} \cdot h/(2\pi)]^2}{r_{n_{\mathrm{orb,H}}}^2} = \frac{m_{\mathrm{ele}} \cdot e^2}{4\pi \cdot \epsilon_0 \cdot r_{n_{\mathrm{orb,H}}}} \;; \quad n_{\mathrm{orb,H}} = 1, 2, 3, \ldots \,, \tag{8.24}$$

or in a simplified form

$$\frac{n_{\mathrm{orb,H}}^2 \cdot h^2}{\pi \cdot r_{n_{\mathrm{orb,H}}}} = \frac{m_{\mathrm{ele}} \cdot e^2}{\epsilon_0} \;; \quad n_{\mathrm{orb,H}} = 1, 2, 3, \ldots \tag{8.25}$$

This yields the electron's permitted orbital radii

$$r_{n_{\mathrm{orb,H}}} = n_{\mathrm{orb,H}}^2 \cdot \frac{\epsilon_0 \cdot h^2}{\pi \cdot m_{\mathrm{ele}} \cdot e^2} \;; \quad n_{\mathrm{orb,H}} = 1, 2, 3, \ldots \tag{8.26}$$

The allowed orbit with the smallest radius is ($n_{\mathrm{orb}} = 1$)

$$r_{n_{\mathrm{orb}}=1,\mathrm{H}} = \frac{\epsilon_0 \cdot h^2}{\pi \cdot m_{\mathrm{ele}} \cdot e^2} = 0.529 \times 10^{-10} \,\mathrm{m} = 0.529 \,\text{Å} \,. \tag{8.27}$$

This orbital radius is called Bohr's radius (ångström = Å = 10^{-10} m), and thus, the diameter of a hydrogen atom is roughly 1 Å. The electron's potential energy for the orbital quantum number of the hydrogen atom ($n_{\mathrm{orb,H}}$th orbit) is shown in Eqs. (8.16) and (8.26), that is,

$$\begin{aligned} E_{\mathrm{pot},n_{\mathrm{orb,H}}} &= F_{\mathrm{Col},n_{\mathrm{orb,H}}} \cdot r_{n_{\mathrm{orb,H}}} \\ &= -\frac{e^2}{4\pi \cdot \epsilon_0 \cdot r_{n_{\mathrm{orb,H}}}} = -\frac{m_{\mathrm{ele}} \cdot e^4}{4\epsilon_0^2 \cdot n_{\mathrm{orb,H}}^2 \cdot h^2} \,. \end{aligned} \tag{8.28}$$

Using Eq. (8.22), the kinetic energy is given by

$$E_{\text{kin},n_{\text{orb,H}}} = \frac{m_{\text{ele}} \cdot v_{\text{ele}}^2}{2} = \frac{e^2}{8\pi \cdot \epsilon_0 \cdot r_{n_{\text{orb,H}}}}$$

$$= -\frac{1}{2} \cdot E_{\text{pot},n_{\text{orb,H}}} \; . \tag{8.29}$$

Thus, the electron's total energy in orbit $n_{\text{orb,H}}$ (orbital quantum number of hydrogen atom) is given by

$$E_{n_{\text{orb,H}}} = E_{\text{pot},n_{\text{orb,H}}} + E_{\text{kin},n_{\text{orb,H}}} = \frac{1}{2} E_{\text{pot},n_{\text{orb,H}}}$$

$$= -\frac{m_{\text{ele}} \cdot e^4}{8\epsilon_0^2 \cdot h^2} \cdot \frac{1}{n_{\text{orb,H}}^2} \; ;$$

$$n_{\text{orb,H}} = 1, 2, 3, \ldots \tag{8.30}$$

Bohr's second postulate states that for emission or absorption of photons, the energy difference between the two states $n_{\text{orb,H}}$ and $m_{\text{orb,H}} > n_{\text{orb,H}}$ is required to match the energy of the emitted or absorbed photon given by

$$h \cdot v_{n_{\text{orb,H}}, m_{\text{orb,H}}} = h \cdot \frac{c}{\lambda_{n_{\text{orb,H}}, m_{\text{orb,H}}}}$$

$$= \Delta E_{n_{\text{orb,H}}, m_{\text{orb,H}}}$$

$$= E_{m_{\text{orb,H}}} - E_{n_{\text{orb,H}}} \; ;$$

$$n_{\text{orb,H}}, m_{\text{orb,H}} = 1, 2, 3, \ldots \tag{8.31}$$

Equation (8.30) leads to the allowed frequencies for atomic spectra of a hydrogen atom with the orbital quantum numbers $n_{\text{orb,H}} < m_{\text{orb,H}}$,

$$\frac{1}{\lambda_{n_{\text{orb,H}}, m_{\text{orb,H}}}} = v_{n_{\text{orb,H}}, m_{\text{orb,H}}} = \frac{m_{\text{ele}} \cdot e^4}{8 \cdot \epsilon_0^2 \cdot h^3 \cdot c} \cdot \left(\frac{1}{n_{\text{orb,H}}^2} - \frac{1}{m_{\text{orb,H}}^2} \right)$$

$$= R_{\text{orb,H}} \cdot \left(\frac{1}{n_{\text{orb,H}}^2} - \frac{1}{m_{\text{orb,H}}^2} \right) \; ;$$

$$n_{\text{orb,H}}, m_{\text{orb,H}} = 1, 2, 3, \ldots, \tag{8.32}$$

where $R_{\text{orb,H}}$ is the Rydberg constant, given by

$$R_{\text{orb,H}} = \frac{m_{\text{ele}} \cdot e^4}{8\epsilon_0^2 \cdot h^3 \cdot c} = 1.0974 \times 10^7 \text{m}^{-1} \; . \tag{8.33}$$

Depending on the subscript $n_{\text{orb,H}}$ (the orbital quantum number of H) of the lower energy state, the terminology for the spectral lines of the hydrogen atom are, see Figure 8.7,

$$
\begin{array}{lll}
n_{\text{orb,H}} = 1 : & \text{Lyman series,} & m_{\text{orb,H}} = 2, 3, 4, \ldots \\
n_{\text{orb,H}} = 2 : & \text{Balmer series,} & m_{\text{orb,H}} = 3, 4, 5, \ldots \\
n_{\text{orb,H}} = 3 : & \text{Paschen series,} & m_{\text{orb,H}} = 4, 5, 6, \ldots \\
n_{\text{orb,H}} = 4 : & \text{Brackett series,} & m_{\text{orb,H}} = 5, 6, 7, \ldots
\end{array}
$$

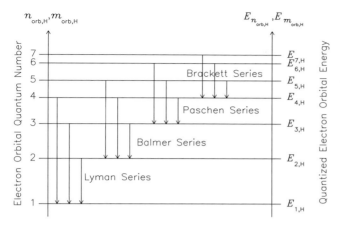

Figure 8.7 Possible transitions between electron orbital energy states of the hydrogen atom and the resulting spectral series. Redrawn from Jain (2007) with modification.

Table 8.1 The energy difference $\Delta E_{m_{orb,H}, n_{orb,H}}$ and allowed frequencies $\nu_{m_{orb,H}, n_{orb,H}}$ for atomic spectra of hydrogen atoms and the respective wavelengths $\lambda_{m_{orb,H}, n_{orb,H}}$ of the two states $n_{orb,H}$ and $m_{orb,H}$ ($> n_{orb,H}$).

$n_{orb,H}$	$m_{orb,H}$	Series		$\Delta E_{m_{orb,H}, n_{orb,H}}$ (eV)	$\nu_{m_{orb,H}, n_{orb,H}}$ (s^{-1})	$\lambda_{m_{orb,H}, n_{orb,H}}$ (µm)
1	2	Lyman	α	10.204 27	2.466 21 × 10^{15}	0.121 56
1	3		β	12.093 95	2.924 27 × 10^{15}	0.102 52
1	4		γ	12.755 34	3.084 19 × 10^{15}	0.097 20
2	3	Balmer	α	1.889 68	4.569 17 × 10^{14}	0.656 12
2	4		β	2.551 07	6.168 38 × 10^{14}	0.486 02
2	5		γ	2.857 20	6.908 58 × 10^{14}	0.433 94
3	4	Paschen	α	0.661 39	1.599 21 × 10^{14}	1.874 63
3	5		β	0.967 52	2.339 41 × 10^{14}	1.281 49
3	6		γ	1.133 81	2.741 50 × 10^{14}	1.093 53
4	5	Brackett	α	0.306 13	7.402 05 × 10^{13}	4.050 13
4	6		β	0.472 42	1.142 29 × 10^{14}	2.624 48
4	7		γ	0.572 69	1.384 74 × 10^{14}	2.164 98
5	6			0.166 29	4.020 87 × 10^{13}	7.455 92
5	7			0.266 56	6.445 32 × 10^{13}	4.651 32
5	8			0.331 64	8.018 89 × 10^{13}	3.738 58
6	7			0.100 27	2.424 46 × 10^{13}	12.365 35
6	8			0.165 35	3.998 02 × 10^{13}	7.498 52
6	9			0.209 96	5.076 85 × 10^{13}	5.905 09
7	8			0.065 08	1.573 57 × 10^{13}	19.051 80
7	9			0.109 70	2.652 40 × 10^{13}	11.302 70
7	10			0.141 61	3.424 08 × 10^{13}	8.755 42

The Lyman-α emission line at the 121.56 nm wavelength is used for humidity measurements. Additional numbers are given in Table 8.1.

8.4
Molecular Absorption/Emission Line Spectra

Since molecules, compared to simple atoms, have several ways to distribute their internal energy, the emission spectrum of a molecule is much more complicated than that of a single atom. In addition to their quantized electron orbital energy, $E_{n_{orb}}$, molecules exhibit quantized rotational energy, $E_{n_{rot}}$, and vibrational energy, $E_{n_{vib}}$, and nonquantized translational energy, E_{trl}. The various molecular rotations and vibrations are only optically active if the molecule possesses either a permanent or oscillating electric or magnetic dipole moment. Molecular energy transitions and the related frequencies or wavelengths for emission and/or absorption are linearly composed of the electron orbital, rotational, vibrational, and translational energy parts, see Eq. (8.1). Additionally, photons may be absorbed and can destroy a molecular bond (photodissociation), which is particularly relevant in the UV spectral region, but is not addressed here.

8.4.1
Molecular Rotational Spectra

Quantum restrictions prevent a molecule from rotating with arbitrary circular rotational frequency. Let $\mathcal{I}$ be the moment of rotational inertia of a molecule, and the molecule rotates with a quantized circular rotational frequency,

$$\omega_{c,n_{rot}} = 2\pi \cdot \nu_{n_{rot}} = \frac{2\pi}{T_{p,n_{rot}}} \quad ; \quad n_{rot} = 1, 2, 3, \dots \, , \tag{8.34}$$

where $T_{p,n_{rot}}$ is the period of the rotation of the molecule around the molecule's axis of symmetry. The angular momentum of the rotating molecule is given by

$$\mathcal{L}_{n_{rot}} = \mathcal{I} \cdot \omega_{c,n_{rot}} \quad ; \quad n_{rot} = 1, 2, 3, \dots \tag{8.35}$$

Bohr's first postulate requires the angular momentum to be an integer multiple of $\hbar = h/(2\pi)$,

$$\mathcal{L}_{n_{rot}} = \mathcal{I} \cdot \omega_{c,n_{rot}} = n_{rot} \cdot \hbar \quad ; \quad n_{rot} = 1, 2, 3, \dots \, , \tag{8.36}$$

where n_{rot} is the rotational quantum number. This leads to

$$\omega_{c,n_{rot}} = \frac{n_{rot} \cdot \hbar}{\mathcal{I}} \quad ; \quad n_{rot} = 1, 2, 3, \dots \tag{8.37}$$

The allowed quantized rotational energy is given by

$$E_{n_{rot}} = \frac{1}{2} \cdot \mathcal{I} \cdot \omega_{c,n_{rot}}^2 = \frac{1}{2} \cdot \frac{n_{rot}^2 \cdot \hbar^2}{\mathcal{I}} \quad ; \quad n_{rot} = 1, 2, 3, \dots \tag{8.38}$$

This is not the strict quantum mechanical result. Furthermore, quantum theory requires that the square of the rotational quantum number n_{rot}^2 in Eq. (8.38) be substituted by $n_{rot} \cdot (n_{rot} + 1)$, which leads to the following expression for the allowable quantized levels for the rotational energy of molecules:

$$
\begin{aligned}
E_{n_{rot}} &= \frac{1}{2} \cdot \frac{n_{rot} \cdot (n_{rot} + 1) \cdot \hbar^2}{\mathcal{I}} \\
&= \frac{n_{rot} \cdot (n_{rot} + 1) \cdot h^2}{8\pi^2 \cdot \mathcal{I}} ; \quad n_{rot} = 1, 2, 3, \dots ,
\end{aligned}
\tag{8.39}
$$

where the $\mathcal{I}$ represents the moment of rotational inertia about the molecule's rotating axis. Bohr's second postulate states that for photon emission or absorption, the energy differences between the two states $n_{rot,H}$ and $m_{rot} > n_{rot}$ (absorption) or $m_{rot} < n_{rot}$ (emission), see Eq. (8.31), are required to match the energy of the photon, that is,

$$
\nu_{n_{rot}, m_{rot}} = \frac{1}{h} \cdot (E_{m_{rot}} - E_{n_{rot}}); \qquad n_{rot}, m_{rot} = 1, 2, 3, \dots
\tag{8.40}
$$

This yields

$$
\begin{aligned}
&\nu_{n_{rot}, m_{rot}} = \frac{h}{8\pi^2 \cdot \mathcal{I}} \cdot [m_{rot} \cdot (m_{rot} + 1) - n_{rot} \cdot (n_{rot} + 1)] ; \\
&n_{rot}, m_{rot} = 1, 2, 3, \dots
\end{aligned}
\tag{8.41}
$$

Quantum mechanics allows transitions between neighboring rotational states, that is,

$$
m_{rot} = n_{rot} + 1 .
\tag{8.42}
$$

We have

$$
\begin{aligned}
\nu_{n_{rot}} &= \frac{h}{8\,\pi^2 \cdot \mathcal{I}} \cdot [(n_{rot} + 1) \cdot (n_{rot} + 1 + 1) - n_{rot} \cdot (n_{rot} + 1)] \\
&= \frac{h}{8\pi^2 \cdot \mathcal{I}} \cdot \left(n_{rot}^2 + 2n_{rot} + n_{rot} + 2 - n_{rot}^2 - n_{rot} \right) \\
&= \frac{h}{4\pi^2 \cdot \mathcal{I}} \cdot (n_{rot} + 1) ; \quad n_{rot} = 1, 2, 3, \dots
\end{aligned}
\tag{8.43}
$$

We define the rotational constant

$$
B = \frac{h}{8\pi^2 \cdot \mathcal{I}} ,
\tag{8.44}
$$

which gives

$$
\nu_{n_{rot}} = 2B \cdot (n_{rot} + 1) ; \quad n_{rot} = 1, 2, 3, \dots
\tag{8.45}
$$

The $\nu_{n_{rot}}$ in Eq. (8.45) represents the allowed quantized frequencies for rotational transitions in diatomic molecules. The frequencies $\nu_{n_{rot}}$ of the absorbed or emitted radiation increase proportionally to n_{rot}, and the spectral emission lines are equidistant, see Figure 8.10. Their mutual distance $2B = h/(4\pi^2 \cdot \mathcal{I})$ can be used to determine the molecule's moment of inertia $\mathcal{I}$. The respective allowed quantized energies for rotational transitions in diatomic molecules are given by

$$
E_{n_{rot}} = h \cdot B \cdot n_{rot} \cdot (n_{rot} + 1) ; \quad n_{rot} = 1, 2, 3, \dots
\tag{8.46}
$$

8.4.2
Ratio of Molecular Electron Orbital and Rotational Energies

The energy changes accompanying a quantum rotational transition are given by Eq. (8.39). For the first excited state, that is, $n_{rot} = 1$, we obtain

$$E_{n_{rot}=1} = \frac{2h^2}{8\pi^2 \cdot \mathcal{I}} = \frac{\hbar^2}{\mathcal{I}} \,. \tag{8.47}$$

In general,

$$\mathcal{I} = \sum_i m_i \cdot r_i^2 \,. \tag{8.48}$$

For a diatomic molecule, we have

$$\mathcal{I} = \sum_{i=1}^2 m_i \cdot r_i^2 = m_1 \cdot r_1^2 + m_2 \cdot r_2^2 \,. \tag{8.49}$$

If the diatomic molecule is symmetric, that is, $m_1 = m_2$ and $r_1 = r_2 = r_{mol}$, the following relation holds:

$$\mathcal{I} = (m_1 + m_2) \cdot r_{mol}^2 = m_{mol} \cdot r_{mol}^2 \,. \tag{8.50}$$

Therefore, we obtain $\mathcal{I} = m_{mol} \cdot r_{mol}^2$; whereas, m_{mol} is the mass of the molecule and r_{mol} is the distance of the molecule's atoms from their center of gravity, that is,

$$E_{n_{rot}=1} \approx \frac{\hbar^2}{m_{mol} \cdot r_{mol}^2} \,. \tag{8.51}$$

The energy associated with an electron orbital jump from the ground state to an excited state is, see Eqs. (8.28) and (8.30),

$$|E_{n_{orb}=1,H}| \approx \frac{e^2}{8\pi \cdot \epsilon_0 \cdot r_{n_{orb}=1,H}} \,, \tag{8.52}$$

which follows from the discussion of the hydrogen atom. $r_{n_{orb}=1,H}$ represents the Bohr radius and from the expressions, we obtain

$$\frac{|E_{n_{orb}=1,H}|}{E_{n_{rot}=1,H}} \approx \left(\frac{e^2}{8\pi \cdot \epsilon_0 \cdot r_{n_{orb}=1,H}} \right) \Big/ \left(\frac{\hbar^2}{m_{mol} \cdot r_{mol}^2} \right) \,. \tag{8.53}$$

In a simple molecule, the typical distance between the nuclei is approximately equal to the Bohr radius $r_{n_{orb}=1,H} = (\epsilon_0 \cdot h^2)/(\pi \cdot m_{ele} \cdot e^2)$, see Eq. (8.27). Therefore, we approximate r_{mol} by the Bohr radius and obtain

$$r_{mol} \approx \frac{\epsilon_0 \cdot h^2}{\pi \cdot m_{ele} \cdot e^2} \approx r_{n_{orb}=1,H} \,. \tag{8.54}$$

From Eq. (8.53), we get

$$\frac{|E_{n_{\text{orb}}=1,\text{H}}|}{E_{n_{\text{rot}}=1,\text{H}}} \approx \frac{e^2 \cdot m_{\text{mol}} \cdot r_{\text{mol}}^2}{8\pi \cdot \epsilon_0 \cdot r_{n_{\text{orb}}=1,\text{H}} \cdot \hbar^2} \ . \tag{8.55}$$

With $r_{\text{mol}} \approx r_{\text{orb},\text{H},1}$ and using Eq. (8.54),

$$\frac{|E_{n_{\text{orb}}=1,\text{H}}|}{E_{n_{\text{rot}}=1,\text{H}}} \approx \frac{e^2 \cdot m_{\text{mol}} \cdot r_{\text{mol}}}{8\pi \cdot \epsilon_0 \cdot \hbar^2} = \frac{e^2 \cdot m_{\text{mol}}}{8\pi \cdot \epsilon_0 \cdot \hbar^2} \cdot \frac{\epsilon_0 \cdot h^2}{\pi \cdot m_{\text{ele}} \cdot e^2} \ , \tag{8.56}$$

and yields

$$\frac{|E_{n_{\text{orb}}=1,\text{H}}|}{E_{n_{\text{rot}}=1,\text{H}}} \approx \frac{m_{\text{mol}}}{m_{\text{ele}}} \cdot \frac{1}{8\pi^2} \cdot (2\pi)^2 = \frac{m_{\text{prot}}}{m_{\text{ele}}} \ , \tag{8.57}$$

where m_{prot} is the proton mass of the hydrogen atom. We know that $m_{\text{prot}}/m_{\text{ele}} \approx 1836$, and thus,

$$\frac{|E_{n_{\text{orb}}=1,\text{H}}|}{E_{n_{\text{rot}}=1,\text{H}}} \approx 1836 \ . \tag{8.58}$$

For the O_2 molecule, we would obtain $32 \times 1836 = 58\,752$. It is clear that the molecular rotational energy is significantly smaller than the typical energy associated with an electron orbit.

8.4.3
Vibrational Spectra of Diatomic Molecules

Let us consider a diatomic molecule for which the two atoms are allowed to vibrate along the axis connecting the atoms. To a good degree of approximation, we can assume the atoms are bound elastically, that is, they prescribe a harmonic oscillation around the position of rest, x_0,

$$x(t) = x_0 \cdot \cos(\omega_{\text{c},n_{\text{vib}}} \cdot t) \ , \tag{8.59}$$

with

$$\omega_{\text{c},n_{\text{vib}}} = 2\pi \cdot \nu_{n_{\text{vib}}} = \frac{2\pi}{T_{\text{p},n_{\text{vib}}}} \ , \tag{8.60}$$

where $T_{\text{p},n_{\text{vib}}}$ is the period of the harmonic vibrating oscillation. Considering Schrödinger's equation for a harmonic oscillator (Zdunkowski et al., 2007), we find the quantized energy levels of vibration to be

$$E_{n_{\text{vib}}} = \left(n_{\text{vib}} + \frac{1}{2} \right) \cdot h \cdot \nu'_{\text{vib},i} \ ;$$

$$n_{\text{vib}} = 0, 1, 2, \ldots \ ; \quad i = 1, 2, 3, \ldots \ , \tag{8.61}$$

where n_{vib} is the vibrational quantum number and $\nu'_{\text{vib},i}$ is the normal mode frequency or the resonant frequency of the harmonic oscillator. Water vapor (H_2O)

has three resonant frequencies ($i = 1, 2, 3$) given here as wavenumbers: $\tilde{\nu}'_{vib,1} = 3657.1 \, cm^{-1}$; $\tilde{\nu}'_{vib,2} = 1594.8 \, cm^{-1}$; and, $\tilde{\nu}'_{vib,3} = 3755.9 \, cm^{-1}$. CO_2 has one resonance frequency at $\tilde{\nu}'_{vib,1} = 2143.3 \, cm^{-1}$. N_2 has neither an electric nor a magnetic dipole moment, and, therefore, no vibrational absorption or emission lines. O_2 has no electric dipole moment, but a permanent magnetic dipole moment, therefore, vibrational absorption or emission lines occur at 60 and 118 GHz.

The quantized line spectra associated with the vibrational energy are given by

$$\nu_{n_{vib},m_{vib}} = \frac{\Delta E_{n_{vib},m_{vib}}}{h} = \frac{E_{m_{vib}} - E_{n_{vib}}}{h} \ ;$$

$$n_{vib} = 0, 1, 2, \dots \ ; \quad m_{vib} = 1, 2, 3, \dots \tag{8.62}$$

The selection rule names the quantized frequencies which are actually realizable. For vibrational spectra, the possible transitions are obtained from the following selection rule, namely,

$$m_{vib} - n_{vib} = \Delta n_{vib} . \tag{8.63}$$

Δn_{vib} represents an integer number of the difference between vibrational quantum numbers. Transitions with $\Delta n_{vib} = \pm 1$ usually correspond to the strongest vibrational bands, referred to as the fundamentals, whereas the overtone bands occur with $\Delta n_{vib} = \pm 2, \pm 3, \dots$ (Goody and Yung, 1989; Thomas and Stamnes, 1999). From Eq. (8.61), we obtain

$$\nu_{n_{vib},m_{vib}} = \frac{E_{m_{vib}} - E_{n_{vib}}}{h}$$

$$= \left(m_{vib} + \frac{1}{2} \right) \cdot \nu'_{vib,i} - \left(n_{vib} + \frac{1}{2} \right) \cdot \nu'_{vib,i} . \tag{8.64}$$

For ($m_{vib} - n_{vib} = \Delta n_{vib}$), we obtain ($m_{vib} = n_{vib} + \Delta n_{vib}$), that is,

$$\nu_{n_{vib},i} = \left(n_{vib} + \Delta n_{vib} + \frac{1}{2} \right) \cdot \nu'_{vib,i} - \left(n_{vib} + \frac{1}{2} \right) \cdot \nu'_{vib,i}$$

$$= \Delta n_{vib} \cdot \nu'_{vib,i} . \tag{8.65}$$

8.4.4
Combined Molecular Vibration-Rotation Spectra

Let us assume a mean thermal energy of ($k_B \cdot T$) $\approx 2.5 \times 10^{-2}$ eV corresponding to $T = 300$ K. The typical transitions between vibrational states are $\Delta E_{n_{vib},n_{vib}+1} \approx (5-100) \times 10^{-2}$ eV. For a mean thermal energy of ($k_B \cdot T$) $\approx 2.5 \times 10^{-2}$ eV, an extremely low number $N_{mol,n}$ of all possible vibrational energy levels are populated, and we have sparsely arranged vibrational frequencies.

In contrast, the rotational energy levels are in the range of $\Delta E_{n_{rot},n_{rot}+1} \approx (10^{-3})$ eV. Sufficient energy levels exist at $T = 300$ K ($\approx 2.5 \times 10^{-2}$ eV), which can contribute to the absorption and emission, to have many densely arranged rotational frequencies.

The number of energy levels $N_{mol,n}$ available for both vibrational and rotational energy level changes can be estimated as

$$\left(\frac{N_{mol,n+1}}{N_{mol,n}} \right)_{vib} \ll 1 \; ; \qquad \left(\frac{N_{mol,n+1}}{N_{mol,n}} \right)_{rot} \approx 1 \, , \tag{8.66}$$

with n the respective quantum numbers for rotation (n_{rot}) and vibration (n_{vib}). However, a strong interaction between vibrations and rotations exists, and for each vibrational transition $\Delta n_{vib} = \pm 1$, there is a corresponding rotational transition $\Delta n_{rot} = \pm 1$. From the densely arranged rotational states ($\Delta E_{n_{rot},n_{rot}+1} \approx 10^{-3}$ eV), each vibrational state is further split into a large number of rotational transitions.

Considering the rotation and vibration contributions to the energy of a molecule, a simple additive combination of both effects allows determination of the vibration-rotation spectrum

$$\begin{aligned} E_{n_{vib},n_{rot}} &= E_{n_{vib}} + E_{n_{rot}} \\ &= \left(n_{vib} + \frac{1}{2} \right) \cdot h \cdot \nu'_{vib,i} + B \cdot h \cdot n_{rot} \cdot (n_{rot} + 1) \, , \end{aligned} \tag{8.67}$$

where B represents the rotational constant, introduced in Eq. (8.44). The moment of inertia is different at different vibrational levels. For the ground vibrational state of carbon monoxide, the rotational constant is (Bohren and Clothiaux, 2006)

$$B = 1.9563 \, \text{cm}^{-1} \, . \tag{8.68}$$

Equation (8.67) only represents a simplified treatment; in reality, several additional terms must be considered. Electronic transitions cause a change in the molecular bonds, and the molecule's moment of inertia changes between the initial and final electronic state. Second-order interaction terms for an anharmonic oscillator lead to correction terms, which are quadratic both in ($n_{vib} + 1/2$) and $n_{rot} \cdot (n_{rot} + 1)$. We will not consider such complications. For further details, the reader may consult a textbook on molecular spectroscopy, for example, Penner (1959). From Eq. (8.67), we get

$$\nu_{n_{vib},n_{rot}} = \frac{\Delta E_{m_{vib},m_{vib}} - \Delta E_{n_{rot},n_{rot}}}{h} \, . \tag{8.69}$$

Three types of combined rotational-vibrational absorption and emission lines for atmospheric molecules exist; the R-branch, the Q-branch, and the P-branch, and are explained in detail in the following discussion.

R-branch For $\Delta n_{rot} = \Delta n_{vib} = +1$ (spectral neighbors), we obtain (for $n_{vib} \longrightarrow n_{vib} + 1$ and $n_{rot} \longrightarrow n_{rot} + 1$) the R-branch of the spectrum with the frequencies

$$\begin{aligned} \nu_{n_{vib},n_{rot}} &= \frac{E_{m_{vib},m_{rot}} - E_{n_{vib},n_{rot}}}{h} \\ &= \frac{E_{n_{vib}+1,n_{rot}+1} - E_{n_{vib},n_{rot}}}{h} \\ &= \frac{E_{n_{vib}+1} + E_{n_{rot}+1} - (E_{n_{vib}} + E_{n_{rot}})}{h} \, . \end{aligned} \tag{8.70}$$

We apply Eq. (8.61),

$$E_{n_{\mathrm{vib}}} = \left(n_{\mathrm{vib}} + \frac{1}{2}\right) \cdot h \cdot v'_{\mathrm{vib},i} \, , \tag{8.71}$$

and Eq. (8.46),

$$E_{n_{\mathrm{rot}}} = h \cdot B \cdot n_{\mathrm{rot}} \cdot (n_{\mathrm{rot}} + 1) \, , \tag{8.72}$$

to obtain

$$
\begin{aligned}
v_{n_{\mathrm{vib}}, n_{\mathrm{rot}}} = {} & \left(n_{\mathrm{vib}} + 1 + \frac{1}{2}\right) \cdot v'_{\mathrm{vib},i} \\
& + B \cdot (n_{\mathrm{rot}} + 1) \cdot (n_{\mathrm{rot}} + 2) \\
& - \left[\left(n_{\mathrm{vib}} + \frac{1}{2}\right) \cdot v'_{\mathrm{vib},i} + B \cdot n_{\mathrm{rot}} \cdot (n_{\mathrm{rot}} + 1)\right],
\end{aligned}
\tag{8.73}
$$

and as a result we get

$$
\begin{aligned}
v_{n_{\mathrm{vib}}, n_{\mathrm{rot}}} = {} & n_{\mathrm{vib}} \cdot v'_{\mathrm{vib},i} + \frac{3}{2} \cdot v'_{\mathrm{vib},i} - n_{\mathrm{vib}} \cdot v'_{\mathrm{vib},i} - \frac{1}{2} \cdot v'_{\mathrm{vib},i} \\
& + (n_{\mathrm{rot}} + 1) \cdot (B \cdot n_{\mathrm{rot}} + 2B - B \cdot n_{\mathrm{rot}}) \\
& \approx v'_{\mathrm{vib},i} + 2 \cdot B \cdot (n_{\mathrm{rot}} + 1) \, ; \\
& n_{\mathrm{rot}} = 1, 2, 3 \dots \, ,
\end{aligned}
\tag{8.74}
$$

where we assume that the quantity B is the same for different vibrational and rotational energy levels.

P-branch   For $\Delta n_{\mathrm{vib}} = 1$; $\Delta n_{\mathrm{rot}} = -1$ (spectral neighbors), we have (for $n_{\mathrm{vib}} \longrightarrow n_{\mathrm{vib}} + 1$ and $n_{\mathrm{rot}} \longrightarrow n_{\mathrm{rot}} - 1$) the P-branch of the spectrum with the frequencies

$$
\begin{aligned}
v_{n_{\mathrm{vib}}, n_{\mathrm{rot}}} = {} & \left(n_{\mathrm{vib}} + 1 + \frac{1}{2}\right) \cdot v'_{\mathrm{vib},i} \\
& + B \cdot (n_{\mathrm{rot}} - 1) \cdot n_{\mathrm{rot}} \\
& - \left[\left(n_{\mathrm{vib}} + \frac{1}{2}\right) \cdot v'_{\mathrm{vib},i} + B \cdot n_{\mathrm{rot}} \cdot (n_{\mathrm{rot}} + 1)\right],
\end{aligned}
\tag{8.75}
$$

and as a result we get

$$
\begin{aligned}
v_{n_{\mathrm{vib}}, n_{\mathrm{rot}}} = {} & n_{\mathrm{vib}} \cdot v'_{\mathrm{vib},i} + \frac{3}{2} \cdot v'_{\mathrm{vib},i} - n_{\mathrm{vib}} \cdot v'_{\mathrm{vib},i} - \frac{1}{2} \cdot v'_{\mathrm{vib},i} \\
& + B \cdot n_{\mathrm{rot}}^2 - B \cdot n_{\mathrm{rot}} - B \cdot n_{\mathrm{rot}}^2 - B \cdot n_{\mathrm{rot}} \\
& \approx v'_{\mathrm{vib},i} - 2 \cdot B \cdot n_{\mathrm{rot}} \, ; \\
& n_{\mathrm{rot}} = 1, 2, 3, \dots \, .
\end{aligned}
\tag{8.76}
$$

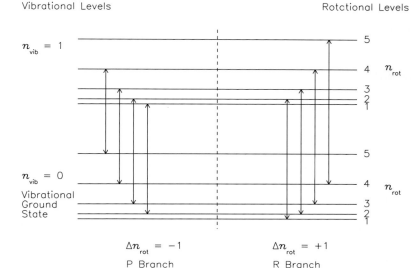

Figure 8.8 Transitions for the vibration-rotation spectrum of a diatomic molecule. Adapted from Zdunkowski et al. (2007), after Houghton and Smith (1966). Courtesy of A. Bott and T. Trautman.

Q-branch In the simple cases discussed above, we have not permitted electron orbital transitions, and if these are allowed, the Q-branch occurs for the particular case $\Delta n_{rot} = 0$. Important atmospheric gases having Q-branches in the IR spectral region are CO_2 and N_2O.

Polyatomic molecules, such as, CO_2 and CH_4, allow more complex fundamental vibrational modes compared to the diatomic molecules. For rotating molecules, the three moments of inertia with respect to their major axes of rotation must be determined from four types of molecules. The asymmetric top molecular structures have three different moments of inertia (e.g., H_2O, O_3). The symmetric top molecular structures have two different moments of inertia (e.g., NH_3, CH_3Cl, $CFCl_3$). The spherical symmetric top molecular structures have three equal moments of inertia (e.g., CH_4). The linear molecules have two equal moments of inertia and a third moment which is practically negligible (e.g., CO_2, CO, O_2, and N_2O).

Figure 8.8 displays the transitions of the vibration-rotation spectrum of a diatomic molecule. The combined vibration-rotation energy levels are shown. Furthermore, the possible transitions for the P- and R-branches between the vibrational ground state ($n_{vib} = 0$) and the first vibrational state ($n_{vib} = 1$) are illustrated.

8.5
Examples of Atmospheric Gas Spectra

8.5.1
Three General Types of Spectra

The three general types of molecular absorption and emission spectra from the overlap of the different absorption and emission lines are individual absorption or emission spectral lines, band spectra, and continuum absorption spectra. Some examples of line spectra are given in Section 8.5.2. A spectral band consists of a series of individual spectral absorption or emission lines which overlap and form bands. Continuum absorption spectra extend over a broad wavelength region and are smooth functions of wavelength. For example, the water vapor continuum absorption spectrum spreads over a wavenumber region between 200–1200 cm^{-1} and remains largely unexplained. It has been suggested that the continuum absorption results from the accumulated absorption of the far wings of distant lines in the FIR. The continuum absorption may also be caused by collision broadening of the absorption/emission lines between water vapor molecules (self-broadening) or between water vapor and nonabsorbing N_2 molecules (foreign broadening).

8.5.2
Infrared (IR) – Combined Vibrational and Rotational Transitions

The most important absorbing and emitting gases in the infrared (IR) spectral region are H_2O, CO_2, O_3, CH_4, N_2O, CO, and CFCs. Due to the particular molecular structure, each of these gases has its own specific absorption and emission spectrum. Some examples follow.

Combined vibration-rotational lines of N_2O: 1220–1330 cm^{-1}   The central region of the strongest vibration-rotational absorption/emission lines of N_2O ($\nu'_{vib,1}$, P, R) are depicted in Figure 8.9. Here, the P- and R-branches can be seen as the regular groups of lines to the left and right of the band center at 1285.6 cm^{-1} (7.78 µm). A second, weaker band – an upper state band, see Goody and Yung (1989) – with a slightly different band center is superimposed.

Purely rotational (i.e., Equidistant) lines of CO: 2096–2226 cm^{-1}   Like nitrous oxide (N_2O), a carbon monoxide (CO) molecule has a linear structure. The band center of the purely rotational spectrum of CO near 4.67 µm shows a very simple structure, see Figure 8.10. As is typical for purely rotational spectra, the lines appear equally spaced. Moreover, the lines are more widely spaced for the CO molecule than for N_2O, and the spacing can be attributed to the fact that CO has a smaller moment of inertia than N_2O. See Section 8.4.1 for an explanation of the mutual distance of the rotational lines and the relation to the molecule's moment of inertia.

Other examples: CH_4, CO_2, H_2O, O_2   From the other line spectra depicted in Figure 8.11, the absorption spectra can be seen for the following gases: CH_4, CO_2, H_2O, and O_2.

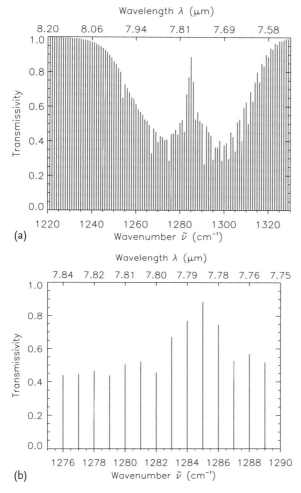

Figure 8.9 Spectrum of N_2O near 7.78 µm. The transmissivity was computed with MOD-TRAN4 (Version 3 Revision 1, MOD4v3r1) for a NO_2 layer between 0 and 100 km for US standard atmospheric N_2O concentration (Anderson et al., 1986) and for an observational zenith angle of 30°. (a) and (b) cover different wavelength-wavenumber regions.

8.5.3
Near Infrared (NIR) to Visible (VIS)

Table 8.2 lists the most important absorption bands for several atmospheric gases in the NIR and the VIS spectral regions. The main absorbing atmospheric trace gases are H_2O, CO_2, O_3, and O_2, whereas, N_2O, CH_4, CO, and NO_2 are of lesser importance.

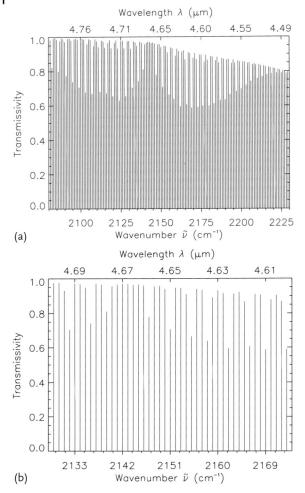

Figure 8.10 Spectrum of atmospheric CO near 4.67 μm. The transmissivity was computed with MODTRAN4 (Version 3 Revision 1, MOD4v3r1), for a CO layer between 0 and 100 km for US standard atmospheric CO concentration (Anderson et al., 1986) and for an observational zenith angle of 30°. (a) and (b) cover different wavelength-wavenumber regions.

8.5.4
Visible (VIS) to Ultraviolet (UV) – Electron Orbital Transitions

Absorption of photons with UV to VIS wavelengths is caused by several atmospheric gases. N_2 absorbs at wavelengths less than 0.1 μm and in the Vegard–Kaplan band (0.23–0.34 μm). O_2 and O_3 have several absorption bands which are outlined later in more detail. Water vapor (H_2O) has major absorption lines in the VIS and UV range for wavelengths less than 0.21 μm and between 0.6–0.72 μm. Other trace gases also absorb photons at VIS and UV wavelengths, such as, H_2O_2

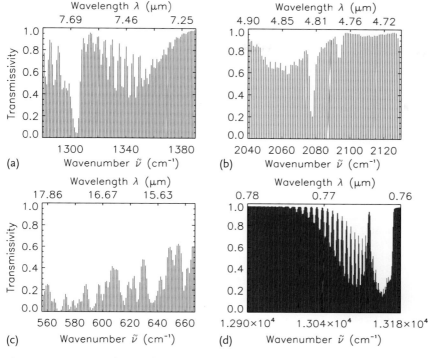

Figure 8.11 Spectrum of atmospheric gases: (a) CH_4, (b) CO_2, (c) H_2O, and (d) O_2. The transmissivity was computed with MOD-TRAN4 (Version 3 Revision 1, MOD4v3r1), for a respective gas layer between 0 and 100 km for US standard atmospheric gas concentration (Anderson et al., 1986), and for an observational zenith angle of 30°.

(hydrogen peroxide) ($< 0.35\,\mu m$), NO_2 (nitrogen oxide) ($< 0.6\,\mu m$), N_2O (nitrous oxide) ($< 0.24\,\mu m$), NO_3 (nitrate radical) (0.41–$0.67\,\mu m$), HONO (nitrous acid) ($< 0.4\,\mu m$), HNO_3 (nitric acid) ($< 0.33\,\mu m$), CH_3Br (methyl bromide) ($< 0.26\,\mu m$), $CFCl_3$ (CFC–11) ($< 0.23\,\mu m$), and HCHO (formaldehyde) (0.25–$0.36\,\mu m$).

The absorption of UV and VIS radiation is primarily caused by molecular oxygen (O_2) and ozone (O_3). Figure 8.12 depicts the spectrum of the absorption cross-sections of O_3 and O_2. We can distinguish lines and bands, see Table 8.3.

Note, the electron orbital transitions are responsible for the bands of O_3 and O_2 at wavelengths shorter than 1 µm. Although the number of ozone molecules in the Earth's atmosphere is small, the molecules have large absorption cross-sections in the UV, and very little UV radiation can reach the Earth's surface at wavelengths less than 305–310 nm.

The UV photons have sufficiently high energy to change orbital levels of electrons circling the nucleus. The associated electron orbital transitions are responsible for the absorption of solar radiation in the UV to visible spectrum. Photons of high energy are emitted (released) if an electron lowers its orbital level. Electron orbital transitions from high-energy photons may cause various photochemical and photophysical reactions.

Table 8.2 Atmospheric gases absorbing solar radiation in the NIR and VIS spectral region. Data taken from Liou (2002, Table 3.3).

Gas	Center $\tilde{v}$ (cm^{-1})/λ (μm)	Band interval $\tilde{v}$ (cm^{-1})
H_2O	3703/2.7	2500–4500
	5348/1.87	4800–6200
	7246/1.38	6400–7600
	9090/1.1	8200–9400
	10 638/0.94	10 100–11 300
	12 195/0.82	11 700–12 700
	13 888/0.72	13 400–14 600
	Visible	15 000–22 600
CO_2	2326/4.3	2000–4000
	3703/2.7	3400–3850
	5000/2.0	4700–5200
	6250/1.6	6100–6450
	7143/1.4	6850–7000
O_3	2110/4.74	2000–2300
	3030/3.3	3000–3100
	Visible	10 600–22 600
O_2	6329/1.58	6300–6350
	7874/1.27	7700–8050
	9433/1.06	9350–9400
	13 158 &0.76	12 850–13 200
	14 493/0.69	14 300–14 600
	15 873/0.63	14 750–15 900
N_2O	2222/4.5	2100–2300
	2463/4.06	2100–2800
	3484/2.87	3300–3500
CH_4	3030/3.3	2500–3200
	4420/2.26	4000–4600
	6005/1.66	5850–6100
CO	2141/4.67	2000–2300
	4273/2.34	4150–4350
NO_2	Visible	14 400–50 000

8.6
Approximations of Absorption/Emission Line Shapes

From experimental investigations, we know that the individual absorption and emission lines (orbital, rotational, vibrational) are broadened over a narrow frequency region. The broadening of spectral lines results from three processes; natural, collision, and Doppler broadening. At normal pressures and temperatures, the natural broadening can be neglected, and the collision broadening is normally more important than the Doppler broadening in the lower atmosphere.

Table 8.3 Absorption of UV and VIS radiation is primarily caused by molecular oxygen (O_2) and ozone (O_3). Data taken from Liou (2002, Table 3.2).

Gas	Name	Absorption wavelengths (μm)
Oxygen (O_2)	Runge-bands	< 0.1
	Schuman–Runge continuum	0.10–0.175
	Schuman–Runge bands	0.175–0.20
	Herzberg continuum	0.20–0.242
	O_2-A band	0.76
	O_2-B band	0.69
Ozone (O_3)	Hartley bands	0.20–0.31
	Huggins bands	0.31–0.40
	Chappuis bands	0.40–0.85

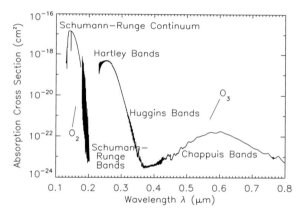

Figure 8.12 Wavelength dependence of the absorption cross sections of O_2 and O_3. 1 Å corresponds to 0.1 nm (1 nm is equivalent to 10 Å); the horizontal scale covers up to 0.8 μm. Data for O_3 from Voigt et al. (2001); O_2 data for the Schumann–Runge bands at 300 K from Yoshino et al. (1992); O_2 data for the Schumann–Runge continuum at 295 K from Yoshino et al. (2005).

8.6.1
Lorentz Line Shape of the Absorption Coefficient – Collision Broadening

The Lorentz line shape describes the line broadening that is fundamentally caused by collisions of molecules. We omit the derivation and give the absorption coefficient $b_{abs,L}(\tilde{\nu})$ (subscript "L" for Lorentz) associated with the collision broadening as a function of wavenumber $\tilde{\nu}$ by

$$b_{abs,L}(\tilde{\nu}) = \frac{1}{\pi} \cdot \frac{S_L \cdot \alpha_L}{\alpha_L^2 + (\tilde{\nu} - \tilde{\nu}_0)^2} \ , \tag{8.77}$$

where S_L indicates the absorption/emission line intensity (in units of cm^{-2}) given by

$$S_L = \int_{-\infty}^{\infty} b_{\text{abs,L}}(\tilde{\nu}) \, d\tilde{\nu} \; . \tag{8.78}$$

It can be shown (Zdunkowski et al., 2007) that the half-width α_L in units of cm^{-1} is a function of temperature T and pressure p,

$$\alpha_L(p, T) = \alpha_{L,0} \cdot \frac{p}{p_0} \cdot \sqrt{\frac{T_0}{T}} \; , \tag{8.79}$$

where $\alpha_{L,0}$ is the half-width at reference pressure and temperature, p_0 and T_0, given by

$$\alpha_{L,0} = \alpha_L(p_0, T_0) = \text{const} \cdot \frac{p_0}{\sqrt{T_0}} \; , \tag{8.80}$$

and provided by HITRAN (Rothman et al., 1992), also see Goody (1964a).

8.6.2
Thermal Doppler Line Shape

At low pressures, the Doppler broadening becomes important which is due to the Doppler effect associated with molecular motion. The Doppler effect refers to a phenomenon that the frequency of radiation emitted by a moving object is increased (decreased) if the object is moving towards (away from) the observer. Without derivation, we give the absorption coefficient of the Doppler line as

$$b_{\text{abs,D}}(\tilde{\nu}) = S_D \cdot \frac{c}{\tilde{\nu}_0} \cdot \sqrt{\frac{m_{\text{mol}}}{2\pi \cdot k_B \cdot T}} \cdot \exp\left[-\frac{m_{\text{mol}} \cdot (\tilde{\nu} - \tilde{\nu}_0)^2 \cdot c^2}{2 \, k_B \cdot T \cdot \tilde{\nu}_0^2}\right], \tag{8.81}$$

where the subscript "D" has been used to make a distinction between the absorption coefficients of the Lorentz and Doppler lines. Let us determine the particular frequency $\tilde{\nu} = \tilde{\nu}_0 \pm \alpha_D$ for which

$$b_{\text{abs,D}}(\tilde{\nu}) = \frac{1}{2} \cdot b_{\text{abs,D,0}} \; , \tag{8.82}$$

where

$$b_{\text{abs,D,0}} = S_D \cdot \frac{c}{\tilde{\nu}_0} \cdot \sqrt{\frac{m_{\text{mol}}}{2\pi \cdot k_B \cdot T}} \; . \tag{8.83}$$

This yields the half-width α_D of the Doppler line, which is only a function of temperature, that is,

$$\alpha_D = \sqrt{\ln 2} \cdot \frac{\tilde{\nu}_0}{c} \cdot \sqrt{\frac{2 \, k_B \cdot T}{m_{\text{mol}}}} \; . \tag{8.84}$$

8.6.3
Voigt Line Shape – Combined Collision and Doppler Broadening

Both absorption-line broadening effects (due to collisions and caused by the Doppler effect) are combined in the Voigt line shape. Before we determine this line shape, let us give the two relative shapes f_L and f_D (in units of cm) of the Lorentz and the Doppler absorption lines, namely,

$$
f_L(\tilde{\nu} - \tilde{\nu}_0) = \frac{1}{\pi} \cdot \frac{\alpha_L}{\alpha_L^2 + (\tilde{\nu} - \tilde{\nu}_0)^2} \, , \tag{8.85}
$$

$$
f_D(\tilde{\nu} - \tilde{\nu}_0) = \frac{1}{\alpha_D} \cdot \sqrt{\frac{\ln 2}{\pi}} \cdot \exp\left[-\frac{(\tilde{\nu} - \tilde{\nu}_0)^2 \cdot \ln 2}{\alpha_D^2} \right]. \tag{8.86}
$$

To include both the Lorentz and the Doppler effect, we have to convolve the two relative shapes, f_L and f_D, over all wavenumbers $\tilde{\nu}_0'$, that is,

$$
b_{abs,V}(\tilde{\nu} - \tilde{\nu}_0) = S_V \int_{-\infty}^{\infty} f_L(\tilde{\nu} - \tilde{\nu}_0') \cdot f_D(\tilde{\nu}_0' - \tilde{\nu}_0) \, d\tilde{\nu}_0' \, . \tag{8.87}
$$

This verifies that the signal of the product of two spectra corresponds to the convolution of the individual signals. Thus, the preceding expression defines the relative shape of the Voigt line which can be formulated as

$$
f_V(\zeta) = \int_{-\infty}^{\infty} f_L(\zeta - x) \cdot f_D(x) \, dx \, , \tag{8.88}
$$

where $\zeta = \tilde{\nu} - \tilde{\nu}_0$, $x = \tilde{\nu}_0' - \tilde{\nu}_0$, and $\zeta - x = \tilde{\nu} - \tilde{\nu}_0'$. The preceding convolution integral reads

$$
f_V(\zeta) = \frac{\alpha_L}{\alpha_D} \cdot \sqrt{\frac{\ln 2}{\pi^3}} \cdot \frac{\ln 2}{\alpha_D^2}
$$
$$
\times \int_{-\infty}^{\infty} \frac{\exp\left[-\ln 2 \cdot (x/\alpha_D)^2 \right]}{(\sqrt{\ln 2} \cdot \alpha_L/\alpha_D)^2 + (\zeta \cdot \sqrt{\ln 2}/\alpha_D - x \cdot \sqrt{\ln 2}/\alpha_D)^2} \, dx \, . \tag{8.89}
$$

We introduce the quantities

$$
A = \sqrt{\ln 2} \cdot \frac{\alpha_L}{\alpha_D} \, , \tag{8.90}
$$

$$
\eta = \sqrt{\ln 2} \cdot \frac{\zeta}{\alpha_D} \, , \tag{8.91}
$$

$$
y = \sqrt{\ln 2} \cdot \frac{x}{\alpha_D} \, . \tag{8.92}
$$

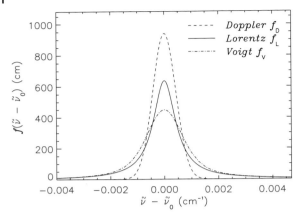

Figure 8.13 Comparison of relative line shapes: Lorentz line (solid) $\alpha_L = 5 \times 10^{-4}$ cm^{-1}; Doppler line (dashed) $\alpha_D = 5 \times 10^{-4}$ cm^{-1}; Voigt profile (dash-dot) as convolution of Lorentz and Doppler profiles with $\alpha_L = \alpha_D = 5 \times 10^{-4}$ cm^{-1}. Redrawn after Zdunkowski et al. (2007) with modification, see also Huang and Yung (2004). Courtesy of A. Bott and T. Trautmann.

In this way, the relative shape of the Voigt absorption line can be expressed in compact notation as

$$f_V(\eta) = \frac{A}{\alpha_D} \cdot \sqrt{\frac{\ln 2}{\pi^3}} \int\limits_{-\infty}^{\infty} e^{-y^2} \cdot \frac{1}{A^2 + (\eta - y)^2} \, dy \ . \tag{8.93}$$

The absorption coefficient for the Voigt profile is given by

$$b_{abs,V}(\eta) = S \cdot f_V(\eta) \ . \tag{8.94}$$

Figure 8.13 compares the three types of relative line shapes. For $\alpha_L = \alpha_D$, the collision line broadening (Lorenz line shape, solid line) is stronger in the line wings as compared to the Doppler line shape.

8.7
Spectral Transmissivity and Absorptivity

Spectral transmissivity is an important quantity in remote sensing applications and radiative budget studies. For simplicity, we consider one species of absorbing gas along a homogeneous atmospheric path whose path length (also called absorbing mass) u is given by

$$u(z_1, z_2) = \int\limits_{z_1}^{z_2} \varrho_{abs}(z') \, dz' \ , \tag{8.95}$$

where ϱ_{abs} represents the mass density of the absorbing gas in units of $kg\,m^{-3}$. The absorption path length u has the units of $kg\,m^{-2}$. The dimensionless vertical transmissivity $\mathcal{T}(u)$ in the wavenumber interval $\Delta\tilde{\nu}$ is defined as

$$\mathcal{T}(u) = \frac{1}{\Delta\tilde{\nu}} \int\limits_{\Delta\tilde{\nu}} \mathcal{T}_{\tilde{\nu}}\,d\tilde{\nu} = \frac{1}{\Delta\tilde{\nu}} \int\limits_{\Delta\tilde{\nu}} \exp[-k_{abs}(\tilde{\nu})\cdot u]\,d\tilde{\nu}\,, \qquad (8.96)$$

where k_{abs} in units of $m^2\,kg^{-1}$ is the mass absorption coefficient. k_{abs} is related to the volume absorption coefficient b_{abs} in units of m^{-1} via the density of the absorbing medium ϱ_{abs} in the form

$$k_{abs}(\tilde{\nu}, z) = \frac{b_{abs}(\tilde{\nu}, z)}{\varrho_{abs}(z)}\,. \qquad (8.97)$$

The corresponding spectral absorptivity is

$$\mathcal{A}(u) = 1 - \mathcal{T}(u) = \frac{1}{\Delta\tilde{\nu}} \int\limits_{\Delta\tilde{\nu}} \left\{1 - \exp\left[-k_{abs}(\tilde{\nu})\cdot u\right]\right\}\,d\tilde{\nu}\,. \qquad (8.98)$$

8.7.1
Weak-Line and Strong-Line Approximations

In the case of a single absorption line, we introduce a quantity referred to as the equivalent width, $W(u)$, and defined as

$$W(u) = \mathcal{A}(u)\cdot\Delta\tilde{\nu}\,. \qquad (8.99)$$

The equivalent width $W(u)$ is the width of an infinitely strong hypothetical line, which gives the same amount of absorption as that of the actual line of interest.

Two cases, the weak-line and strong-line conditions, provide interesting insight into the dependence of the equivalent width on the amount of absorber, or more specifically, the path length. The weak-line condition indicates a special case when $k_{abs}(\tilde{\nu})\cdot u \ll 1$. Thus, using Taylor expansion and Eq. (8.97), we obtain

$$\mathcal{A}_{weak}(u) \approx \frac{1}{\Delta\tilde{\nu}} \int\limits_{-\infty}^{\infty} \left[1 - 1 + k_{abs}(\tilde{\nu})\cdot u\right]\,d\tilde{\nu}$$

$$= \frac{1}{\Delta\tilde{\nu}} \int\limits_{-\infty}^{\infty} k_{abs}(\tilde{\nu})\cdot u\,d\tilde{\nu}$$

$$= \frac{1}{\Delta\tilde{\nu}} \int\limits_{-\infty}^{\infty} \frac{b_{abs}(\tilde{\nu})}{\varrho_{abs}}\cdot u\,d\tilde{\nu}\,.$$

From Eq. (8.94), we have

$$
\mathcal{A}_{\text{weak}}(u) \approx \frac{1}{\Delta \tilde{\nu} \cdot \varrho_{\text{abs}}} \int_{-\infty}^{\infty} S \cdot f(\tilde{\nu} - \tilde{\nu}_0) \cdot u \, d\tilde{\nu}
$$

$$
= \frac{S \cdot u}{\Delta \tilde{\nu} \cdot \varrho_{\text{abs}}}
$$

$$
= \frac{S_k \cdot u}{\Delta \tilde{\nu}} \; , \tag{8.100}
$$

where $S_k = S/\varrho_{\text{abs}}$ indicates the mass-specific line intensity. In Eq. (8.100), the normalization condition for the line profile $f(\tilde{\nu} - \tilde{\nu}_0)$ is implied. Accordingly, we have

$$
W_{\text{weak}}(u) = \mathcal{A}_{\text{weak}}(u) \cdot \Delta \tilde{\nu} \approx S_k \cdot u \; . \tag{8.101}
$$

Thus, under the weak-line approximation, the equivalent width is linearly proportional to the line intensity S_k and the absorption path length u and is independent of the detailed line profile (e.g., the line half-width).

To illustrate the strong-line approximation, we consider the Lorentz profile. Under the strong-line condition, the absorption is saturated at the line center. Thus, the contribution to transmissivity is primarily from the line wing regions, that is, $|\tilde{\nu} - \tilde{\nu}_0| \gg \alpha_{\text{L}}$. Note that in the line wing regions, the shape of an absorption line can be well approximated by the Lorentz profile. Thus, we have the expression, following Eqs. (8.98) and (8.77),

$$
W_{\text{L}}(u) = \mathcal{A}_{\text{L}}(u) \cdot \Delta \tilde{\nu}
$$

$$
= \int_{-\infty}^{\infty} \left\{ 1 - \exp\left[-\frac{b_{\text{abs,L}}(\tilde{\nu})}{\varrho_{\text{abs}}} \cdot u \right] \right\} d\tilde{\nu}.
$$

$$
= \int_{-\infty}^{\infty} \left\{ 1 - \exp\left[-\frac{S_{k,\text{L}} \cdot u}{\pi} \cdot \frac{\alpha_{\text{L}}}{\alpha_{\text{L}}^2 + (\tilde{\nu} - \tilde{\nu}_0)^2} \right] \right\} d\tilde{\nu} \; . \tag{8.102}
$$

Introducing the strong-line approximation $|\tilde{\nu} - \tilde{\nu}_0| \gg \alpha_{\text{L}}$ yields

$$
W_{\text{strong,L}}(u) = \int_{-\infty}^{\infty} \left\{ 1 - \exp\left[-\frac{S_{k,\text{L}} \cdot u}{\pi} \cdot \frac{\alpha_{\text{L}}}{(\tilde{\nu} - \tilde{\nu}_0)^2} \right] \right\} d\tilde{\nu}
$$

$$
= 2 \int_{0}^{\infty} \left\{ 1 - \exp\left[-\frac{S_{k,\text{L}} \cdot u}{\pi} \cdot \frac{\alpha_{\text{L}}}{(\tilde{\nu} - \tilde{\nu}_0)^2} \right] \right\} d\tilde{\nu} \; . \tag{8.103}
$$

Furthermore, we substitute

$$
\eta = \sqrt{\frac{S_{k,\text{L}} \cdot u \cdot \alpha_{\text{L}}}{\pi}} \cdot \frac{1}{(\tilde{\nu} - \tilde{\nu}_0)} \tag{8.104}
$$

into Eq. (8.103) and get

$$d\tilde{\nu} = -\sqrt{\frac{S_{k,L} \cdot u \cdot \alpha_L}{\pi}} \cdot \frac{d\eta}{\eta^2} .$$

(8.105)

Substituting Eq. (8.105) into Eq. (8.103) yields

$$W_{\text{strong,L}}(u) = 2\sqrt{\frac{S_{k,L} \cdot u \cdot \alpha_L}{\pi}} \int_0^\infty \left(1 - e^{-\eta^2}\right) \cdot \frac{1}{\eta^2} \, d\eta .$$

(8.106)

Integrating by parts in the preceding expression, we obtain

$$W_{\text{strong,L}}(u) = 4\sqrt{\frac{S_{k,L} \cdot u \cdot \alpha_L}{\pi}} \int_0^\infty e^{-\eta^2} \, d\eta .$$

(8.107)

To evaluate the integral in Eq. (8.107), let us consider

$$\gamma = \int_0^\infty e^{-x^2} dx = \int_0^\infty e^{-y^2} \, dy .$$

(8.108)

From Eq. (8.108), we have

$$\gamma^2 = \int_0^\infty e^{-(x^2+y^2)} \, dx \, dy .$$

(8.109)

If we use a polar coordinate system to conduct the integration in Eq. (8.109), we obtain

$$\gamma^2 = \int_0^{\pi/2} \int_0^\infty e^{-R^2} \cdot R \, dR \, d\varphi = \frac{\pi}{4} .$$

(8.110)

Thus, it follows that

$$\gamma = \frac{\sqrt{\pi}}{2} .$$

(8.111)

Consequently,

$$W_{\text{strong,L}}(u) = 2\sqrt{S_{k,L} \cdot u \cdot \alpha_L} .$$

(8.112)

The equivalent width $W_{\text{strong,L}}$ is proportional to the square root of the absorption path length u under the strong-line condition, assuming a Lorentz profile. Because of this feature, the strong-line condition is also often known as the square root regime. Note that a more rigorous analysis of the strong-line absorptivity based on the use of the Ladenburg and Reiche function can be found in Goody and Yung (1989) and Liou (2002).

8.7.2
Line-By-Line Method (LBLM)

The Line-By-Line Method (LBLM) provides a rigorous approach to accurately compute the spectral transmissivity. To illustrate the basic principle of this method, let us consider N species of absorbing gases. For each species, J_n absorption lines contribute to the spectrum of interest and the subscript n indicates the species. Thus, the spectral transmissivity can be formulated as

$$\mathcal{T} = \frac{1}{\Delta\tilde{\nu}} \int_{\Delta\tilde{\nu}} \mathcal{T}_{\tilde{\nu}} \, d\tilde{\nu}$$

$$= \frac{1}{\Delta\tilde{\nu}} \int_{\Delta\tilde{\nu}} \exp\left\{ -\sum_{n=1}^{N} u_n \cdot \left[-k_{\text{abs,c},n}(\tilde{\nu}) + \sum_{j=1}^{J_n} S_{k,n,j} \cdot f_{n,j}(\tilde{\nu} - \tilde{\nu}_{0,nj}) \right] \right\} d\tilde{\nu} ,$$

$$(8.113)$$

where $k_{\text{abs,c},n}$ is the continuum mass absorption coefficient in units of $\text{m}^2\,\text{kg}^{-1}$, u_n indicates the absorption path length of the nth species in units of $\text{kg}\,\text{m}^{-2}$, $S_{k,n,j}$ indicates the mass-specific line intensity, and $f_{n,j}$ denotes the line profile. The LBLM uses high-spectral resolution to resolve the rapid variation of absorption coefficients versus wavenumber. In practice, the LBLM may require a substantial amount of computational effort.

8.7.3
Band Models

Many applications practically prohibit conducting LBLM calculations. To mitigate the computational burden, various band models have been developed. Although band models seem to be largely obsolete nowadays with the advent of fast computers and the increasing popularity of the correlated k-distribution method (see Section 8.7.5), band models have been discussed in essentially all existing radiative transfer texts from historical and pedagogical perspectives. Here, we briefly describe a well-known band model, the Goody statistical model (Goody, 1952).

Upon scrutinizing the water-vapor rational spectrum reported by Randall et al. (1937), Goody (1952) found "the only common feature of these ranges is a nearly random distribution of the line intensities and positions" and suggested the use of the Poisson function to specify the probability distribution function (PDF) of the line intensities. The function is given by

$$P(S) = \frac{1}{S_{\text{m}}} \cdot \exp\left(-\frac{S}{S_{\text{m}}} \right). \tag{8.114}$$

Note that the PDF in Eq. (8.114) is normalized such that

$$\int_0^\infty P(S) \, dS = \int_0^\infty \frac{1}{S_{\text{m}}} \cdot \exp\left(-\frac{S}{S_{\text{m}}} \right) dS = 1 . \tag{8.115}$$

To understand the physical meaning of the term S_m in Eq. (8.114), we consider the mean line intensity associated with the PDF given by

$$\overline{S} = \int_0^\infty S \cdot P(S)\, dS$$

$$= \int_0^\infty S \cdot \frac{1}{S_m} \cdot \exp\left(-\frac{S}{S_m}\right)\, dS$$

$$= S_m \cdot \int_0^\infty \frac{1}{S_m} \cdot \exp\left(-\frac{S}{S_m}\right)\, dS$$

$$= S_m\,, \tag{8.116}$$

and S_m is the mean line intensity.

Let us consider the spectral transmissivity associated with a spectral band comprised of n absorption lines with a mean line spacing of $\delta\tilde{\nu}$. If we postulate that all the line positions are equally probable, we can formulate spectral transmissivity in the form of

$$\mathcal{T} \approx \left\{ 1 - \frac{1}{n \cdot \delta\tilde{\nu}} \int_{-\infty}^{\infty} \int_0^\infty P(S) \cdot \left(1 - \exp\left[-u \cdot S \cdot f(\tilde{\nu}, \alpha_L)\right]\right)\, dS\, d\tilde{\nu} \right\}^n\,, \tag{8.117}$$

where we assume n to be large. Under this condition, the following approximation is valid (Liou, 2002),

$$\left(1 - \frac{t}{n}\right)^n \approx e^{-t}\,, \quad \text{for a large } n\,. \tag{8.118}$$

Thus, applying the relation in Eq. (8.118) to Eq. (8.117) yields

$$\mathcal{T} \approx \exp\left\{ -\frac{1}{n \cdot \delta\tilde{\nu}} \int_{-\infty}^{\infty} \int_0^\infty P(S) \cdot \left(1 - \exp\left[-u \cdot S \cdot f(\tilde{\nu}, \alpha_L)\right]\right)\, dS\, d\tilde{\nu} \right\}\,. \tag{8.119}$$

If the Lorentz profile is assumed for the line shape in Eq. (8.119), the spectral transmissivity is given by

$$\mathcal{T}_L \approx \exp\left[-\frac{1}{n \cdot \delta\tilde{\nu}} \cdot \frac{\sqrt{\pi} \cdot u \cdot S_m \cdot \alpha_L}{\left(\pi \cdot \alpha_L^2 + u \cdot S_m \cdot \alpha_L\right)^{1/2}} \right]\,. \tag{8.120}$$

Goody (1952) compared the transmissivities computed from the Goody model with those reported by Cowling (1950) and observed remarkably good agreement between the two results. Cowling (1950) used a more rigorous method with spectral resolutions of $\delta\tilde{\nu} = 2.00$, 1.48, 2.27, and 3.50 cm^{-1}, and with various values of the mean line intensity and semiwidth.

Similar statistical models to describe the distribution of line intensities, such as, the Godson inverse power law (Godson, 1955) and the Malkmus model (Malkmus, 1967), will not be described here. For more details, see Zdunkowski et al. (2007) and the references cited therein.

8.7.4
Scaling Techniques for Inhomogeneous Path

Band models were developed for homogeneous atmospheres. To account for an inhomogeneous atmosphere, various scaling techniques, such as, the *Curtis–Godson technique* (Curtis, 1952; Godson, 1953) and the three-parameter scaling approaches (Fu and Liou, 1992a; Goody, 1964b), were introduced. To illustrate the basic principle of the scaling approaches, we present a one-parameter method in the manner of Chou and Arking (1980).

Consider the transmissivity associated with water vapor. The half-widths of the absorption are usually smaller than the mean line spacing that lead to the spectral dominance of the line wings, that is, in the region with $|\tilde{\nu} - \tilde{\nu}_0| \gg \alpha_L$. Furthermore, the absorption at line centers is usually saturated because of an abundance of water vapor; thus, an inaccurate treatment of the absorption coefficient at line centers has little impact on the resultant transmissivity. For these reasons, with respect to a reference pressure p_r and a reference level of temperature T_r, the absorption coefficient can be written as

$$b_{\text{abs}}(\tilde{\nu}, p, T) = b_{\text{abs}}(\tilde{\nu}, p_r, T_r) \cdot \frac{p}{p_r} \cdot \sqrt{\frac{T_r}{T}} \,. \tag{8.121}$$

To derive Eq. (8.121), we apply Eq. (8.79), which gives

$$\alpha_L(p, T) = \alpha_L(p_r, T_r) \cdot \frac{p}{p_r} \cdot \sqrt{\frac{T_r}{T}} \,. \tag{8.122}$$

Furthermore, we assume

$$\frac{\sum_i S_i(T) \cdot \alpha_{L,i}(p_r, T_r) \cdot (\tilde{\nu} - \tilde{\nu}_{0,i})^{-1}}{\sum_i S_i(T_r) \cdot \alpha_{L,i}(p_r, T_r) \cdot (\tilde{\nu} - \tilde{\nu}_{0,i})^{-1}} \approx 1 \,. \tag{8.123}$$

It should be noted that the relation in Eq. (8.123) is an inaccurate approximation. To improve the accuracy, Chou and Arking (1980) used the mean value of the following ratio over a wide spectral region, that is,

$$\varsigma_{\tilde{\nu}}(T, T_r) = \sqrt{\frac{T_r}{T}} \cdot \frac{\sum_i S_i(T) \cdot \alpha_{L,i}(p_r, T_r) \cdot (\tilde{\nu} - \tilde{\nu}_{0,i})^{-1}}{\sum_i S_i(T_r) \cdot \alpha_{L,i}(p_r, T_r) \cdot (\tilde{\nu} - \tilde{\nu}_{0,i})^{-1}} \,, \tag{8.124}$$

and expressed the absorption coefficient in the form

$$k_{\text{abs}}(\tilde{\nu}, p, T) = k_{\text{abs}}(\tilde{\nu}, p_r, T_r) \cdot \frac{p}{p_r} \cdot \overline{\varsigma}_{\tilde{\nu}}(T, T_r) \,. \tag{8.125}$$

Thus, the spectral transmissivity can be expressed using Eq. (8.96) as

$$
\begin{aligned}
\mathcal{T}(u) &= \frac{1}{\Delta\tilde{\nu}} \int\limits_{\Delta\tilde{\nu}} \exp\left[-\int\limits_u k_{\mathrm{abs}}(\tilde{\nu}, p, T)\mathrm{d}u\right] \mathrm{d}\tilde{\nu} \\
&= \frac{1}{\Delta\tilde{\nu}} \int\limits_{\Delta\tilde{\nu}} \exp\left[-k_{\mathrm{abs}}(\tilde{\nu}, p_{\mathrm{r}}, T_{\mathrm{r}}) \cdot \tilde{u}\right] \mathrm{d}\tilde{\nu} ,
\end{aligned}
\tag{8.126}
$$

where $\tilde{u}$ is the scaled absorption path length given by

$$
\tilde{u} = \int\limits_u \frac{p}{p_{\mathrm{r}}} \cdot \sqrt{\frac{T_{\mathrm{r}}}{T}} \, \mathrm{d}u
\tag{8.127}
$$

if validity of Eq. (8.121) is assumed, or

$$
\tilde{u} = \int\limits_u \frac{p}{p_{\mathrm{r}}} \cdot \overline{\varsigma}_{\tilde{\nu}}(T, T_{\mathrm{r}}) \, \mathrm{d}u
\tag{8.128}
$$

if validity of Eq. (8.125) is assumed. Equation (8.127) proves that the band models can be extended to an inhomogeneous absorption path by using the reference absorption coefficient and scaled path length.

8.7.5
The *k*-Distribution Method

The *k*-distribution method provides an efficient and accurate approach to compute the spectral transmissivity associated with a homogeneous path.

The absorption coefficient, if viewed in the wavenumber domain, may be highly repetitive within a given spectral region. If the spectral region is divided into a number of subintervals and the absorption coefficient is evaluated at the centers of all the subintervals, the corresponding spectral transmissivity is independent of the ordering of the spectral subintervals. Thus, the coefficient values can be regrouped to improve the computational efficiency. The origin of this innovative idea can be traced to the studies by Ambartzumian (1936) and Lebedinsky (1939).

To illustrate the basic principle of the *k*-distribution method, we consider the following transformation in calculating the spectral transmissivity $\mathcal{T}(u)$ defined in Eq. (8.96):

$$
\begin{aligned}
\mathcal{T}(u) &= \frac{1}{\Delta\tilde{\nu}} \int\limits_{\tilde{\nu}_1}^{\tilde{\nu}_2} \exp\left[-k_{\mathrm{abs}}(\tilde{\nu}) \cdot u\right] \mathrm{d}\tilde{\nu} \\
&= \int\limits_{k_{\mathrm{abs,min}}}^{k_{\mathrm{abs,max}}} \exp\left(-k_{\mathrm{abs}} \cdot u\right) \cdot F(k_{\mathrm{abs}}) \, \mathrm{d}k_{\mathrm{abs}} ,
\end{aligned}
\tag{8.129}
$$

where $\Delta\tilde{\nu} = \tilde{\nu}_2 - \tilde{\nu}_1$ is the width of the spectrum of interest and $k_{abs,min}$ and $k_{abs,max}$ indicate the minimum and maximum values of the absorption coefficient within $\Delta\tilde{\nu}$. The term $F(k_{abs})$ is the probability distribution function of the absorption coefficient and denotes the fraction of $\Delta\tilde{\nu}$ when the absorption coefficient value is between $k_{abs} - (1/2)dk_{abs}$ and $k_{abs} + (1/2)dk_{abs}$. If the spectral region $\Delta\tilde{\nu}$ is divided into N subregions and the absorption coefficient monotonically increases or decreases in each subregion, the probability distribution function can be expressed mathematically in the form of (Lacis and Oinas, 1991; Thomas and Stamnes, 1999)

$$
F(k_{abs}) = \frac{1}{\Delta\tilde{\nu}} \cdot \left(\left| \frac{d\tilde{\nu}}{dk_{abs}} \right|_{k_{abs,min,1} < k_{abs} < k_{abs,max,1}} \right.
$$
$$
+ \left| \frac{d\tilde{\nu}}{dk_{abs}} \right|_{k_{abs,min,2} < k_{abs} < k_{abs,max,2}} + \cdots +
$$
$$
\left. + \left| \frac{d\tilde{\nu}}{dk_{abs}} \right|_{k_{abs,min,N} < k_{abs} < k_{abs,max,N}} \right) . \tag{8.130}
$$

If the absorption coefficient is beyond the absorption coefficient limits, that is, $k_{abs} < k_{abs,min,i}$ or $k_{abs} > k_{abs,max,i}$ where i indicates the subspectral region, this subspectral region does not contribute to $F(k_{abs})$. After the probability distribution function is defined, we can compute the corresponding cumulative probability function G as

$$
G(k_{abs}) = \int_0^{k_{abs}} F(k'_{abs}) \, dk'_{abs} . \tag{8.131}
$$

With an inversion of the cumulative probability function, the mass absorption coefficient can be expressed in the form of

$$
k_{abs} = k_{abs}(G) . \tag{8.132}
$$

Thus, the spectral transmissivity can be written as

$$
\mathcal{T} = \int_0^1 \exp\left[-k_{abs}(G) \cdot u\right] dG . \tag{8.133}
$$

A straightforward approach to compute $F(k_{abs})$ and $G(k_{abs})$ is to divide $\Delta\tilde{\nu}$ into a number of uniform subspectral intervals with spectral spacing of $\delta\tilde{\nu}$ ($\delta\tilde{\nu} \ll \Delta\tilde{\nu}$), to divide $\Delta k_{abs} = k_{abs,max} - k_{abs,min}$ into a number of intervals with spacing of δk_{abs} ($\delta k_{abs} \ll \Delta k_{abs}$), and to evaluate the absorption coefficient values at the centers of all the subspectral intervals, that is, at $\tilde{\nu}_1 + (i - 1/2) \cdot \delta\tilde{\nu}$, $i = 1, 2, 3, \ldots, N$. If n points among the N points of the evaluated mass absorption coefficient lie between $[k_{abs,min} + (i - 1) \cdot \delta k_{abs}]$ and $[k_{abs,min} + i \cdot \delta k_{abs}]$, then we have

$$
F\left[k_{abs,min} + (i - 1/2) \cdot \delta k_{abs}\right] = \frac{n}{N} . \tag{8.134}
$$

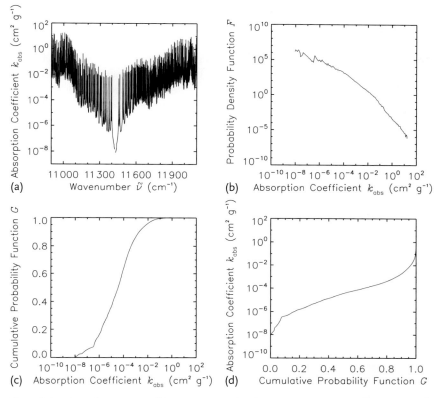

Figure 8.14 Mass absorption coefficients associated with water vapor. (a) Mass absorption coefficients associated with water vapor, (b) $F(k_{abs})$, (c) $G(k_{abs})$, (d) $k_{abs}(G)$. Courtesy of S. Ding.

Furthermore, if m points among the N points of the evaluated mass absorption coefficient are smaller than $[k_{abs,min} + i \cdot \delta k_{abs}]$, we have

$$G[k_{abs,min} + i \cdot \delta k_{abs}] = \frac{m}{N} . \tag{8.135}$$

As an example of the numerical computation, Figure 8.14a shows the mass absorption coefficients associated with water vapor between 10 902.75 and 12 205.25 cm^{-1} at a temperature of 250 K and a pressure of 398 hPa. The mass absorption coefficient in the wavenumber domain significantly oscillates with the variation of the wavenumber. Figure 8.14b shows $F(k_{abs})$, Figure 8.14c shows $G(k_{abs})$, and Figure 8.14d shows $k_{abs}(G)$. Because the absorption is a smooth function of G, we can use fewer points in comparison with the LBLM to compute the spectral transmissivity based on Eq. (8.133).

8.7.6

The Correlated *k*-Distribution Method (CKDM)

The preceding *k*-distribution method is applicable to a homogeneous path, but the Correlated *k*-Distribution Method (CKDM) is used for an inhomogeneous path. The method assumes the *k*-distribution is correlated in the frequency domain, that is, the ordering of the absorption coefficient in terms of a monotonic increase or decrease of the line strength leads to the same spectral alignment of the absorptions lines at various atmospheric heights (Lacis and Oinas, 1991). Fu and Liou (1992b) proved that the validation of the CKDM requires two conditions: (i) if $k_{abs}(\tilde{\nu}_1) = k_{abs}(\tilde{\nu}_2)$ at p_r and T_r, where p_r and T_r are reference pressure and temperature, then $k_{abs}(\tilde{\nu}_1) = k_{abs}(\tilde{\nu}_2)$ at any random pressure and temperature; and, (ii) If $k_{abs}(\tilde{\nu}_1) > k_{abs}(\tilde{\nu}_2)$ at p_r and T_r, then $k_{abs}(\tilde{\nu}_1) > k_{abs}(\tilde{\nu}_2)$ at any random pressure and temperature.

Under the preceding conditions, the radiative transfer simulation involving an inhomogeneous absorption path can be achieved on a layer-by-layer basis with the correlated assumption between different layers. Unlike the band models, the CKDM can incorporate, in a straightforward manner, the multiple-scattering by clouds and aerosol particles. Fu and Liou (1992b) have proved that the CKDM is accurate at single-, weak-, and strong-line limits. However, the correlated assumption is not always perfectly true under realistic atmospheric conditions. Specifically, a satisfactory sorting of $k_{abs}(G)$ at one pressure level may not be consistently fulfilled at another level. If the two pressure levels are relatively close, the correlation may be approximately valid under normal atmospheric conditions (Fu and Liou, 1992b). The errors from applications of the CKDM to realistic atmospheres have been investigated by Fu and Liou (1992b) and will not be recaptured here.

For practical applications, it is an essential to calculate the total mean transmission of multiple gases within one spectral band. Generally, two approaches can be used to treat overlap absorption in the G space in the CKDM. In the first approach, it is assumed that the gases' absorptions are independent and interactions do not occur between the gases. For example, the spectral mean transmittance of two overlapping gases for a given band can be calculated in the form of

$$
\begin{aligned}
\mathcal{T}_{\Delta\tilde{\nu}}(1,2) &= \mathcal{T}_{\Delta\tilde{\nu}}(1) \times \mathcal{T}_{\Delta\tilde{\nu}}(2) \\
&= \sum_{i=1}^{N} \exp(-k_{abs,1,i} \cdot u_1) \cdot \Delta G_{1,i} \cdot \sum_{j=1}^{M} \exp(-k_{abs,2,j} \cdot u_2) \cdot \Delta G_{2,j} \\
&= \sum_{i=1}^{N} \Delta G_{1,i} \cdot \sum_{j=1}^{M} \Delta G_{2,j} \cdot \exp(-k_{abs,1,i} \cdot u_1 - k_{abs,2,j} \cdot u_2) ,
\end{aligned}
$$

(8.136)

where u_1 and u_2 are the absorber amounts of overlapping gases one and two, and M and N are the total number of the G-values for the two gases. Based on Eq. (8.136), we perform $M \times N$ operations to obtain the spectral mean transmittance in each band of the two overlapping gases.

The same approach can be applied to bands with three or more overlapping gases. However, the calculation will increase N times for each additional absorbing gas given that the total number of the G-points is N for the gas. Although the approach can produce good results, it involves a significant amount of calculations, which makes the method inefficient, particularly with more than three types of gas.

To overcome the complications and inefficiencies of treating gaseous overlapping absorptions, a second choice is the amount-weighted approach (Fu and Liou, 1992b; Li and Barker, 2005; Shi et al., 2009). The underlying concept of this approach is to consider a gaseous mixture as a "single gas" by combining the absorption coefficients of all the gases. The absorption coefficients are weighted by absorber amounts and the correlated assumptions for each individual gas are required. Further details can be found in Fu and Liou (1992b) and Shi et al. (2009).

8.7.7
Application of the CKDM to Satellite Remote Sensing

For remote sensing applications with the CKDM, the spectral response function of the measuring instrument must be considered. Kratz (1995) and Kratz and Rose (1999) attempted an approximation by subdividing the spectral interval of interest into subintervals in which the instrument response could be considered constant.

Many instrument spectral response functions show considerable spectral structures. Edwards and Francis (2000) presented a more feasible approach to include the instrument spectral response function based in the CKDM. In the later method, the band transmissivity for each atmospheric layer is given by

$$\mathcal{T}_\Phi = \frac{1}{R_\Phi} \int_{\Delta\tilde{\nu}} t(\tilde{\nu}) \cdot \Phi(\tilde{\nu}) \, d\tilde{\nu} , \tag{8.137}$$

where $t(\tilde{\nu})$ is monochromatic transmittance, $\Phi(\tilde{\nu})$ is the spectral response function of specified instrument, and the integrated spectral response is

$$R_\Phi = \int_{\Delta\tilde{\nu}} \Phi(\tilde{\nu}) \, d\tilde{\nu} . \tag{8.138}$$

In practice, the instrument spectral channel is divided into a number of subintervals within which the product of the response function and the absorption coefficient monotonically increases and decreases. The cumulative distribution function can be calculated via

$$G_\Phi(k'_{\text{abs}}) = \frac{1}{R_\Phi} \sum_{m=1}^{M} \Phi_m \cdot \delta\tilde{\nu}_m(k_{\text{abs}} < k'_{\text{abs}})$$
$$\times W_m(k_{\text{abs},m,\text{min}} \leq k_{\text{abs}} \leq k_{\text{abs},m,\text{max}}) , \tag{8.139}$$

where W_m is the window function associated with the mth subinterval. The function is unity when the absorption coefficient is between the minimum and maximum absorption coefficient values within the subinterval, and zero otherwise. $\delta\tilde{\nu}_m$

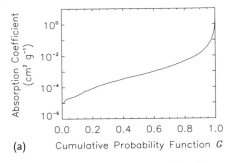
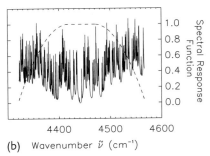

(a) Cumulative Probability Function G (b) Wavenumber $\tilde{\nu}$ (cm⁻¹)

Figure 8.15 *k*-distribution including the effect of the instrument spectral response function. (a) The absorption coefficient as a function of the cumulative probability, which includes the effect of the instrument spectral response; (b) H_2O mass absorption coefficient at 300 K for levels at 631 hPa (solid line), and spectral response function of the GOES-R ABI channel 6 (dashed line). Courtesy of S. Ding.

indicates the fraction of the mth spectral subinterval, corresponding to coefficients smaller than k'_{abs}. In this way, the spectral response function of the instrument is included in the k-distribution and incorporated into radiative transfer calculations. As an example, Figure 8.15 shows the k-distribution numerical computation including the effect of the instrument spectral response function. Figure 8.15b shows the mass absorption coefficients associated with water vapor between 4321.55 and 4566.55 cm⁻¹, at a temperature of 300 K, a pressure of 631 hPa, and the spectral response function of the GOES-R Advanced Baseline Imager (ABI) channel 6. Figure 8.15a presents the variation of the absorption coefficient versus the cumulated probability function.

Problems

Problem 8.1 Molecular Rotational Spectra

a) Calculate the moment of inertia $\mathcal{I}$ for the linear molecules of carbon dioxide CO_2 and carbon monoxide CO. The distance between the C and O atoms is 116.32×10^{-12} m in CO_2 and 112.8×10^{-12} m in CO. The mass of a C atom is 1.9926×10^{-26} kg and of an O atom is 2.6569×10^{-26} kg.

b) Determine the rotational quantum numbers of a CO molecule involved in the rotational transition that produces an emission line at a wavelength of 4.67 μm. Compare the rotational energy of this state to the thermal energy of an atom at room temperature.

Problem 8.2 Rotational Spectra of Isotopes

a) The molecule of hydrochloric acid HCl consists of one atom each of hydrogen and chlorine. However, natural chlorine contains two chlorine isotopes ^{35}Cl and ^{37}Cl, and two different "versions" of the HCl molecule exist. The mass of the two chlorine isotopes is 35 and 37 times the mass m_H of the hydrogen atom. The distance r_{HCl} between the two atoms is practically the same in both molecules (128 pm). Determine the frequency resolution $\Delta\tilde{\nu}/\tilde{\nu}$ of a spectrometer required to distinguish the rotational emission lines of both molecules. Which property of the emission spectrum is determined by the fact that ^{35}Cl and ^{37}Cl naturally occur at a number ratio of 3 : 1?

b) In a similar fashion, determine the frequency resolution required to distinguish the emission lines of carbon monoxide with the isotopes ^{12}C and ^{13}C. Why is the effect larger than for HCl?

Problem 8.3 Vibrational Energy of Nitrogen Monoxide (NO)

a) The fundamental wavenumber of the NO molecule is $1904.0\,\text{cm}^{-1}$. Calculate the wavenumbers of the emission lines that correspond to transitions from the first and the second vibrational energy state to the ground state.

b) The emission lines are actually measured at 1876.1 and $3724.2\,\text{cm}^{-1}$ because the description of the vibration as a harmonic oscillator is not entirely accurate. The description is improved by the addition of a quadratic term into the formula of the vibrational energy levels,

$$E_{n_{vib}} = \left(n_{vib} + \frac{1}{2} \right) \cdot h \cdot c \cdot \tilde{\nu}'_{vib,i} + \chi \cdot \left(n_{vib} + \frac{1}{2} \right)^2 \cdot h \cdot c \cdot \tilde{\nu}'_{vib,i} \ . \quad (8.140)$$

Determine the anharmonic constant χ.

Problem 8.4 Doppler Shift

a) Express the Doppler shift in terms of wavelength.

b) Many cosmic objects, for example, stars and clouds of hydrogen, emit a strong Lyman-α line. These emissions are an important tool in astronomy. For example, the Doppler shift of this line indicates the velocity of an object relative to an observer on Earth. The discovery that the emission lines of all distant galaxies are shifted toward longer (not shorter) wavelengths led to the now broadly accepted theory of an expanding universe (although a distinction has to be made between the actual motion of objects and the expansion of space itself). Calculate the relative velocity of the Andromeda galaxy from the observed wavelength of its Lyman-α line of 121.44 nm. What does this mean for the theory of an expanding universe?

9
Terrestrial Radiative Transfer

The atmospheric energy budget is largely determined by solar insolation at the TOA, and by emission and absorption in the terrestrial (thermal) spectral region. In Section 3.1.5, the EM radiation spectrum in the atmosphere was divided into the solar region ($\lambda = 0.2–5$ µm) and into the terrestrial region ($\lambda = 5–100$ µm), which includes the TIR ($\lambda = 5–50$ µm) and the FIR ($\lambda > 50$ µm). In general, scattering, absorption, and emission modify the EM radiation within the atmosphere. However, in the TIR spectral region, scattering by molecules, aerosol particles, cloud droplets, ice crystals, and precipitation particles is much less important than absorption at wavelengths longer than 4 µm. Exceptions are found in the atmospheric window region (roughly 8–13 µm) where the atmosphere is mostly transparent to TIR radiation and particle scattering and absorption cannot be neglected.

In the troposphere, net diabatic heating is determined by emission from the surface and from atmospheric gas molecules. Latent heat processes are important near the surface, but atmospheric gas emission is more important in the troposphere. In the stratosphere, net heating mainly depends on the imbalance between ozone (O_3) absorption of solar radiation in the UV spectral region and carbon dioxide (CO_2) emission affecting thermal radiative cooling. In the stratosphere, O_3 is the most important solar absorber and CO_2 is the most important TIR emitter.

The Sun is an external source of atmospheric radiation, but the sources of terrestrial radiation are internal. By combination, the Earth's land, ocean, ice, and atmosphere produce terrestrial radiation, which can be described by Planck's function. The solar and terrestrial spectral regions have little overlap and can be treated separately. The internal terrestrial sources and the boundary radiation fields are mostly isotropic. Thus, the terrestrial radiation field is axially symmetric with respect to the vertical direction, the z axis, and the radiative transfer equation does not depend on the azimuth angle φ. In contrast to the solar spectral region, the radiance expansion in a Fourier cosine series is not necessary.

We begin with the Schwarzschild–Emden equation, see Eq. (6.68), in the form with the vertical coordinate z. We neglect scattering processes (i.e., $J_{\mathrm{dir},\lambda} = J_{\mathrm{diff},\lambda} = 0$, $\tilde{\omega} = 0$, $b_{\mathrm{ext}} \equiv b_{\mathrm{abs}}$) and introduce Eq. (6.8) as

$$d\tau(\lambda, z) = -b_{\mathrm{abs}}(\lambda, z)\, dz .\qquad (9.1)$$

Theory of Atmospheric Radiative Transfer, First Edition. Manfred Wendisch and Ping Yang
© 2012 WILEY-VCH Verlag GmbH & Co. KGaA. Published 2012 by WILEY-VCH Verlag GmbH & Co. KGaA.

Applying these assumptions, we obtain the concise 1D RTE valid in the terrestrial spectral range and omitting scattering processes,

$$\frac{d I_{\text{diff},\lambda}(z,\mu)}{dz} = -\frac{b_{\text{abs}}(\lambda,z)}{\mu} \cdot \{I_{\text{diff},\lambda}(z,\mu) - B_\lambda[T(z)]\}.\tag{9.2}$$

9.1
Downward Spectral Radiation

9.1.1
Diffuse Downward Radiance $I^{\downarrow}_{\text{diff},\lambda}$

For the downward terrestrial spectral radiance $I^{\downarrow}_{\text{diff},\lambda}(z,-\mu)$, the directional coordinate $\cos\theta = -\mu$ where $\mu > 0$, thus Eq. (9.2) becomes

$$\frac{d I^{\downarrow}_{\text{diff},\lambda}(z,-\mu)}{dz} = \frac{b_{\text{abs}}(\lambda,z)}{\mu} \cdot \left\{I^{\downarrow}_{\text{diff},\lambda}(z,-\mu) - B_\lambda[T(z)]\right\}.\tag{9.3}$$

Equation (9.3) is an ordinary differential equation of the first kind, and we present two approaches for the solution.

First, we use the "standard" method that gives the solution in terms of the summation of a general solution to the homogeneous counterpart of Eq. (9.3) and a special solution to the inhomogeneous equation, subject to an appropriate boundary condition. The homogeneous equation corresponding to Eq. (9.3) is given as

$$\frac{d I^{\downarrow}_{\text{diff},\lambda,\text{hom}}(z,-\mu)}{dz} = \frac{b_{\text{abs}}(\lambda,z)}{\mu} \cdot I^{\downarrow}_{\text{diff},\lambda,\text{hom}}(z,-\mu).\tag{9.4}$$

The formal solution for Eq. (9.4) is in the form of

$$I^{\downarrow}_{\text{diff},\lambda,\text{hom}}(z,-\mu) = C_0 \cdot \exp\left[-\frac{1}{\mu}\int_z^\infty b_{\text{abs}}(\lambda,\zeta)\,d\zeta\right],\tag{9.5}$$

where C_0 is a constant to be determined by the boundary condition applied to the ultimate solution for Eq. (9.3).

We assume the special solution for the inhomogeneous Eq. (9.3) to be in a form similar to Eq. (9.5). However, for the special solution, the constant factor C_0 in Eq. (9.5) is replaced by a function of the independent coordinate z, that is,

$$I^{\downarrow}_{\text{diff},\lambda,\text{spec}}(z,-\mu) = C_1(z) \cdot \exp\left[-\frac{1}{\mu}\int_z^\infty b_{\text{abs}}(\lambda,\zeta)\,d\zeta\right].\tag{9.6}$$

By substituting Eq. (9.6) into Eq. (9.3), we obtain

$$\frac{\mathrm{d}C_1(z)}{\mathrm{d}z} \cdot \exp\left[-\frac{1}{\mu}\int_z^\infty b_{\mathrm{abs}}(\lambda, \zeta)\,\mathrm{d}\zeta\right]$$

$$+ C_1(z) \cdot \frac{b_{\mathrm{abs}}(\lambda, z)}{\mu} \cdot \exp\left[-\frac{1}{\mu}\int_z^\infty b_{\mathrm{abs}}(\lambda, \zeta)\,\mathrm{d}\zeta\right]$$

$$= C_1(z) \cdot \frac{b_{\mathrm{abs}}(\lambda, z)}{\mu} \cdot \exp\left[-\frac{1}{\mu}\int_z^\infty b_{\mathrm{abs}}(\lambda, \zeta)\,\mathrm{d}\zeta\right]$$

$$- \frac{b_{\mathrm{abs}}(\lambda, z)}{\mu} \cdot B_\lambda[T(z)]\,. \tag{9.7}$$

Two terms on the left and right of the equal sign cancel, and we obtain

$$\frac{\mathrm{d}C_1(z)}{\mathrm{d}z} \cdot \exp\left[-\frac{1}{\mu}\int_z^\infty b_{\mathrm{abs}}(\lambda, \zeta)\,\mathrm{d}\zeta\right] = -\frac{b_{\mathrm{abs}}(\lambda, z)}{\mu} \cdot B_\lambda[T(z)]\,. \tag{9.8}$$

By simplifying Eq. (9.8), we obtain

$$\frac{\mathrm{d}C_1(z)}{\mathrm{d}z} = -\frac{b_{\mathrm{abs}}(\lambda, z)}{\mu} \cdot B_\lambda[T(z)] \cdot \exp\left[\frac{1}{\mu}\int_z^\infty b_{\mathrm{abs}}(\lambda, \zeta)\,\mathrm{d}\zeta\right]\,. \tag{9.9}$$

The solution of Eq. (9.9) is given by

$$C_1(z) = A - \frac{1}{\mu}\int_0^z b_{\mathrm{abs}}(\lambda, z') \cdot B_\lambda[T(z')] \cdot \exp\left[\frac{1}{\mu}\int_{z'}^\infty b_{\mathrm{abs}}(\lambda, \zeta)\,\mathrm{d}\zeta\right]\,\mathrm{d}z'\,, \tag{9.10}$$

where A is an integration constant.

After the factor $C_1(z)$ is specified by Eq. (9.10), the solution of Eq. (9.3) can be obtained by adding the general homogeneous solution given by Eq. (9.5) and the special solution given by Eq. (9.6), that is,

$$I_{\mathrm{diff},\lambda}^\downarrow(z, -\mu) = I_{\mathrm{diff},\lambda,\mathrm{hom}}^\downarrow(z, -\mu) + I_{\mathrm{diff},\lambda,\mathrm{spec}}^\downarrow(z, -\mu)$$

$$= \exp\left[-\frac{1}{\mu}\int_z^\infty b_{\mathrm{abs}}(\lambda, \zeta)\,\mathrm{d}\zeta\right] \cdot \{C_0 + C_1(z)\}$$

$$= \exp\left[-\frac{1}{\mu}\int_z^\infty b_{\mathrm{abs}}(\lambda, \zeta)\,\mathrm{d}\zeta\right] \cdot \left\{C_0 + A - \frac{1}{\mu}\int_0^z b_{\mathrm{abs}}(\lambda, z') \cdot B_\lambda[T(z')]\right.$$

$$\left. \times \exp\left[\frac{1}{\mu}\int_{z'}^\infty b_{\mathrm{abs}}(\lambda, \zeta)\,\mathrm{d}\zeta\right]\,\mathrm{d}z'\right\}\,. \tag{9.11}$$

As the temperature of the cosmic background is on the order of 2.7 K, there is practically no downward radiation at the TOA in the terrestrial spectrum. Thus, the boundary condition for the downward radiance is given by

$$I_{\text{diff},\lambda}^{\downarrow}(\infty, -\mu) = 0 . \tag{9.12}$$

By applying the preceding boundary condition to Eq. (9.11), we obtain

$$C_0 + A = \frac{1}{\mu} \int\limits_0^\infty b_{\text{abs}}(\lambda, z') \cdot B_\lambda[T(z')] \cdot \exp\left[\frac{1}{\mu} \int\limits_{z'}^\infty b_{\text{abs}}(\lambda, \zeta)\,d\zeta\right] dz' . \tag{9.13}$$

Substituting Eq. (9.13) into Eq. (9.11) gives

$$I_{\text{diff},\lambda}^{\downarrow}(z, -\mu) = \exp\left[-\frac{1}{\mu} \int\limits_z^\infty b_{\text{abs}}(\lambda, \zeta)\,d\zeta\right]$$

$$\times \left\{ \frac{1}{\mu} \int\limits_0^\infty b_{\text{abs}}(\lambda, z') \cdot B_\lambda[T(z')] \cdot \exp\left[\frac{1}{\mu} \int\limits_{z'}^\infty b_{\text{abs}}(\lambda, \zeta)\,d\zeta\right] dz' \right.$$

$$\left. - \frac{1}{\mu} \int\limits_0^z b_{\text{abs}}(\lambda, z') \cdot B_\lambda[T(z')] \cdot \exp\left[\frac{1}{\mu} \int\limits_{z'}^\infty b_{\text{abs}}(\lambda, \zeta)\,d\zeta\right] dz' \right\} . \tag{9.14}$$

We apply a transformation as

$$\int\limits_0^\infty \ldots dz' - \int\limits_0^z \ldots dz' = \int\limits_z^\infty \ldots dz' , \tag{9.15}$$

yielding

$$I_{\text{diff},\lambda}^{\downarrow}(z, -\mu) = \exp\left[-\frac{1}{\mu} \int\limits_z^\infty b_{\text{abs}}(\lambda, \zeta)\,d\zeta\right] \cdot \left\{ \frac{1}{\mu} \cdot \int\limits_z^\infty b_{\text{abs}}(\lambda, z') \cdot B_\lambda[T(z')] \right.$$

$$\left. \times \exp\left[\frac{1}{\mu} \int\limits_{z'}^\infty b_{\text{abs}}(\lambda, \zeta)\,d\zeta\right] dz' \right\} . \tag{9.16}$$

Another transformation is applied, namely,

$$-\int\limits_z^\infty \ldots d\zeta + \int\limits_{z'}^\infty \ldots d\zeta = -\int\limits_z^{z'} \ldots d\zeta , \tag{9.17}$$

and we combine the two exponential terms to obtain

$$I_{\text{diff},\lambda}^{\downarrow}(z, -\mu) = \frac{1}{\mu} \int\limits_z^\infty b_{\text{abs}}(\lambda, z') \cdot B_\lambda[T(z')] \cdot \exp\left[-\frac{1}{\mu} \int\limits_z^{z'} b_{\text{abs}}(\lambda, \zeta)\,d\zeta\right] dz' .$$

$$\tag{9.18}$$

In Eq. (9.18), the terms with their units are: the spectral volume absorption coefficient $b_{abs}(\lambda)$ in units of $\mathrm{m^{-1}\,nm^{-1}}$; the spectral radiance Planck function $B_\lambda[T(z')]$ in units of $\mathrm{W\,m^{-2}\,sr^{-1}\,nm^{-1}}$; the absolute temperature T in units of kelvin; and, the dimensionless zenith distance $\mu = \cos\theta$.

Alternatively, to solve for the downward diffuse radiance, we rewrite Eq. (9.3) as

$$\frac{d I_{\mathrm{diff},\lambda}^{\downarrow}(z',-\mu)}{dz'} - I_{\mathrm{diff},\lambda}^{\downarrow}(z',-\mu)\cdot\frac{b_{abs}(\lambda,z')}{\mu} = -\frac{b_{abs}(\lambda,z')}{\mu}\cdot B_\lambda[T(z')]\,. \quad (9.19)$$

Using the chain rule of differentiation, we can prove the following relation:

$$\frac{d\left\{ I_{\mathrm{diff},\lambda}^{\downarrow}(z',-\mu)\cdot\exp\left[-\int_{z''}^{z'} b_{abs}(\lambda,\zeta)/\mu\,d\zeta\right]\right\}}{dz'}$$

$$= \frac{d I_{\mathrm{diff},\lambda}^{\downarrow}(z',-\mu)}{dz'}\cdot\exp\left[-\int_{z''}^{z'}\frac{b_{abs}(\lambda,\zeta)}{\mu}\,d\zeta\right]$$

$$- I_{\mathrm{diff},\lambda}^{\downarrow}(z',-\mu)\cdot\frac{b_{abs}(\lambda,z')}{\mu}\cdot\exp\left[-\int_{z''}^{z'}\frac{b_{abs}(\lambda,\zeta)}{\mu}\,d\zeta\right]\,, \quad (9.20)$$

where z'' indicates a certain reference altitude in the atmosphere. Thus, if we multiply Eq. (9.19) with

$$\exp\left[-\int_{z''}^{z'}\frac{b_{abs}(\lambda,\zeta)}{\mu}\,d\zeta\right]\,,$$

we can rewrite Eq. (9.19) in the form of

$$\frac{d\left\{ I_{\mathrm{diff},\lambda}^{\downarrow}(z',-\mu)\cdot\exp\left[-\int_{z''}^{z'} b_{abs}(\lambda,\zeta)/\mu\,d\zeta\right]\right\}}{dz'}$$

$$= -\frac{b_{abs}(\lambda,z')}{\mu}\cdot B_\lambda[T(z')]\cdot\exp\left[-\int_{z''}^{z'}\frac{b_{abs}(\lambda,\zeta)}{\mu}\,d\zeta\right]\,. \quad (9.21)$$

Integrating the preceding equation over z' from z to ∞ gives

$$I_{\mathrm{diff},\lambda}^{\downarrow}(\infty,-\mu)\cdot\exp\left[-\int_{z''}^{\infty}\frac{b_{abs}(\lambda,\zeta)}{\mu}\,d\zeta\right]$$

$$- I_{\mathrm{diff},\lambda}^{\downarrow}(z,-\mu)\cdot\exp\left[-\int_{z''}^{z}\frac{b_{abs}(\lambda,\zeta)}{\mu}\,d\zeta\right]$$

$$= -\int_{z}^{\infty}\frac{b_{abs}(\lambda,z')}{\mu}\cdot B_\lambda[T(z')]\cdot\exp\left[-\int_{z''}^{z'}\frac{b_{abs}(\lambda,\zeta)}{\mu}\,d\zeta\right]dz'\,. \quad (9.22)$$

Applying the boundary condition, Eq. (9.12), to the preceding equation, results in

$$I_{\text{diff},\lambda}^{\downarrow}(z,-\mu) \cdot \exp\left[-\int_{z''}^{z} \frac{b_{\text{abs}}(\lambda,\zeta)}{\mu} \, d\zeta\right]$$

$$= \int_{z}^{\infty} \frac{b_{\text{abs}}(\lambda,z')}{\mu} \cdot B_{\lambda}[T(z')] \cdot \exp\left[-\int_{z''}^{z'} \frac{b_{\text{abs}}(\lambda,\zeta)}{\mu} \, d\zeta\right] dz' , \qquad (9.23)$$

from which follows

$$I_{\text{diff},\lambda}^{\downarrow}(z,-\mu) = \int_{z}^{\infty} \frac{b_{\text{abs}}(\lambda,z')}{\mu} \cdot B_{\lambda}[T(z')]$$

$$\times \exp\left[-\int_{z''}^{z'} \frac{b_{\text{abs}}(\lambda,\zeta)}{\mu} \, d\zeta + \int_{z''}^{z} \frac{b_{\text{abs}}(\lambda,\zeta)}{\mu} \, d\zeta\right] dz' . \qquad (9.24)$$

We apply the transformation

$$-\int_{z''}^{z'} \ldots d\zeta + \int_{z''}^{z} \ldots d\zeta = -\int_{z}^{z'} \ldots d\zeta , \qquad (9.25)$$

which gives the same results as found in Eq. (9.18), that is,

$$I_{\text{diff},\lambda}^{\downarrow}(z,-\mu) = \frac{1}{\mu} \int_{z}^{\infty} b_{\text{abs}}(\lambda,z') \cdot B_{\lambda}[T(z')] \cdot \exp\left[-\int_{z}^{z'} \frac{b_{\text{abs}}(\lambda,\zeta)}{\mu} \, d\zeta\right] dz' .$$

$$(9.26)$$

In Eq. (9.26), the exponential term under the integral represents the spectral transmissivity of the layer $[z, z']$

$$\mathcal{T}(\lambda, z, z') = \exp\left[-\frac{1}{\mu} \int_{z}^{z'} b_{\text{abs}}(\lambda,\zeta) \, d\zeta\right] . \qquad (9.27)$$

With this definition, we obtain

$$I_{\text{diff},\lambda}^{\downarrow}(z,-\mu) = \frac{1}{\mu} \int_{z}^{\infty} b_{\text{abs}}(\lambda,z') \cdot B_{\lambda}[T(z')] \cdot \mathcal{T}(\lambda, z, z') \, dz' . \qquad (9.28)$$

Figure 9.1 is an illustration of Eq. (9.28). The Planck function emission is weighted with the volume absorption coefficient, and the layer transmissivity determines the number of emitted photons in altitude z' to reach the detector situated at altitude z. The weighting function $\mathcal{W}(\lambda, z, z')$ of the layer $[z, z']$ is introduced, such that

$$\mathcal{W}(\lambda, z, z') = \frac{b_{\text{abs}}(\lambda,z')}{\mu} \cdot \mathcal{T}(\lambda, z, z') . \qquad (9.29)$$

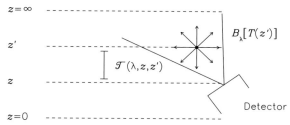

Figure 9.1 Illustration of Eq. (9.28).

Thus,

$$I_{\text{diff},\lambda}^{\downarrow}(z,-\mu) = \int_{z}^{\infty} \mathcal{W}(\lambda,z,z') \cdot B_{\lambda}[T(z')]\,dz' \,. \tag{9.30}$$

The weighting function $\mathcal{W}(\lambda, z, z')$ weights the contributions of Planck's function associated with each atmospheric layer ($z \rightarrow z'$), see Figure 9.1. If the atmosphere is opaque ($\mathcal{T} \rightarrow 0$), the measured radiance is mainly determined by the thermal emission near the receiver. If the atmosphere is less opaque, or transparent ($\mathcal{T} \rightarrow 1$), the relevant weighting function will include emission from greater distances. In the case of distant radiances, the receiver will see through a "window."

The radiance is converted to the brightness temperature by inverting Planck's function, see Eq. (3.58). If the atmosphere is opaque at a particular wavelength, the brightness temperature gives a reasonable estimate of the physical temperature at the level where the weighting function reaches a peak.

A receiver on the ground receives a radiance of

$$I_{\text{diff},\lambda}^{\downarrow}(0,-\mu) = \frac{1}{\mu} \int_{z=0}^{\infty} b_{\text{abs}}(\lambda,z') \cdot B_{\lambda}[T(z')] \cdot \mathcal{T}(\lambda, z=0, z')\,dz'$$

$$= \int_{0}^{\infty} \mathcal{W}(\lambda, z=0, z') \cdot B_{\lambda}[T(z')]\,dz' \,. \tag{9.31}$$

Figure 9.2 illustrates two typical spectra of $I_{\text{diff},\lambda}^{\downarrow}(z=0, \mu=-1)$.

There are two major differences between the model atmospheres assumed for the simulations of the spectra shown in Figure 9.2. The "Tropical" atmosphere is much warmer and more humid (especially in lower altitudes) compared to the "Subarctic Winter" atmosphere. Because the strong water vapor concentrations near the surface are highly absorbing and emitting, most of the downward radiance received near the ground originates from low altitudes, particularly those in wavelength regions where water vapor absorption dominates ($\lambda > 14\,\mu m$, wavenumber $\tilde{\nu} < 730\,cm^{-1}$, and $\lambda < 8\,\mu m$, wavenumber $\tilde{\nu} > 1270\,cm^{-1}$). Therefore, the Planck function corresponding to the near-surface temperature of 300 K closely matches the "Tropical" spectrum in the wavelength regions of high water vapor

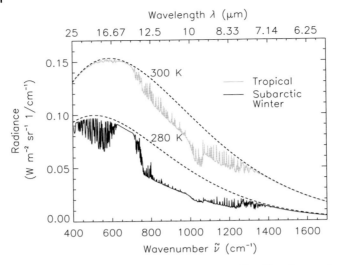

Figure 9.2 Two examples of simulated spectra of downward spectral radiances at the surface. The dashed lines indicate the spectral variations at the approximate surface temperatures. "Tropical" and "Subarctic Win- ter" profiles of atmospheric parameters were assumed (Anderson et al., 1986). The simulations were performed with MODTRAN4 (Version 3 Revision 1, MOD4v3r1).

absorption and emission. This also holds for the second spectrum shown in Figure 9.2. However, the water vapor amount is much lower for the "Subarctic Winter" atmosphere and causes major fluctuations in the spectrum for wavelengths larger than about 16 μm. In the "Subarctic Winter" spectrum, the high absorption due to CO_2 is not covered by water vapor absorption as in the case of the "Tropical" atmosphere.

In the atmospheric window wavelength region (8–13 μm), the spectra in both cases drops because radiance originates from higher, colder atmospheric layers. In particular for the more humid "Tropical" atmosphere, certain line structures appear and are commonly referred to as the "dirty window." For the very dry "Subarctic Winter" atmosphere, the atmospheric window is much cleaner and less absorption lines appear. The atmospheric window is also disturbed by emission due to O_3 at the 9.6 μm wavelength.

9.1.2
Diffuse Downward Irradiance $F_{\text{diff},\lambda}^{\downarrow}$

To obtain the downward irradiance $F_{\text{diff},\lambda}^{\downarrow}$, we integrate the radiance $I_{\text{diff},\lambda}^{\downarrow}$ as defined in Eq. (3.49)

$$F_{\text{diff},\lambda}^{\downarrow}(z) = -2\,\pi \int_{\pi/2}^{\pi} I_{\text{diff},\lambda}^{\downarrow}(z, \theta') \cdot \cos\theta' \cdot \sin\theta' \mathrm{d}\theta' \, . \tag{9.32}$$

If we substitute $\theta = \pi - \theta'$, we get

$$\mathrm{d}\theta' = -\mathrm{d}\theta \, , \tag{9.33}$$

$$\cos \theta' = \cos(\pi - \theta) = -\cos \theta \, , \tag{9.34}$$

$$\sin \theta' = \sin(\pi - \theta) = \sin \theta \, , \tag{9.35}$$

meaning

$$
\begin{aligned}
F^{\downarrow}_{\mathrm{diff},\lambda}(z) &= -2\,\pi \int\limits_{\pi/2}^{0} I^{\downarrow}_{\mathrm{diff},\lambda}(z, \pi - \theta) \cdot \cos \theta \cdot \sin \theta \, \mathrm{d}\theta \\
&= 2\,\pi \int\limits_{0}^{\pi/2} I^{\downarrow}_{\mathrm{diff},\lambda}(z, \pi - \theta) \cdot \cos \theta \cdot \sin \theta \, \mathrm{d}\theta \, .
\end{aligned}
\tag{9.36}
$$

Note that $I^{\downarrow}_{\mathrm{diff},\lambda}(z, \pi - \theta)$ is the same as $I^{\downarrow}_{\mathrm{diff},\lambda}(z, -\mu)$ in Eq. (9.18), where $\mu = \cos \theta$. We substitute Eq. (9.18) into the preceding Eq. (9.36) to obtain

$$
\begin{aligned}
F^{\downarrow}_{\mathrm{diff},\lambda}(z) = 2\pi \int\limits_{0}^{\pi/2} &\left\{ \frac{1}{\cos \theta} \int\limits_{z}^{\infty} b_{\mathrm{abs}}(\lambda, z') \cdot B_{\lambda}[T(z')] \right. \\
&\left. \times \exp\left[-\frac{1}{\cos \theta} \int\limits_{z}^{z'} b_{\mathrm{abs}}(\lambda, \zeta)\mathrm{d}\zeta \right] \mathrm{d}z' \right\} \cdot \cos \theta \cdot \sin \theta \, \mathrm{d}\theta \, ,
\end{aligned}
\tag{9.37}
$$

or in a slightly simplified form

$$
\begin{aligned}
F^{\downarrow}_{\mathrm{diff},\lambda}(z) = 2\,\pi \int\limits_{0}^{\pi/2} &\left\{ \int\limits_{z}^{\infty} b_{\mathrm{abs}}(\lambda, z') \cdot B_{\lambda}[T(z')] \right. \\
&\left. \times \exp\left[-\frac{1}{\cos \theta} \int\limits_{z}^{z'} b_{\mathrm{abs}}(\lambda, \zeta)\mathrm{d}\zeta \right] \mathrm{d}z' \right\} \cdot \sin \theta \, \mathrm{d}\theta \, .
\end{aligned}
\tag{9.38}
$$

Exchange of integration yields

$$
\begin{aligned}
F^{\downarrow}_{\mathrm{diff},\lambda}(z) = 2\pi \int\limits_{z}^{\infty} & b_{\mathrm{abs}}(\lambda, z') \cdot B_{\lambda}[T(z')] \\
&\times \int\limits_{0}^{\pi/2} \exp\left[-\frac{1}{\cos \theta} \int\limits_{z}^{z'} b_{\mathrm{abs}}(\lambda, \zeta)\mathrm{d}\zeta \right] \cdot \sin \theta \, \mathrm{d}\theta \, \mathrm{d}z' \, .
\end{aligned}
\tag{9.39}
$$

Gold's, or Exponential Integral Function

We use $\chi = 1/\cos\theta$ to get

$$\frac{d\chi}{d\theta} = -\frac{1}{\cos^2\theta} \cdot (-\sin\theta)\,, \tag{9.40}$$

$$\sin\theta\,d\theta = \cos^2\theta\,d\chi = \frac{1}{\chi^2}\,d\chi\,. \tag{9.41}$$

This gives

$$F_{\mathrm{diff},\lambda}^{\downarrow}(z) = 2\pi \int\limits_{z}^{\infty} b_{\mathrm{abs}}(\lambda, z') \cdot B_\lambda[T(z')]$$

$$\times \int\limits_{1}^{\infty} \exp\left[-\chi \cdot \int\limits_{z}^{z'} b_{\mathrm{abs}}(\lambda, \zeta)\,d\zeta\right] \cdot \chi^{-2}\,d\chi\,dz'\,. \tag{9.42}$$

We will use Gold's function $G_n(x)$, also called the exponential integral of nth order, defined as (Bronstein and Semendjajev, 1985)

$$G_n(x) = \int\limits_{1}^{\infty} e^{-\xi \cdot x} \cdot \xi^{-n}\,d\xi\,. \tag{9.43}$$

In the case of Eq. (9.42), we have $n = 2$, $\xi = \chi$, and

$$x = \int\limits_{z}^{z'} b_{\mathrm{abs}}(\lambda, \zeta)\,d\zeta\,. \tag{9.44}$$

Furthermore, we introduce the isotropic blackbody irradiance $F_{\mathrm{BB},\lambda}(z)$, see Eq. (3.83), by

$$F_{\mathrm{BB},\lambda}(z) = \pi \cdot B_\lambda[T(z)]. \tag{9.45}$$

From this, we have

$$F_{\mathrm{diff},\lambda}^{\downarrow}(z) = 2 \int\limits_{z}^{\infty} b_{\mathrm{abs}}(\lambda, z') \cdot F_{\mathrm{BB},\lambda}(z') \cdot G_2\left[\int\limits_{z}^{z'} b_{\mathrm{abs}}(\lambda, \zeta)\,d\zeta\right]\,dz'\,. \tag{9.46}$$

For Gold's function, the following recursion holds, that is,

$$G_{i-1}(x) = -\frac{dG_i(x)}{dx}\,; \quad \text{for} \quad i = 1, 2, 3, \ldots \tag{9.47}$$

In our case, we get

$$G_2\left[\int\limits_{z}^{z'} b_{\mathrm{abs}}(\lambda, \zeta)\,d\zeta\right] = -\frac{dG_3\left[\int_{z}^{z'} b_{\mathrm{abs}}(\lambda, \zeta)\,d\zeta\right]}{d\left[\int_{z}^{z'} b_{\mathrm{abs}}(\lambda, \zeta)\,d\zeta\right] = b_{\mathrm{abs}}(\lambda, z')\,dz'}$$

$$= -\frac{1}{b_{\mathrm{abs}}(\lambda, z')}\frac{dG_3\left[\int_{z}^{z'} b_{\mathrm{abs}}(\lambda, \zeta)\,d\zeta\right]}{dz'}\,. \tag{9.48}$$

Hence, we can rewrite Eq. (9.46) as

$$F^{\downarrow}_{\text{diff},\lambda}(z) = -2 \int\limits_{z}^{\infty} F_{\text{BB},\lambda}(z') \, dG_3 \left[\int\limits_{z}^{z'} b_{\text{abs}}(\lambda, \zeta) \, d\zeta \right] . \tag{9.49}$$

Absorbing Mass

As an equivalent to the geometric altitude z, we use a more suitable vertical co-ordinate, the absorbing mass $u(\lambda, z)$. In the terrestrial spectral region where absorption is of most importance, the absorbing mass can be used as a surrogate of optical thickness. The volume absorption coefficient $b_{\text{abs}}(\lambda, z)$ in units of m^{-1} is related to the mass absorption coefficient $k_{\text{abs}}(\lambda, z)$ in units of $m^2 \, kg^{-1}$ via the density of the absorbing medium $\varrho_{\text{abs}}(z)$ in units of $kg \, m^{-3}$ by Eq. (8.97). We define the differential absorbing mass, also called the differential absorbing path within the atmospheric column $du(\zeta)$ in units of $kg \, m^{-2}$ by

$$du(z) = -\varrho_{\text{abs}}(z) \, dz . \tag{9.50}$$

The minus sign indicates that the absorbing mass decreases with increasing altitude. The differential absorbing mass is another type of vertical coordinate, equivalent to the optical thickness, see Eq. (6.2), and is defined as

$$u(z) = \int\limits_{z}^{\infty} \varrho_{\text{abs}}(z') \, dz' . \tag{9.51}$$

For comparison, the optical thickness was defined by Eq. (6.2), also see the definition of the absorption path length u given by Eq. (8.95). With this definition, the inner integral from Eq. (9.39) gives

$$\int\limits_{z}^{z'} b_{\text{abs}}(\lambda, \zeta) d\zeta = \int\limits_{z}^{z'} k_{\text{abs}}(\lambda, \zeta) \cdot \varrho_{\text{abs}}(\zeta) \, d\zeta$$

$$\approx -k_{\text{abs},0}(\lambda) \int\limits_{u(z)}^{M'} du(\zeta)$$

$$= k_{\text{abs},0}(\lambda) \cdot [u(z) - M'] , \tag{9.52}$$

with the definition included in Eq. (9.52)

$$M' = u(z') . \tag{9.53}$$

Thus, we have transformed the height-dependence of $k_{\text{abs}}(\lambda, z)$ into the absorbing mass, where $k_{\text{abs},0}(\lambda)$ is related to standard atmospheric conditions. We now apply Gold's function to obtain

$$dG_3 \left[\int\limits_{z}^{z'} b_{\text{abs}}(\lambda, \zeta) \, d\zeta \right] = dG_3 \left\{ k_{\text{abs},0}(\lambda) \cdot [u(z) - M'] \right\} . \tag{9.54}$$

Water vapor is the major atmospheric absorber in the terrestrial spectral region. The water vapor absorbing mass $w(z)$ within the atmospheric layer from z to ∞ is defined by

$$w(z) = \int_z^\infty \varrho_{wv}(\zeta) \, d\zeta = u_{wv}(z) \,, \tag{9.55}$$

with ϱ_{wv} being the water vapor density. The columnar (total) water vapor absorbing mass w_∞ is given by

$$w_\infty = \int_0^\infty \varrho_{wv}(z) \, dz \,. \tag{9.56}$$

The columnar water vapor absorbing mass w_∞ is the total (columnar) water vapor mass per unit area and lies between 0.3 and $3 \, \mathrm{g \, cm^{-2}}$ under realistic atmospheric conditions. Because water vapor is the major absorber, we may replace $u(z)$ by $w(z)$. $u(z)$ corresponds to the optical thickness in the solar spectral region and is a type of vertical coordinate in the terrestrial, nonscattering atmosphere.

Resulting Irradiance
With these definitions, we obtain the spectral downward irradiance from Eq. (9.49), namely,

$$F_{\mathrm{diff},\lambda}^\downarrow[u(z)] = - \int_{u(z)}^{u(z\longrightarrow\infty)} F_{\mathrm{BB},\lambda}[T(M')] \frac{d\mathcal{T}_{\mathrm{F}}[\lambda, u(z) - M']}{dM'} \, dM' \,. \tag{9.57}$$

The blackbody irradiance is weighted using the spectral transmissivity

$$\mathcal{W}_{\mathrm{F}} = \frac{d\mathcal{T}_{\mathrm{F}}[\lambda, u(z) - M']}{dM'} \,. \tag{9.58}$$

The quantity $\mathcal{W}_{\mathrm{F}}$ is called the weighting function. The spectral transmissivity for flux densities $\mathcal{T}_{\mathrm{F}}(\lambda)$ is related to the previous equations by

$$d\mathcal{T}_{\mathrm{F}}[\lambda, u(z) - M'] = 2d\,G_3 \left\{ k_{\mathrm{abs},0}(\lambda) \cdot [u(z) - M'] \right\} \,. \tag{9.59}$$

9.2
Upward Terrestrial Spectral Radiation

9.2.1
Diffuse Upward Radiance $I^{\uparrow}_{\text{diff},\lambda}$

To find the upward radiance, the approach is similar to the downward radiation with the following results:

$$
I^{\uparrow}_{\text{diff},\lambda}(z,\mu) = \varepsilon(\lambda, z = 0) \cdot B_\lambda[T(z = 0)] \cdot \exp\left[-\frac{1}{\mu}\int_0^z b_{\text{abs}}(\lambda, \zeta)\,d\zeta\right]
$$

$$
+ \frac{1}{\mu}\int_0^z b_{\text{abs}}(\lambda, z') \cdot B_\lambda[T(z')] \cdot \exp\left[-\frac{1}{\mu}\int_{z'}^z b_{\text{abs}}(\lambda, \zeta)\,d\zeta\right] dz' .
$$

$$(9.60)$$

The exponential term under the integral in Eq. (9.60) represents the spectral transmissivity, $\mathcal{T}(\lambda)$. Applying Eq. (9.50) to Eq. (9.60), we obtain

$$
I^{\uparrow}_{\text{diff},\lambda}(z,\mu) = \varepsilon(\lambda, z = 0) \cdot B_\lambda[T(z = 0)] \cdot \mathcal{T}(\lambda, 0, z)
$$

$$
+ \frac{1}{\mu}\int_0^z b_{\text{abs}}(\lambda, z') \cdot B_\lambda[T(z')]\mathcal{T}(\lambda, z', z)\,dz' .
$$

$$(9.61)$$

Figure 9.3 illustrates the configuration associated with Eq. (9.61). The two terms in Eq. (9.61) represent the contributions from the surface and the atmosphere. Using the weighting function $\mathcal{W}(\lambda)$ defined in Eq. (9.29), we obtain

$$
I^{\uparrow}_{\text{diff},\lambda}(z,\mu) = \varepsilon(\lambda, z = 0) \cdot B_\lambda[T(z = 0)] \cdot \mathcal{T}(\lambda, 0, z)
$$

$$
+ \int_0^z \mathcal{W}(\lambda, z', z) \cdot B_\lambda[T(z')]\,dz' .
$$

$$(9.62)$$

The physical meaning of Eq. (9.61) is schematically illustrated in Figure 9.3.

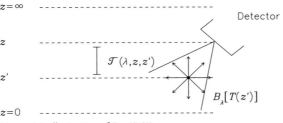

Figure 9.3 Illustration of Eq. (9.62).

As an example of the application of Eq. (9.61), the radiance received by an infrared sensor aboard a satellite situated at the TOA is given by

$$
I_{\text{diff},\lambda}^{\uparrow}(\infty, \mu) = \varepsilon(\lambda, z = 0) \cdot B_{\lambda}[T(z = 0)] \cdot \mathcal{T}(\lambda, 0, \infty)
$$

$$
+ \int_{0}^{\infty} \mathcal{W}(\lambda, z', \infty) \cdot B_{\lambda}[T(z')] \, \mathrm{d}z' . \tag{9.63}
$$

In Chapter 1, Figure 1.4 and the associated discussion, we presented a concise overview of typical spectral patterns of the radiance signal received by a satellite, $I_{\text{diff},\lambda}^{\uparrow}(z \longrightarrow \infty, \mu)$. In the atmospheric window region ($\approx 8{-}13\,\mu\text{m}$), the two spectra in Figure 1.4 approximately indicate the surface temperature of $0\,°\text{C}$ for the "Subarctic Winter" and $15\,°\text{C}$ for the "Tropical" atmospheres. At $15\,\mu\text{m}$, the strong absorption due to CO_2 is obvious as CO_2 is well mixed throughout the troposphere and stratosphere.

9.2.2
Diffuse Upward Irradiance $F_{\text{diff},\lambda}^{\uparrow}$

For upward spectral irradiance, we obtain

$$
F_{\text{diff},\lambda}^{\uparrow}[u(z)] = \varepsilon(\lambda, z = 0) \cdot F_{\text{BB},\lambda}[T(z = 0)] \cdot \mathcal{T}_{\text{F}}[\lambda, u(z)]
$$

$$
+ \int_{0}^{u(z)} F_{\text{BB},\lambda}[T(M')]\frac{\mathrm{d}\mathcal{T}_{\text{F}}[\lambda, u(z) - M']}{\mathrm{d}M'} \, \mathrm{d}M' . \tag{9.64}
$$

This result is similar to Eq. (9.57) with an additional surface term.

9.3
Example of Simulated Spectra

9.3.1
Downward and Upward Radiances

Figure 9.4a shows the downward radiances $I_{\text{diff},\lambda}^{\downarrow}(z = 0, -\mu = -1)$ simulated at the surface. In Figure 9.4b, the corresponding upward radiances $I_{\text{diff},\lambda}^{\uparrow}(z \longrightarrow \infty, \mu = 1)$ simulated at the TOA are shown. Respective Planck functions are indicated by dashed lines. In the simulations, different gaseous absorbers, such as, N_2O, CH_4, O_2, O_3, CO_2, H_2O, and all other gases are shown. In the figure, the effect of gaseous emissions at different altitudes is evident.

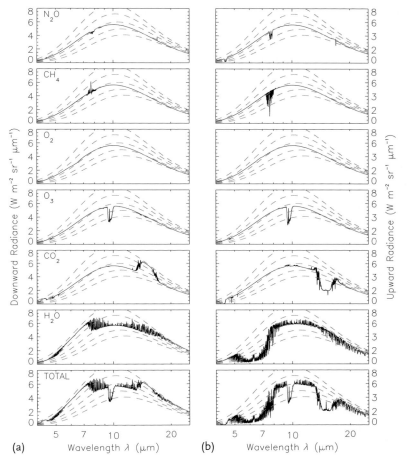

Figure 9.4 Zenith radiances of a cloudless and aerosol-free atmosphere simulated at the surface (a) and the TOA (b). Only gas absorption is considered here. The dashed lines indicate Planck's function for temperatures from 253 K (lowest dashed curve) to 293 K (upper dashed curve) in 10 K intervals.

9.3.2
Influence of Cirrus on Terrestrial Spectral Irradiance

Figure 9.5 and parts of the discussion in Section 9.3.2 are adapted from Wendisch et al. (2007); for details please refer to the original paper. Figure 9.5 shows results of simulations under cirrus cloudy conditions. The spectra of the downward (Figure 9.5a,c) and upward (Figure 9.5b,d) TIR irradiance for a high, optically thin cirrus are displayed. The vertical lines mark the spectral location of the absorption bands of CO_2 (4.3 and 15 μm), H_2O (6.3 μm), and O_3 (9.6 μm). Asterisks indicate the blackbody radiation following Planck's law assuming the temperature at the re-

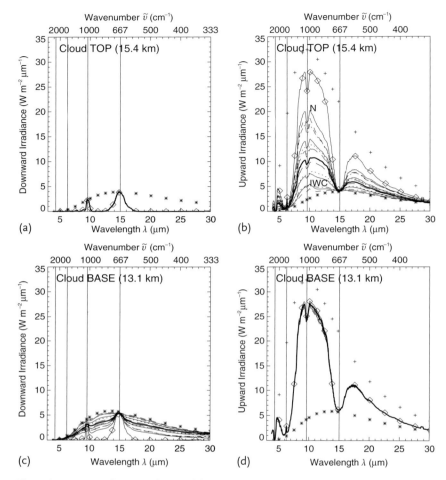

Figure 9.5 Downward (a, c) and upward (b, d) spectral TIR irradiance at cirrus top (a, b) and base (c, d) for the high (cold), optically thin cirrus. The different curves represent simulation results assuming several crystal shapes, for details and curve notation see Table 1 and Figure 3 in Wendisch et al. (2007). Reproduced with permission of AGU © AGU.

spective altitude. The plus signs (Figure 9.5b,d) represent the blackbody radiation corresponding to the surface temperature. Open diamonds indicate the simulations in cloudless conditions without the presence of cirrus.

Downward spectral thermal infrared irradiance The downward TIR irradiance at the cirrus top (see Figure 9.5a) is generally small. In the centers of the O_3 and CO_2 absorption bands, the TIR downward radiation approximately corresponds to the Planck emission (asterisks) at the cirrus top level. However, the Planck emission is low because of the low temperatures at this height. The cirrus emits radiation at its base and, thus, slightly increases downward irradiances emitted outside the O_3 and CO_2 absorption bands, see Figure 9.5b. Outside the gas absorption bands, the optically thin cirrus does not behave like a blackbody. Only in the centers of absorption bands does the emissivity approach unity.

Upward spectral terrestrial irradiance   Figure 9.5d shows the upward irradiance at the cirrus base. Between 8–13 µm (atmospheric window), most of the irradiance emitted by the surface reaches the cirrus base; part of the surface-emitted irradiance (roughly 15%) is reduced due to continuum gas absorption, mainly H_2O, and the respective emission at lower temperatures. Note that the water vapor content is enhanced in the subtropical environment of this special case. The reduction due to absorption and emission at lower temperatures is most obvious in the 9.6 µm O_3 absorption band.

The emission at the major gas absorption bands is dominated by the Planck function curve at the corresponding temperature. The upward irradiance below the cirrus does not depend on ice crystal shape. However, at the cirrus top, see Figure 9.5b, the upward irradiance outside the gas absorption bands is decreased compared to the cloudless case.

9.4
Broadband Terrestrial Radiative Transfer

The techniques to calculate integrated, broadband radiation quantities, for example, radiance and irradiance from spectral radiation, are tricky, although the LBLM integration approach is straightforward. Instead of detailed calculation techniques, we will show and discuss results of terrestrial broadband radiative quantities. First, we investigate atmospheric broadband terrestrial irradiance profiles in the presence of cirrus (Section 9.4.1). Second, we will look at terrestrial radiative cooling rates derived from the broadband terrestrial irradiances (Section 9.4.2). We will also touch on profiles of solar heating rates, although most of the section is concerned with terrestrial radiation.

9.4.1
Impact of Cirrus on Irradiance

Cirrus clouds are an important component of the Earth's climate system (Liou, 1986). The annually averaged global cirrus cover is estimated at 13–27% by Sassen and Wang (2008). 17% cirrus cover is given as a minimum value for the polar latitudes. In the tropics, the maximum of cirrus cloud cover can be as large as 45%.

Cirrus clouds interact (scatter and absorb) with solar radiation at wavelengths between 0.2–5 µm, absorb and emit terrestrial radiation between 5 and 50 µm, and, thus, influence the radiative energy budget of the Earth. Cirrus may either warm or cool the atmospheric layer below the cloud, depending on the vertical position and the thickness, the microphysical and optical properties, especially optical thickness. If the cirrus is optically thin, it generally acts similar to a greenhouse gas and warms the underlying atmosphere. If the cirrus is optically thick, it reflects a considerable portion of the incoming solar radiation which leads to a cooling effect.

Figure 9.6 illustrates the warming and cooling effects of cirrus in the terrestrial spectral region. This figure and parts of the discussion in this Section 9.4.1 are adapted from Wendisch et al. (2007). Figure 9.6 shows two typical examples of broadband TIR upward and downward irradiance profiles as simulated and observed during the Cirrus Regional Study of Tropical Anvils and Cirrus Layers – Florida Area Cirrus Experiment (CRYSTAL-FACE) (Jensen et al., 2004). Figure 9.6a,c are for an optically thin cirrus; Figure 9.6b,d show the simulation

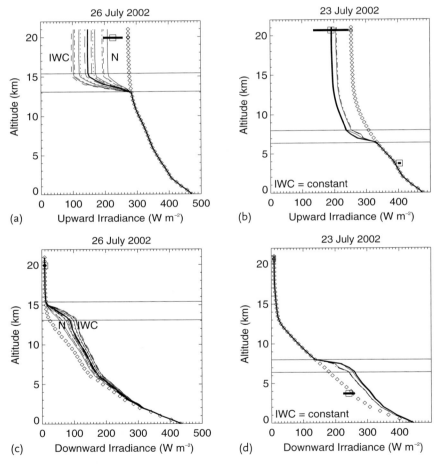

Figure 9.6 Profiles of simulated broadband TIR upward (a,b) and downward (c,d) irradiance. The horizontal lines mark the cirrus top and bottom heights respectively. The different curves represent simulation results assuming several crystal shapes. For details and curve notation see Table 1 and Figure 3 in Wendisch et al. (2007). Adapted from Wendisch et al. (2007). Reproduced with permission of AGU © AGU.

results for an optically thick cirrus. The open diamonds indicate the simulations without cirrus in cloudless conditions. The open squares show the average of the respective measured irradiance, and the horizontal bars represent the two-sigma standard deviations along the flight track.

The simulated profiles of the broadband TIR irradiance are displayed as different lines, which represent the results for different crystal shape assumptions. The results for two different ice crystal shape assumptions are marked by N (first scenario: N = constant) and IWC (second scenario: IWC = constant). For details of the scenarios and the curve notation see Wendisch et al. (2007).

Upward irradiance As evident from Figure 9.6a,c, the cirrus decreases the upward broadband TIR irradiance above its base height, compared to the cloudless case. If the cirrus is situated in a higher and colder environment, this greenhouse effect of the cirrus is more pronounced.

The optically thin cirrus, see Figure 9.6a, was located at a higher altitude and in a colder environment compared to the optically thick cirrus presented in Figure 9.6c. Therefore, the decrease of the upward irradiance from the optically thin cirrus is less pronounced because the competing effects of high altitude and small optical thickness are balanced. The optically thick cirrus, see Figure 9.6c, was situated at a lower altitude and in a warmer environment, and the compensating effects of lower altitude and large optical thickness are partially canceled.

The various lines in Figure 9.6a,c represent the simulation results for different ice crystal shapes. Below the cirrus, no crystal shape effects become evident. Above the cirrus base, there are significant shape effects on the upward irradiance if the optical thickness is small, see Figure 9.6a. For the cirrus with large optical thickness, the impact of crystal shape can be neglected, see Figure 9.6c.

Downward irradiance Figure 9.6b,d shows that the cirrus increases the downward broadband TIR irradiance compared to the cloudless curve due to emission. The magnitude of the emitted irradiance depends on the cirrus altitude, temperature, and optical thickness. The cirrus has almost no impact on the downward broadband TIR irradiance at the surface, and the shape effects are diminished near the surface.

In summary, the impact of crystal shape is important for the high, optically thin cirrus only.

9.4.2
Radiative Cooling and Heating

In this section, the expressions for the divergence of the net flux density will be presented first and will be followed by an equation for the local temperature change due to radiative processes of solar absorption and terrestrial emission. The spectral net flux density is defined by Eq. (3.50). For the broadband net flux density F_{net} in a horizontally homogeneous atmosphere, we have

$$F_{net} = F^\downarrow - F^\uparrow .$$ (9.65)

To calculate the effect of the radiation field on the local and temporal temperature change at a certain altitude z, we use the divergence of the flux density, that is,

$$\left.\frac{\partial T}{\partial t}\right|_z = \frac{1}{\varrho \cdot c_p} \frac{\partial F_{net}(z)}{\partial z} = \frac{1}{\varrho \cdot c_p} \frac{\partial}{\partial z}\left[F^\downarrow(z) - F^\uparrow(z)\right]$$

$$\approx \frac{1}{\varrho \cdot c_p} \cdot \left\{\frac{\left[F^\downarrow(z_2) - F^\downarrow(z_1)\right] - \left[F^\uparrow(z_2) - F^\uparrow(z_1)\right]}{z_2 - z_1}\right\},$$ (9.66)

where ϱ and c_p stand for the air density (water vapor plus dry air) and the specific heat at constant air pressure p. This equation basically means a local heating effect if

$$\left[F^\downarrow(z_2) - F^\downarrow(z_1)\right] > \left[F^\uparrow(z_2) - F^\uparrow(z_1)\right],$$ (9.67)

or in other words, local heating occurs if

$$F_{net}(z_2) > F_{net}(z_1) .$$ (9.68)

In general, Eq. (9.66) holds in both the solar and terrestrial spectral regions. In the solar wavelength region, we either have absorption causing a convergence of radiant energy in the considered local volume or solar heating. In the terrestrial spectral region, emission causes either a negative convergence (i.e., divergence) of radiant energy within the considered volume or terrestrial cooling.

Since the flux densities $F^\downarrow$ and $F^\uparrow$ in the terrestrial spectral region are evaluated in terms of the absorbing mass $u(z)$, see Eq. (9.50), we rewrite Eq. (9.66) in the form

$$\left.\frac{\partial T}{\partial t}\right|_z = \frac{1}{\varrho \cdot c_p} \frac{\partial F_{net}[u(z)]}{\partial z} = \frac{1}{\varrho \cdot c_p} \frac{\partial F_{net}}{\partial u}\frac{du}{dz} .$$ (9.69)

Using Eq. (9.50) yields

$$\left.\frac{\partial T}{\partial t}\right|_z = -\frac{1}{\varrho \cdot c_p} \frac{\partial F_{net}}{\partial u} \cdot \varrho_{abs} .$$ (9.70)

The specific humidity q is defined by

$$q = \frac{\varrho_{wv}}{\varrho} = \frac{\varrho_{wv}}{\varrho_{wv} + \varrho_{dry}} = \frac{\varrho_{abs}}{\varrho} ,$$ (9.71)

where ϱ_{wv} and ϱ_{dry} represent the water vapor and dry air densities, respectively. We assume that water vapor is the major absorber ($\varrho_{wv} \equiv \varrho_{abs}$) and thus obtain

$$\left.\frac{\partial T}{\partial t}\right|_z = -\frac{q(z)}{c_p} \frac{\partial F_{net}}{\partial u} .$$ (9.72)

With Eq. (9.65), the local temperature change is given by

$$\frac{\partial T}{\partial t}\bigg|_z = -\frac{q(z)}{c_\mathrm{p}} \cdot \left(\frac{\partial F^\downarrow}{\partial u} - \frac{\partial F^\uparrow}{\partial u}\right). \tag{9.73}$$

Subsequently, we need to know the broadband downward and upward flux densities, $F^\downarrow$ and $F^\uparrow$, which are given by spectral integration of Eqs. (9.57) and (9.64). In Eq. (9.73) a partial derivative $(\partial/\partial u)$ of an integral term is involved, see Eqs. (9.57) and (9.64). To calculate the partial derivative, we need to utilize the Leibniz rule given by

$$\frac{\partial}{\partial \zeta} \int_{u(\zeta)}^{v(\zeta)} \phi(x,\zeta)\,\mathrm{d}x = \int_{u(\zeta)}^{v(\zeta)} \frac{\partial}{\partial \zeta}[\phi(x,\zeta)]\,\mathrm{d}x + \phi(v,\zeta)\frac{\partial v}{\partial \zeta} - \phi(u,\zeta)\frac{\partial u}{\partial \zeta}, \tag{9.74}$$

where $\phi(x,\zeta)$, $u(\zeta)$, and $v(\zeta)$ are continuously differentiable functions with respect to the parameter ζ. In our current problem, we assign the variables as

$$\zeta \longrightarrow u, \tag{9.75}$$

$$x \longrightarrow M', \tag{9.76}$$

$$u(\zeta) \longrightarrow 0, \tag{9.77}$$

$$v(\zeta) \longrightarrow u, \tag{9.78}$$

and

$$\phi(x,\zeta) \longrightarrow F_{\mathrm{BB},\lambda}[T(M')]\frac{\mathrm{d}\mathcal{T}_\mathrm{F}[\lambda, M' - u(z)]}{\mathrm{d}M'}. \tag{9.79}$$

Applying the method of integration by parts to Eq. (9.64),

$$F_{\mathrm{diff},\lambda}^\uparrow[u(z)] = \varepsilon(\lambda, z = 0) \cdot F_{\mathrm{BB},\lambda}[T(z = 0)] \cdot \mathcal{T}_\mathrm{F}[\lambda, u(z)] + F_{\mathrm{BB},\lambda}\{T[u(z)]\}$$
$$- F_{\mathrm{BB},\lambda}\{T[u(z = 0)]\} \cdot \mathcal{T}_\mathrm{F}[\lambda, u(z) - u(z = 0)]$$
$$- \int_{u(z=0)}^{u(z)} \mathcal{T}_\mathrm{F}[\lambda, u(z) - M']\frac{\partial F_{\mathrm{BB},\lambda}[T(M')]}{\partial M'}\,\mathrm{d}M'. \tag{9.80}$$

To derive the partial derivative in Eq. (9.80), we apply the Leibniz rule in Eq. (9.74) to the fourth term on the right side of Eq. (9.80) and obtain

$$\frac{\partial}{\partial u(z)} \int_{u(z=0)}^{u(z)} \mathcal{T}_\mathrm{F}[\lambda, u(z) - M']\frac{\partial F_{\mathrm{BB},\lambda}[T(M')]}{\partial M'}\,\mathrm{d}M'$$
$$= \frac{\partial F_{\mathrm{BB},\lambda}\{T[u(z)]\}}{\partial u(z)}$$
$$+ \int_{u(z=0)}^{u(z)} \frac{\partial \mathcal{T}_\mathrm{F}[\lambda, u(z) - M']}{\partial u(z)}\frac{\partial F_{\mathrm{BB},\lambda}[T(M')]}{\partial M'}\,\mathrm{d}M'. \tag{9.81}$$

Thus,

$$
\frac{\mathrm{d}F^{\uparrow}_{\mathrm{diff},\lambda}[u(z)]}{\mathrm{d}u(z)} =
$$

$$
\pi \cdot \left\{ \varepsilon(\lambda, z = 0) \cdot B^{\uparrow}_{\mathrm{BB},\lambda}[T(z = 0)] - B^{\uparrow}_{\mathrm{BB},\lambda}[T(u(z = 0))] \right\} \frac{\mathrm{d}\mathcal{T}_{\mathrm{F}}[\lambda, u(z)]}{\mathrm{d}u(z)}
$$

$$
- \pi \int_{u(z=0)}^{u(z)} \frac{\partial \mathcal{T}_{\mathrm{F}}(\lambda, u(z) - M')}{\partial u(z)} \frac{\partial B^{\uparrow}_{\mathrm{BB},\lambda}[T(M')]}{\partial M'} \, \mathrm{d}M' .
$$

$$(9.82)$$

Similarly, we have

$$
\frac{\mathrm{d}F^{\downarrow}_{\mathrm{diff},\lambda}[u(z)]}{\mathrm{d}u(z)} = -\pi \cdot B^{\downarrow}_{\mathrm{BB},\lambda} \left\{ T[u(z \longrightarrow \infty)] \right\} \frac{\mathrm{d}\mathcal{T}_{\mathrm{F}}[\lambda, u(z) - u(z \longrightarrow \infty)]}{\mathrm{d}u(z)}
$$

$$
- \pi \int_{u(z \longrightarrow \infty)}^{u(z)} \frac{\partial \mathcal{T}_{\mathrm{F}}(\lambda, u(z) - M')}{\partial u(z)} \frac{\partial B^{\downarrow}_{\mathrm{BB},\lambda}[T(M')]}{\partial M'} \, \mathrm{d}M' .
$$

$$(9.83)$$

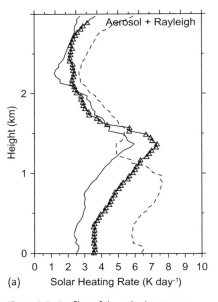
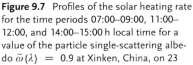
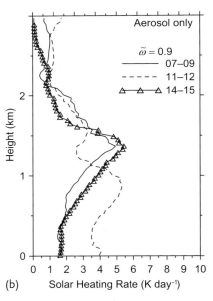

Figure 9.7 Profiles of the solar heating rate for the time periods 07:00–09:00, 11:00–12:00, and 14:00–15:00 h local time for a value of the particle single-scattering albedo $\tilde{\omega}(\lambda) = 0.9$ at Xinken, China, on 23 October 2004. (a) Shows the total solar heating rates, and (b) depicts the heating rates due to aerosol particles only. Reprinted from Wendisch et al. (2008) with permission of Elsevier.

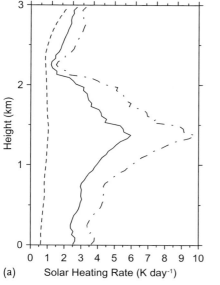

Figure 9.8 Average solar heating (a) and terrestrial cooling (b) rate profiles for the time periods 07:00–09:00 h local time for different values of the particle single-scattering albedo $\tilde{\omega}(\lambda)$ at Xinken, China, on 23 October 2004. Both aerosol and Rayleigh contributions are included. Reprinted from Wendisch et al. (2008) with permission of Elsevier.

After wavelength integration of Eqs. (9.82) and (9.83), we can derive the local temperature change from Eq. (9.73).

As an example, for simulation results, we consider the impact of aerosol particles on solar radiative heating measured in the Pearl River Delta in Southeast China. The two following figures and parts of the text in this section are adapted from Wendisch et al. (2008). In Figure 9.7a, both scattering and absorption by aerosol particles and gas molecules are included, and in Figure 9.7b, the Rayleigh contribution is subtracted. The simulations have been performed for a particle single-scattering albedo of $\tilde{\omega}(\lambda) = 0.9$. The presence of aerosol particles causes substantial warming with maximum values at the top of the aerosol layer of up to $5\,\mathrm{K\,day^{-1}}$. This warming causes the atmospheric stratification to become more stable.

The solar heating due to the absorbing particles is generally not counterbalanced by the terrestrial cooling. Figure 9.8 compares the solar heating rates (Figure 9.8a) with the respective terrestrial cooling rates (Figure 9.8b) for different values of particle single-scattering albedo $\tilde{\omega}$. Most of the solar heating occurs at the top of the Planetary Boundary Layer (PBL), which causes a stabilizing effect. With increasing particulate absorption (decreasing $\tilde{\omega}$), the aerosol stabilizes the PBL stratification further and reduces convective mixing and PBL heights.

Problems

Problem 9.1 Cloud Radiative Forcing

The radiative forcing ΔF of a cloud is defined as the difference of the TOA net irradiance between cloudy and cloud-free conditions, that is,

$$\Delta F = \left(F^\downarrow - F^\uparrow \right)_c - \left(F^\downarrow - F^\uparrow \right)_{no} . \tag{9.84}$$

Here, the index c indicates the atmosphere with the cloud, and the index no indicates the clear atmosphere without the cloud.

a) What does a negative value of ΔF indicate, a warming or a cooling of the Earth-atmosphere system? Use the online radiative transfer model at: snowdog.larc.nasa.gov/cgi-bin/rose/flp200503/flp200503.cgi to estimate the radiative forcing of clouds. The input controls include fields for two clouds and two aerosol layers. Make sure that the optical depth of both aerosol layers is zero. Leave the other fields at their default values. The model is started by a click on COMPUTE (upper left). The results are then printed below the input section. Note the NET TOA Forced value for both solar and terrestrial irradiance in the following scenarios.

b) Increase the optical depth of CLOUD 1 (a cirrus cloud) between 1 and 10. Compare the change in the solar and in the terrestrial range.

c) Now, investigate a water cloud: Set the cloud fraction of CLOUD 1 to zero and that of CLOUD 2 to one. Calculate and plot the radiative forcing of this water cloud for optical depths between 1 and 50. At which optical depth does the cloud block 50% of the downward irradiance at the surface (right part of the table)? Why does the longwave TOA radiative forcing stall at low values, much lower than those of the cirrus cloud?

d) Although the total radiative forcing (the sum of the solar and terrestrial components) of the cloud is always negative, why do clouds still have a warming effect on frosty nights?

e) Change the surface albedo type from "17 Ocean" to "FLAT" and change its numerical value (the default field value is IGBP) between zero and one for a cirrus cloud (CLOUD 1) of optical depth five. From which value of the surface albedo is the total TOA radiative forcing of the cloud positive? Why?

Problem 9.2 Dust Radiative Forcing

The net (solar plus terrestrial) radiative forcing of aerosol depends linearly on the surface albedo. The values in the following Table 9.1 have been determined over the Sahara desert (for an aerosol plume of pure mineral dust with a single-scattering albedo of 0.96):

Table 9.1 Dust radiative forcing observed in the Sahara.

Surface albedo	Dust radiative forcing $(W\,m^{-2})$
0.321	24.35
0.283	18.41
0.254	15.89
0.171	5.53

a) What is the radiative forcing of the same dust plume when it is blown out over the Atlantic Ocean?

b) What would happen qualitatively if the aerosol plume consisted of soot from savannah fires?

c) What would you expect to be the most likely sign of global aerosol radiative forcing?

Appendix A
Abbreviations, Symbols, and Constants

A.1
Acronyms

1D	One-Dimensional
2D	Two-Dimensional
3D	Three-Dimensional
6D	Six-Dimensional
ABI	Advanced Baseline Imager
A-DM	Adding-Doubling Method
AGU	American Geophysical Union
AMS	American Meteorological Society
AU	Astronomical Unit
BMCM	Backward Monte Carlo Method
BOA	Bottom of the Atmosphere
BRDF	Bidirectional Reflectance Distribution Function
CFC	Chlorofluorocarbons
CKDM	Correlated k-Distribution Method
DFM	Delta-Fit Method
DMM	Delta-M Method
DOM	Discrete Ordinate Method
EM	Electromagnetic
FIR	Far Infrared
HITRAN	High-Resolution Transmission Molecular Absorption Database
IGOM	Improved Geometric-Optics Method
IR	Infrared
IWC	Ice Water Content
LBLM	Line-By-Line Method
LBLRTM	Line-By-Line Radiative Transfer Model
LIDAR	Light Detection and Ranging
LIM	Leipzig Institute for Meteorology
LMS	Lorenz–Mie Structure
LWC	Liquid Water Content
MCM	Monte Carlo Method

Theory of Atmospheric Radiative Transfer, First Edition. Manfred Wendisch and Ping Yang
© 2012 WILEY-VCH Verlag GmbH & Co. KGaA. Published 2012 by WILEY-VCH Verlag GmbH & Co. KGaA.

MODIS	Moderate Resolution Imaging Spectroradiometer
MODTRAN	Moderate Spectral Resolution Atmospheric Transmittance Algorithm and Computer Model
MSOS	Method of Successive Order of Scattering
MW	Microwave
NIR	Near Infrared
NSD	Number Size Distribution
OLR	Outgoing Longwave Radiation
OSA	Optical Society of America
PBL	Planetary Boundary Layer
PDF	Probability Distribution Function
RADAR	Radio Wave Detection and Ranging
RTE	Radiative Transfer Equation
SHM	Spherical Harmonics Method
SI	Le Système International d'Unité
TIR	Thermal Infrared
TOA	Top of the Atmosphere
TSA	Two-Stream Approximation
UV	Ultraviolet
VIS	Visible

A.2
Subscripts and Superscripts

$X_{\parallel}$	Parallel with respect to a reference plane (mostly the scattering plane)
$X_{\perp}$	Perpendicular with respect to a reference plane (mostly the scattering plane)
X_{abs}	Absorption
X_{act}	Actinic
X_{ato}	Atomic
X_{B}	Brewster
X_{BB}	Blackbody
X_{BOA}	Bottom of atmosphere
X_{C}	Cloud
X_{cen}	Centrifugal
X_{cir}	Circularly polarized
X_{Col}	Coulomb
X_{dfr}	Diffraction
X_{diff}	Diffuse
X_{dir}	Direct
X_{D}	Doppler
X_{DOM}	Discrete ordinates method
X_{E}	Earth

X_{Edd}	Eddington approximation
X_{eff}	Effective
X_{ele}	Electron
X_{emi}	Emission
X_{ext}	Extinction
X_g	Global
X_{HG}	Henyey–Greenstein
X_{hom}	Homogeneous
X_{ice}	Ice
X_{im}	Imaginary
X_{inc}	Incident
X_{int}	Integrated
X_{iso}	Isotropic
X_{kin}	Kinetic
X_L	Lorentz
X_{lin}	Linearly polarized
X_{lw}	Liquid water
X_m	Medium
X_{max}	Maximum
X_{mol}	Molecule
X_{Mie}	Lorenz–Mie
X_{min}	Minimum
X_{net}	Net
X_{nor}	Normal
X_{orb}	Orbital
X_p	Planetary
X_{phot}	Photon
X_{pot}	Potential
X_{prim}	Primary
X_{prot}	Proton
X_r	Reference
X_{rad}	Radiant
X_{Rayl}	Rayleigh
X_{re}	Real
X_{ref}	Reflected
X_{rot}	Rotational
X_S	Sun
X_{sca}	Scattered
X_{sla}	Slant path
X_{sur}	Surface
X_T	Total reflection
X_{theo}	Theoretical
X_{top}	Top
X_{tot}	Total
X_{tra}	Transmitted/refracted

X_{trl}	Translational
X_{unp}	Unpolarized
X_V	Voigt
X_{vib}	Vibrational
X_{vol}	Volume
X_{wv}	Water vapor
$X^{\uparrow}$	Upward
$X^{\downarrow}$	Downward
X^*	Complex conjugate

A.3
Greek Symbols

Symbol	Unit	Name	Eq.
α	–	Size parameter	(4.75)
α_L	cm^{-1}	Half-width of a Lorentz line	(8.79)
α_D	cm^{-1}	Half-width of a Doppler line	(8.84)
γ	–	Bidirectional reflectance distribution function (BRDF)	(6.77)
γ_p	–	Planetary (local) surface albedo	(6.78)
γ_g	–	Global (spherical) surface albedo	(6.79)
δ_{ij}	–	Kronecker symbol	(2.14)
Δ	–	Deviation of ray	(4.243)
$\delta(x)$	–	Dirac δ-function	(2.28)
ϵ	$A\,s\,V^{-1}\,m^{-1} = F\,m^{-1}$	Complex electric permittivity	(3.5)
ε	–	Emissivity	(3.89)
ε_{ijk}		Permutation symbol	(2.11)
Θ	rad or °	Angle of ray (geometric optics, with respect to normal on surface)	(4.157)
$\Theta_{inc,B}$	rad or °	Brewster incident angle	(4.223)
$\Theta_{inc,T}$	rad or °	Critical incident angle of total reflection	(4.171)
Θ_{inc}	rad or °	Angle of incident ray	(4.157)
Θ_{ref}	rad or °	Angle of reflected ray	(4.157)
Θ_{tra}	rad or °	Angle of transmitted/refracted ray	(4.158)
θ	rad or °	Atmospheric zenith angle	(2.34)
ϑ	rad or °	Scattering angle	(2.41)
κ	$V\,s\,A^{-1}\,m^{-1} = H\,m^{-1}$	Complex magnetic permeability	(3.6)
λ	µm	Wavelength	(3.24)
μ	–	Zenith distance	(3.45)
ν	$Hz = s^{-1}$	Frequency	(3.24)
$\tilde{\nu}$	cm^{-1}	Wavenumber	(3.25)
ξ_λ	$m^{-3}\,sr^{-1}\,µm^{-1}$	Spectra photon distribution function	(6.16)
π_n	–	Angular coefficients of nth order	(4.84)
ρ	$kg\,m^{-3}$	Material density	(5.5)
τ	–	Optical thickness	(6.2)
τ^*	–	Total atmospheric optical thickness	(6.3)
τ_n	–	Angular coefficients of nth order	(4.85)
ϕ		Arbitrary scalar function	
φ	rad or °	Azimuth angle	(2.34)
Φ_λ	$W\,µm^{-1}$	Spectral radiant energy flux	(3.28)

Symbol	Unit	Name	Eq.
ψ	rad or $°$	Rotation angle	(4.18)
Ω	sr	Solid angle	(2.38)
$d^2\Omega$	sr	Differential solid angle	(2.35)
ω_c	$s^{-1} = $ Hz	Circular frequency	(3.24)
$\tilde{\omega}$	–	Single-scattering albedo	(4.50)
$\vec{\nabla}$	m^{-1}	Del operator	(2.5)
∇^2	m^{-2}	Scalar product of two del operators	(2.22)

A.4
Latin Symbols

Symbol	Unit	Name	Eq.
IA	–	Complex amplitude scattering matrix	(4.12)
$A_{ij}, \quad i, j = 1, 2$	–	Complex scattering amplitudes	(4.12)
d^2A	m^2	Differential area element	(2.36)
A_{proj}	m^2	Projected particle cross section	(4.49)
$\mathcal{A}$	–	Absorptivity	(3.57)
$\vec{a}$		Arbitrary vector	
a	m^3	Polarizability	(4.110)
a_n	–	Lorenz–Mie-coefficients of nth order	(4.76)
$\vec{B}$	$V\,s\,m^{-2}$	Complex magnetic induction vector	(3.6)
B	cm^{-1}	Rotational constant	(8.44)
B_λ	$W\,m^{-2}\,sr^{-1}\,\mu m^{-1}$	Spectral Planck function	(3.58)
$\vec{b}$		Arbitrary vector	
b	–	Backscatter fraction	(7.268)
b_n	–	Lorenz–Mie-coefficients of nth order	(4.76)
$b_{abs}, b_{sca}, b_{ext}$	m^{-1}	Volumetric absorption, scattering, and extinction coefficients	(5.16)–(5.18)
$C_{abs}, C_{sca}, C_{ext}$	m^2	Absorption, scattering, and extinction cross sections	(4.47)
C_n	–	Legendre coefficients	(2.68)
C'_n	–	Legendre coefficients of the δ-scaled, truncated phase function	(7.39)
$\vec{c}$		Arbitrary vector	
c_m	$m\,s^{-1}$	Speed of light in a medium	(3.15)

Symbol	Unit	Name	Eq.
c_i	–	Gaussian weights	(2.71)
$\vec{\mathbf{D}}$	$\mathrm{A\,s\,m^{-2}}$	Complex dielectric displacement vector	(3.5)
D	$\mathrm{\mu m}$	Maximum particle dimension	
D_{eff}	$\mathrm{\mu m}$	Effective diameter	(5.9)
$\vec{\mathbf{E}}$	$\mathrm{V\,m^{-1}}$	Complex electric field vector	(3.4), (4.6)
$\vec{\mathbf{E}}_0$	$\mathrm{V\,m^{-1}}$	Complex electric amplitude vector	(4.6)
E	$\mathrm{V\,m^{-1}}$	Complex magnitude of $\vec{\mathbf{E}}$	
E_0	$\mathrm{V\,m^{-1}}$	Complex magnitude of $\vec{\mathbf{E}}_0$	
E_{rad}	$\mathrm{J = W\,s}$	Radiant energy	(3.28)
E_{phot}	$\mathrm{J = W\,s}$	Photon energy	(3.26)
$\hat{\mathbf{e}}_1, \hat{\mathbf{e}}_2, \hat{\mathbf{e}}_3$	–	Euklidic base vectors	(2.1)
F	$\mathrm{W\,m^{-2}}$	Spectrally integrated radiant energy flux density, or irradiance	(3.37)
F_{BB}	$\mathrm{W\,m^{-2}}$	Blackbody irradiance	(3.77)
F_λ	$\mathrm{W\,m^{-2}\,\mu m^{-1}}$	Spectral radiant energy flux density or irradiance	(3.29)
$F_{\mathrm{diff},\lambda}$	$\mathrm{W\,m^{-2}\,\mu m^{-1}}$	Spectral diffuse radiant energy flux density, or irradiance	(7.45), (7.46)
$F_{\mathrm{phot},\lambda}$	$\mathrm{W\,m^{-2}\,\mu m^{-1}}$	Spectral photon flux density	(3.27)
f	–	Scattering function (not normalized)	(4.53)
f_{fws}	–	Fraction of forward-scattering	(7.13), (7.36)
g	–	Asymmetry factor	(4.64)
$\vec{\mathbf{H}}$	$\mathrm{A\,m^{-1}}$	Complex magnetic field vector	(3.2)
I	$\mathrm{W\,m^{-2}\,sr^{-1}}$	Spectrally integrated radiance	(3.36)
I_λ	$\mathrm{W\,m^{-2}\,sr^{-1}\,\mu m^{-1}}$	Spectral radiance	(3.30)
$I_{\mathrm{diff},\lambda}$	$\mathrm{W\,m^{-2}\,sr^{-1}\,\mu m^{-1}}$	Diffuse spectral radiance	(6.47)
$I_{\mathrm{dir},\lambda}$	$\mathrm{W\,m^{-2}\,sr^{-1}\,\mu m^{-1}}$	Direct spectral radiance	(6.47)
IWC	$\mathrm{g\,m^{-3}}$	Ice water content	(5.8)
$J_{\mathrm{diff},\lambda}$	$\mathrm{W\,m^{-2}\,sr^{-1}\,\mu m^{-1}}$	Source function of diffuse radiance	(6.70)
$J_{\mathrm{dir},\lambda}$	$\mathrm{W\,m^{-2}\,sr^{-1}\,\mu m^{-1}}$	Source function of direct radiance	(6.69)
$J_{\mathrm{emi},\lambda}$	$\mathrm{W\,m^{-3}\,sr^{-1}\,\mu m^{-1}}$	Source function of emitted radiance	(6.45)
$j_{\mathrm{emi},\lambda}$	$\mathrm{s^{-1}}$	Source coefficient of emission	(6.41)
k	$\mathrm{m^{-1}}$	Modified wavenumber in a vacuum	(3.21)
k_{m}	$\mathrm{m^{-1}}$	Modified wavenumber in a medium	(4.7)
k_{abs}	$\mathrm{m^2\,kg^{-1}}$	Mass absorption coefficient	(8.97)
LWC	$\mathrm{g\,m^{-3}}$	Liquid water content	(5.7)
$\tilde{m}$	–	Relative refractive index	(4.172)
N	$\mathrm{m^{-3}}$	Number concentration	(5.1)
$N_{\mathrm{phot},\lambda}$	$\mathrm{s^{-1}\,m^{-2}\,\mu m^{-1}}$	Spectral photon number flux density	(3.27)
N_{mol}	$\mathrm{J^{-1}}$	Partition number of energy levels	(8.7)
$\hat{\mathbf{n}}$	–	Normal unit vector of a surface	(3.45)
$n(D)$	$\mathrm{m^{-3}\,\mu m^{-1}}$	Number size distribution	(5.2)
n_{ato}	$\mathrm{m^{-3}}$	Number concentration of atoms	

Symbol	Unit	Name	Eq.
n_{mol}	m^{-3}	Number concentration of air molecules	(4.142)
$n_{phot,\lambda}$	$m^{-3} sr^{-1} \mu m^{-1}$	Spectral photon number density	(3.43)
$d^6 n_{phot}$	–	Differential photon number	(6.15)
$\tilde{n}$	–	Complex refractive index	(4.1)
$\tilde{n}_{re}$	–	Real part of $\tilde{n}$	(4.2)
$\tilde{n}_{im}$	–	Imaginary part of $\tilde{n}$	(4.3)
O		Origin of Cartesian coordinate system	
IP	–	Real phase matrix (normalized Mueller matrix)	(4.33)
$\vec{P}$	A s m	Electric dipole moment vector	(4.108)
$\vec{P}_0$	A s m	Amplitude vector of $\vec{P}$	(4.109)
$P_{ij}, \quad i, j = 1, 2, 3, 4$	–	Real elements of IP	(4.35)
$\mathcal{P}$	–	Phase function normalized to 4π	(4.61)
$\mathcal{P}'$	–	Phase function for δ-scaling	(7.31)
P	–	Degree of polarization	(4.19)
P_n	–	Legendre polynomials	(2.43)
$P_n^{(j)}$	–	Legendre functions (Legendre polynomials of jth order)	(2.52)
$\tilde{P}_n^{(j)}$	–	Renormalized Legendre functions	(2.62)
p	sr^{-1}	Phase function normalized to 1	(4.58)
$Q_{abs}, Q_{sca}, Q_{ext}$	–	Absorption, scattering, and extinction efficiency factors	(4.49)
Q	$W m^{-2}$	Parallel minus perpendicular linear polarized irradiance	(4.15)
$\vec{R}$	m	Radial distance vector	(4.107)
R	m	Radial distance from a scattering particle	(4.11)
$\mathcal{R}$	–	Reflectivity	(4.212), (4.213)
$\vec{r}$	–	Euklidic position vector	(3.28)
r	m	Radius of a sphere	(4.75)
r_{eff}	m	Effective radius	(5.9)
IS	–	Real Mueller matrix	(4.32)
$\vec{S}$	$W m^{-2}$	Stokes vector	(4.13)
S	cm^{-2}	Line intensity	(8.78)
$S_{ij}, \quad i, j = 1, 2, 3, 4$	–	Real elements of IS	(4.41)–(4.46)
$S_{dir,\lambda}$	$W m^{-2} \mu m^{-1}$	Direct spectral irradiance	(6.13)
$S_{dir,\lambda,TOA}$	$W m^{-2} \mu m^{-1}$	Direct spectral irradiance at the TOA	(6.14)
$\hat{s}$	–	Direction unit vector	(2.41)
s	m	Slant path length	(6.10)

Symbol	Unit	Name	Eq.
T	K	Absolute temperature	
$T_{\text{p,vib}}$	s	Period of an vibrating harmonic oscillation	(8.60)
$\mathcal{T}$	–	Transmissivity	(4.204), (4.205)
t	s	Time	
U	W m^{-2}	Linear polarized irradiance under $45°$	(4.16)
u_λ	$\text{J m}^{-3}\,\text{sr}^{-1}\,\mu\text{m}^{-1}$	Spectral radiant energy density	(3.40)
u	kg m^{-2}	Absorption path length, or absorbing mass	(8.95)
V	W m^{-2}	Circularly polarized irradiance	(4.17)
$\text{d}^3 V$	m^3	Differential volume element	(6.15)
v_{eff}	–	Effective variance of size distribution	(5.13)
W	cm^{-1}	Equivalent width	(8.99)
X, Y, Z	–	Axes of Cartesian coordinate system	
x, y, z	m	Cartesian coordinates	
		(z vertical altitude above ground)	
x_i		Gaussian abscissas	(2.70)

A.5
Physical Constants

Greek:

$\epsilon_0 = 8.8542 \times 10^{-12}\,\text{A s V}^{-1}\,\text{m}^{-1}$ Electric permittivity of a vacuum
$\kappa_0 = 1.257 \times 10^{-6}\,\text{V s A}^{-1}\,\text{m}^{-1}$ Magnetic permeability of a vacuum
$\sigma = 5.671 \times 10^{-8}\,\text{W m}^{-2}\,\text{K}^{-4}$ Stefan–Boltzmann constant

Latin:

$c = 2.997\,925 \times 10^8\,\text{m s}^{-1}$ Speed of light in a vacuum
$e = 1.602 \times 10^{-19}\,\text{A s}\ (1\,\text{A s} = 1\,\text{C})$ Electrical elementary charge
$h = 6.6262 \times 10^{-34}\,\text{J s} = 4.138 \times 10^{-15}\,\text{eV s}$ Planck's constant
$\hbar = h/2\pi = 1.054\,59 \times 10^{-34}\,\text{J s}$ Modified Planck's constant
$k_B = 1.3805 \times 10^{-23}\,\text{J K}^{-1} = 8.62 \times 10^{-5}\,\text{eV K}^{-1}$ Boltzmann constant
$k_W = 2897\,\mu\text{m K}$ Constant in Wien's displacement law
$m_{\text{ele}} = 9.108 \times 10^{-31}\,\text{kg}$ Static electron mass
$R_{\text{orb,H}} = 1.0974 \times 10^7\,\text{m}^{-1}$ Rydberg's constant for hydrogen

A.6
Mathematical Constants

$\pi = 3.141\,592\,653\,589$
$e = 2.718\,281\,828$
$i = \sqrt{-1}$

References

Adams, J. and Kattawar, G. (1997) Neutral points in an atmosphere-ocean system. 1. Upwelling light field. *Appl. Opt.*, **36** (9), 1976–1986.

Aden, A. and Kerker, M. (1951) Scattering of electromagnetic waves from two concentric spheres. *J. Appl. Phys.*, **22** (10), 1242–1246.

Ambartzumian, V. (1936) The effect of absorption lines on the radiative equilibrium of the outer layers of stars. *Publ. Astron. Obs., Univ. Leningr.*, **6**, 7–18.

Anderson, G., Clough, S., Kneizys, F., Chetwynd, J., and Shettle, E. (1986) AFGL Atmospheric Constituent Profiles (0–120 km), *Technical Report AFGL-TR-86-0110*, AFGL (OPI), Hanscom AFB, MA 01736.

Arfken, G. and Weber, H. (2005) *Mathematical Methods for Physicists*, 6th edn, Elsevier Academic Press, Burlington, USA.

Asano, S. and Yamamoto, G. (1975) Light scattering by randomly oriented spheroidal particles. *Appl. Opt.*, **14**, 29–49.

Asano, S. and Yamamoto, G. (1995) Light scattering by a spheroidal particle. *Appl. Opt.*, **14** (1), 29–49.

Bates, R. (1975) Analytic constraints on electromagnetic computations. *IEEE Trans. Microwave Theory Tech.*, **MTT-23**, 605–622.

Baum, B., Heymsfield, A., Yang, P., and Thomas, S. (2005) Bulk scattering properties for the remote sensing of ice clouds I: Microphysical data and models. *J. Appl. Meteorol.*, **44**, 1885–1895.

Berk, A., Anderson, G., Bernstein, L., Acharya, P., Dothe, H., Matthew, M., Adler-Golden, S., Chetwynd Jr., J., Richtsmeier, S., Pukall, B., Allred, C., Jeong, L., and Hoke, M. (1999) MODTRAN4 radiative transfer modeling for atmospheric correction. *Proc. SPIE*, **3756**, 348–353.

Bhandari, R. (1985) Scattering coefficients for a multilayered sphere: Analytic expressions and algorithms. *Appl. Opt.*, **24** (13), 1960–1967.

Bi, L., Yang, P., Kattawar, G., and Kahn, R. (2010) Modeling optical properties of mineral aerosol particles by using nonsymmetric hexahedra. *Appl. Opt.*, **49** (3), 334–342.

Bierwirth, E., Wendisch, M., Ehrlich, A., Heese, B., Tesche, M., Althausen, D., Schladitz, A., Müller, D., Otto, S., Trautmann, T., Dinter, T., von Hoyningen-Huene, W., and Kahn, R. (2009) Spectral surface albedo over Morocco and its impact on the radiative forcing of Saharan dust. *Tellus B*, **61**, 252–269.

Bohr, N. (1913) On the constitution of atoms and molecules. *Philos. Mag.*, **26**, 1–25.

Bohren, C. (1974) Light scattering by an optically active sphere. *Chem. Phys. Lett.*, **29** (3), 458–462.

Bohren, C. (1978) Scattering of electromagnetic waves by an optically active cylinder. *J. Colloid Interface Sci.*, **66** (1), 105–109.

Bohren, C. and Clothiaux, E. (2006) *Fundamentals of Atmospheric Radiation*, John Wiley & Sons, Ltd, Chichester.

Bohren, C. and Huffman, D. (1983) *Absorption and Scattering of Light by Small Particles*, John Wiley & Sons, Ltd, New York.

Borghese, F., Denti, P., and Saija, R. (2007) *Scattering from Model Non-Spherical Particles – Theory and Applications to Environmental Physics*, Springer, Berlin.

Born, M. and Wolf, E. (2003) *Principles of Optics*, 7th edn, Cambridge University Press.

Bronstein, I. and Semendjajev, K. (1985) *Taschenbuch der Mathematik*, 22nd edn, Verlag Harri Deutsch, Thun, Frankfurt am Main.

Chandrasekhar, S. (1950) *Radiative Transfer*, Oxford University Press, UK.

Chou, M.D. and Arking, A. (1980) Computation of infrared cooling rates in the water vapor bands. *J. Atmos. Sci.*, **37**, 855–867.

Chwolson, O. (1889) Grundzüge einer mathematischen Theorie der inneren Diffusion des Lichtes. *Bull. Acad. Imp. Sci. St. Petersburg*, **33**, 221–256.

Chylek, P. (1977) Light scattering by small particles in an absorbing medium. *J. Opt. Soc. Am.*, **67**, 561–563.

Clough, S., Iacono, M., and Moncet, J.L. (1992) Line-by-line calculation of atmospheric fluxes and cooling rates: application to water vapor. *J. Geophys. Res.*, **97**, 15761–15785.

Coburn, N. (1955) *Vector and Tensor Analysis*, The Macmillan Company, New York.

Collins, D., Blättner, W., Wells, M. and Horak, H. (1972) Backward Monte Carlo calculations of the polarization characteristics of the radiation emerging from spherical-shell atmospheres. *Appl. Opt.*, **11**, 2684–2696.

Cowling, T. (1950) Atmospheric absorption of heat radiation by water vapour. *Philos. Mag.*, **41** 109–123.

Curtis, A. (1952) Contribution to a discussion of "A statistical model for water vapor absorption" by R.M. Goody. *Q. J. R. Meteorol. Soc.*, **78**, 638–640.

Dave, J.V. and Armstrong, B. (1970) Computations of high-order associated Legendre polynomials. *J. Quant. Spectrosc. Radiat. Transf.*, **10**, 557–562.

Debye, P. (1909) Der Lichtdruck auf Kugeln von beliebigem Material. *Ann. Phys.*, **335**, 57–136.

Eddington, A. (1916) On the radiative equilibrium of the stars. *Mon. Not. R. Astron. Soc.*, **77**, 16–35.

Edwards, D. and Francis, G. (2000) Improvements to the correlated-k radiative transfer method: application to satellite remote sensing. *J. Geophys. Res.*, **105** (D14), 18135–18156.

Evans, K. (1998) The spherical harmonics discrete ordinate method for three-dimensional atmospheric radiative transfer. *J. Atmos. Sci.*, **55**, 429–446.

Evans, K. (2007) SHDOMPPDA, a radiative transfer model for cloudy sky data assimilation. *J. Atmos. Sci.*, **64**, 3854–3864.

Fu, Q. and Liou, K. (1992a) A three-parameter approximation for radiative transfer in non-homogeneous atmospheres: application to the O_3 9.6-μm band. *J. Geophys. Res.*, **97**, 13051–13058.

Fu, Q. and Liou, K. (1992b) On the correlated k-distribution method for radiative transfer in nonhomogeneous atmospheres. *J. Atmos. Sci.*, **49**, 2139–2156.

Fu, Q. and Sun, W. (2001) Mie theory for light scattering by a spherical particle in an absorbing medium. *Appl. Opt.*, **40**, 1354–1361.

Gao, B.C. and Kaufman, Y. (1995) Selection of the 1.375-μm MODIS channel for remote sensing of Cirrus clouds and stratospheric aerosols from space. *J. Atmos. Sci.*, **52**, 4231–4237.

Godson, W. (1953) The evaluation of infrared radiative fluxes due to atmospheric water vapor. *Q. J. R. Meteorol. Soc.*, **79**, 367–379.

Godson, W. (1955) The computation of infrared transmission by atmospheric water vapor. *J. Atmos. Sci.*, **12**, 272–284.

Goody, R. (1952) A statistical model for water vapor absorption. *Q. J. R. Meteorol. Soc.*, **78**, 165–169.

Goody, R. (1964a) *Atmospheric Radiation I: Theoretical Basis*, Clarendon Press, Oxford.

Goody, R. (1964b) The transmission of radiation through an inhomogeneous atmosphere. *J. Atmos. Sci.*, **21**, 575–581.

Goody, R. and Yung, Y. (1989) *Atmospheric Radiation – Theoretical Basis*, Oxford University Press, Oxford.

Greenler, R. (1990) *Rainbows, Halos, and Glories*, Cambridge University Press, New York.

Hansen, J. (1971) Multiple scattering of polarized light in planetary atmospheres. Part II. Sunlight reflected by terrestrial water clouds. *J. Atmos. Sci.*, **28**, 1400–1426.

Hansen, J. and Travis, L. (1974) Light scattering in planetary atmospheres. *Space Sci. Rev.*, **16**, 527–610.

Hildebrand, E. (1974) *Introduction to Numerical Analysis*, Dover, Mineola, New York.

Houghton, J.T. and Smith, S.D. (1966) *Infrared Physics*, Clarendon Press, Oxford.

Hovenier, J.W., Van Der Mee, C., and Domke, H. (2004) *Transfer of Polarized Light in Planetary Atmospheres*, Kluwer Academic Publishers, Dordrecht, Netherlands.

Hu, Y.X., Wielicki, B., Lin, B., Gibson, G., Tsay, S.C., Stamnes, K., and Wong, T. (2000) δ-Fit: a fast and accurate treatment of particle scattering phase functions with weighted singular-value decomposition least-squares fitting. *J. Quant. Spectrosc. Radiat. Transf.*, **65**, 681–690.

Huang, X. and Yung, Y. (2004) A common misunderstanding about the Voigt line profile. *J. Atmos. Sci.*, **61**, 1630–1632.

Jackson, J. (1975) *Classical Electrodynamics*, 2nd edn, John Wiley & Sons, Inc., New York.

Jackson, J. (1999) *Classical Electrodynamics*, 3rd edn, John Wiley & Sons, Inc., New York.

Jain, U. (2007) *Introduction to Atomic and Molecular Spectroscopy*, Alpha Science International Ltd., Oxford, UK.

Jensen, E., Starr, D., and Toon, O. (2004) Mission investigates tropical cirrus clouds. *EOS*, **85** (5), 45–50.

Johnson, J. (1954) *Physical Meteorology*, The M.I.T. Press, Cambridge, Massachusetts.

Jones, D. (1957a) Approximate methods in high-frequency scattering. *Proc. R. Soc. A*, **239**, 338–348.

Jones, D. (1957b) High-frequency scattering of electromagnetic waves. *Proc. R. Soc. A*, **240**, 206–213.

Joseph, J., Wiscombe, W., and Weinman, J. (1976) The Delta-Eddington approximation for radiative flux transfer. *J. Atmos. Sci.*, **33**, 2452–2459.

Kahnert, F. (2003) Numerical methods in electromagnetic scattering theory. *J. Quant. Spectrosc. Radiat. Transf.*, **79**, 775–824.

Keeling, C., Bacastow, R., Bainbridge, A., Ekdahl, Jr., C., Guenther, P., and Waterman, L. (1976) Atmospheric carbon dioxide variations at Mauna Loa Observatory, Hawaii. *Tellus*, **28**, 538–551.

Kim, C. and Yeh, C. (1991) Scattering of an obliquely incident wave by a multilayered elliptical lossy dielectric cylinder. *Radio. Sci.*, **26** (5), 1165–1176.

King, M., Kaufman, Y., Menzel, W., and Tanré, D. (1992) Remote sensing of cloud, aerosol, and water vapor properties from the Moderate Resolution Imaging Spectrometer (MODIS). *IEEE Trans. Geosci. Remote Sens.*, **30**, 2–27.

Kokhanovsky, A. (2006) *Light Scattering Review*, Praxis Publishing Ltd, Chichester, UK.

Kourganoff, V. (1963) *Basic Methods in Transfer Problems*, Dover.

Kratz, D. (1995) The correlated *k*-distribution technique as applied to the AVHRR channels. *J. Quant. Spectrosc. Radiat. Transf.*, **53** (5), 501–517.

Kratz, D. and Rose, F. (1999) Accounting for molecular absorption within the spectral range of the CERES window channel. *J. Quant. Spectrosc. Radiat. Transf.*, **61** (1), 83–95.

Lacis, A. and Oinas, V. (1991) A description of the correlated *k*-distribution method for modeling nongray gaseous absorption, thermal emission, and multiple-scattering in vertically inhomogeneous atmospheres. *J. Geophys. Res.*, **96** (D5), 9027–9063.

Lacis, A., Wang, W., and Hansen, J. (1979) Correlated *k*-distribution method for radiative transfer in climate models: application to effects of cirrus clouds on climate. *NASA Conf. Publ.*, **2076**, 309–314.

Lebedinsky, A. (1939) Radiative equilibrium in the Earth's atmosphere. *Proc. Leningr. Univ. Ser. Math.*, **3**, 152–175.

Lenoble, J.E. (1985) *Radiative Transfer in Scattering and Absorbing Atmospheres: Standard Computational Procedures*, A. Deepak Publishing, Hampton, Virginia, USA.

Levoni, C., Cervino, M., Guzzi, R., and Torricella, F. (1997) Atmospheric aerosol optical properties: a database of radiative characteristics for different components and classes. *Appl. Opt.*, **36** (30), 8031–8041.

Li, J. and Barker, H. (2005) A radiation algorithm with correlated *k*-distribution. Part I: local thermal equilibrium. *J. Atmos. Sci.*, **62**, 286–309.

Liou, K.N. (1986) Influence of cirrus clouds on weather and climate processes: a global perspective. *Month. Weather Rev.*, **114**, 1167–1199.

Liou, K.N. (1973) A numerical experiment on Chandrasekhar's discrete-ordinate method for radiative transfer: applications to cloud and hazy atmospheres. *J. Atmos. Sci.*, **30**, 1303–1326.

Liou, K.N. (2002) *An Introduction to Atmospheric Radiation*, International Geophysics Series, vol. 84, Academic Press, New York, Oxford.

Liu, F., Stagg, B., and Snelling, D. (2006) Effects of primary soot particle size distribution on the temperature of soot particles heated by a nanosecond pulsed laser in an atmospheric laminar diffusion flame. *Int. J. Heat Mass Transf.*, **49**, 777–788.

Logan, N. (1965) Survey of some early studies of the scattering of plane waves by a sphere. *Proc. Inst. Electr. Electron, Eng. (IEEE)*, **53** (8), 773–785.

Lommel, E. (1887) Die Photometrie der diffusen Zurückwerfung. *Sitzungsber. Akad. Wiss. München*, **17**, 95–124.

Lorenz, L. (1890) Lysbevegelser i og uden for en af plane Lysbolger belyst Kugle. Det kongelig danske Videnskabernes Selskabs Skrifter. *Naturvidenskab. Math. Afd.*, **VI** (1), 2–62.

Love, A. (1899) Scattering of electric waves by a sphere. *Proc. Lond. Math. Soc.*, **30**, 308–321.

Lucarini, V., Saarinen, J., Peiponen, K.E., and Vartiainen, E. (2005) *Kramers–Kronig Relations in Optical Materials Research*, Springer, Berlin, Germany.

Malkmus, W. (1967) Random Lorentz band model with exponential-tailed S^{-1} line-intensity distribution functions. *J. Opt. Soc. Am.*, **57**, 323–329.

Marshak, A. and Davis, A. (eds) (2005) *3D Radiative Transfer in Cloudy Atmospheres*, Springer, Berlin, Heidelberg, New York.

Meador, W. and Weaver, W. (1980) Two-stream approximations to radiative transfer in planetary atmospheres: a unified description of existing methods and a new improvement. *J. Atmos. Sci.*, **37**, 630–643.

Mie, G. (1908) Beiträge zur Optik trüber Medien, speziell kolloidaler Metallösungen. *Ann. Phys., Vierte Folge*, **25** (3), 377–445.

Mikulski, J. and Murphy, E. (1963) The computation of electromagnetic scattering from concentric spherical structures. *IEEE Trans. Geosci. Remote Sens.*, **March**, 169–177.

Miles, N., Verlinde, J., and Clothiaux, E. (2000) Cloud droplet size distributions in low-level stratiform clouds. *J. Atmos. Sci.*, **57**, 295–311.

Minnaert, M. (1993) *Light and Color in the Outdoors*, Springer Science+Business Media, New York.

Mishchenko, M. (2006) Maxwell's equations, radiative transfer, and coherent backscattering: a general perspective. *J. Quant. Spectrosc. Radiat. Transf.*, **101**, 540–555.

Mishchenko, M. (2008) Multiple scattering, radiative transfer, and weak localization in discrete random media: unified microphysical approach. *Rev. Geophys.*, **46**, doi:10.1029/2007RG000 230.

Mishchenko, M. (2009) Electromagnetic scattering by non-spherical particles: a tutorial review. *J. Quant. Spectrosc. Radiat. Transf.*, **110** (11), 808–832.

Mishchenko, M., Hovenier, J.W., and Lacis, A. (2000) *Light Scattering by Non-spherical Particles: Theory, Measurements, and Applications*, Academic Press, New York, London, Sydney, Tokyo, San Francisco, San Diego.

Mishchenko, M., Rosenbush, V., Kiselev, N., Lupishko, D., Tishkovets, V., Kaydash, V., Belskaya, I., Etimov, Y., and Shakovskoy, N. (2010) *Polarimetric Remote Sensing of Solar System Objects*, Academperiodyka, Kiev.

Mishchenko, M. and Travis, L. (1994a) Light scattering by polydispersions of randomly oriented spheroids with sizes comparable to wavelengths of observations. *Appl. Opt.*, **33**, 7206–7225.

Mishchenko, M. and Travis, L. (1994b) T-matrix computations of light scattering by large spheroidal particles. *Opt. Commun.*, **109**, 16–21.

Mishchenko, M., Travis, L., and Lacis, A. (eds) (2002) *Scattering, Absorption, and Emission of Light by Small Particles*, Cambridge University Press, Cambridge.

Mishchenko, M., Travis, L., and Lacis, A. (2006) *Multiple Scattering of Light by Particles*, Cambridge University Press, Cambridge, New York.

Moncet, J.L., Uymin, G., Lipton, A., and Snell, H. (2008) Infrared radiance modeling by optimal spectral sampling. *J. Atmos. Sci.*, **65**, 3917–3934.

Morse, P. and Feshbach, H. (1953) *Methods of Theoretical Physics, Part I*. McGraw-Hill.

Muinonen, K. (1989) Scattering of light by crystals: a modified Kirchhoff approximation. *Appl. Opt.*, **28**, 3044–3050.

Nagel, M., Quenzel, H., Kweta, W., and Wendling, R. (1978) *Daylight Illumination–Color–Contrast Tables for Full-Form Objects*, Academic Press, New York, San Francisco, London.

Nicolet, M. (1984) On the molecular scattering in the terrestrial atmosphere: an empirical formula for its calculation in the homosphere. *Planet. Space Sci.*, **32**, 1467–1468.

Nuland, S.B. (2000) *Leonardo da Vinci*, Penguin Group, New York.

Nussenzveig, H. and Wiscombe, W. (1980) Efficiency factors in Mie scattering. *Phys. Rev. Lett.*, **18**, 1490–1494.

Nussenzveig, H. and Wiscombe, W. (1991) Complex angular momentum approximation to hard-core scattering. *Phys. Rev. A*, **43**, 2093–2112.

Oguchi, T. (1973) Scattering properties of oblate raindrops and cross polarization of radio waves due to rain: Calculations at 19.3 and 34.8 GHz. *J. Radio Res. Lab. Jpn.*, **20**, 79–118.

Penner, S. (1959) *Quantitative Molecular Spectroscopy and Gas Emissivities*, Addison-Wesley, Reading, MA.

Petty, G. (2006) *A First Course in Atmospheric Radiation*, 2nd edn, Sundog Publishing, Madison, Wisconsin.

Plass, G. and Kattawar, G. (1968) Monte Carlo calculations of light scattering from clouds. *Appl. Opt.*, **7**, 415–419.

Platnick, S., King, M., Ackerman, S., Menzel, W., Baum, B., Riedi, J., and Frey, R. (2003) The MODIS cloud products: Algorithms and examples from TERRA. *IEEE Trans. Geosci. Remote Sens.*, **41**, 459–473.

Pomraning, G. (1973) *The Equations of Radiation Hydrodynamics*, Pergamon Press, Oxford, U.K.

Preisendorfer, R. (1965) *Radiative Transfer on Discrete Spaces*, Pergamon Press, Oxford, London, Edinburgh, New York, Paris, Frankfurt.

Press, W.H., Teukolsky, S.A., Vetterling, W., and Flannery, B. (1992) *Numerical Recipes in C*, 2nd edn, Cambridge University Press, New York.

Purcell, E. and Pennypacker, C. (1973) Scattering and absorption of light by non-spherical dielectric grains. *Astrophys. J.*, **186**, 705–714.

Randall, H., Dennison, D., Ginsburg, N., and Weber, L. (1937) The far infrared spectrum of water vapor. *Phys. Rev.*, **52**, 160–174.

Rothman, L., Gamache, R., Tipping, R., Rinsland, C., Smith, M., Benner, D., Devi, V., Flaud, J.M., Camy-Peyret, C., Perrin, A., Goldman, A., Massie, S., Brown, L., and Toth, R. (1992) The HITRAN molecular database: editions of 1991 and 1992. *J. Quant. Spectrosc. Radiat. Transf.*, **48** (5/6), 469–507.

Rybicki, G. (1996) *Radiative Transfer in "From White Dwarfs to Black Holes – the Legacy of S. Chandrasekhar"* (ed. G. Srinivasan), The University of Chicago Press, Chicago, IL.

Sassen, K. and Wang, Z. (2008) Classifying clouds around the globe with the CloudSat radar: 1-year of results. *Geophys. Res. Lett.*, **25**, L04805.

Schuster, A. (1905) Radiation through a foggy atmosphere. *Astrophys. J.*, **21** (1), 1–22.

Segelstein, D. (1981) *The Complex Refractive Index of Water*, Ph.D. thesis, University of Missouri, Kansas City.

Shi, G., Xu, N., and Wang, B. (2009) An improved treatment of overlapping absorption bands based on the correlated *k*-distribution model for thermal infrared radiative transfer calculations. *J. Quant. Spectrosc. Radiat. Transf.*, **110**, 435–451.

Simmonds, J. (2000) *A Brief on Tensor Analysis*, 2nd edn, Springer-Verlag, New York.

Sobolev, V. (1975) *Light Scattering in Planetary Atmospheres*, Pergamon Press, Oxford, New York, Toronto, Sydney, Braunschweig.

Sorensen, C. and Roberts, G. (1997) The prefactor of fractal aggregates. *J. Colloid Interf. Sci.*, **186**, 447–452.

Stamnes, K., Tsay, S., Wiscombe, W., and Jayaweera, K. (1988) A numerically stable algorithm for discrete-ordinate-method radiative transfer in multiple-scattering and emitting layered media. *Appl. Opt.*, **27** (12), 2502–2509.

Stephens, G. (1994) *Remote Sensing of the Lower Atmosphere*, Oxford University Press, New York.

Stephens, G., Vane, D., Boain, R., Mace, G., Sassen, K., Wang, Z., Illingworth, A., O'Connor, E., Rossow, W., Durden, S., Miller, S., Austin, R., Benedetti, A., Mitrescu, C., and the CloudSat Science Team (2002) The cloudsat mission and the a-train. *Bull. Amer. Meteorol. Soc.*, **83**, 1771–1790.

Stokes, G. (1862) On the intensity of the light reflected from or transmitted through a pile of plates. *Proc. R. Soc. Lond.*, **11**, 545–556.

Sykes, J. (1951) Approximate integration of the equation of transfer. *Month. Not. R. Astronom. Soc.*, **111**, 377–386.

Takano, Y. and Liou, K.N. (1989) Solar radiative transfer in cirrus clouds. Part I: Single-scattering and optical properties of hexagonal ice crystals. *J. Atmos. Sci.*, **46**, 1–19.

Thomas, G. and Stamnes, K. (1999) *Radiative Transfer in the Atmosphere and Ocean*, Cambridge University Press.

Trenberth, K., Fasullo, J., and Kiehl, J. (2009) Earth's global energy budget. *Bull. Amer. Meteorol. Soc.*, **90**, 311–323, doi:10.1175/2008BAMS2634.1.

Tsang, L. and Kong, J. (2001) *Scattering of Electromagnetic Waves: Advanced Topics*, John Wiley & Sons, New York, USA.

van de Hulst, H. (1957) *Light Scattering by Small Particles*, Dover Publications, Inc., New York.

van de Hulst, H. (1980) *Radiation and Cloud Processes in the Atmosphere. Theory, Observation and Modeling, Multiple Light Scattering*, vol. 2, Academic Press, New York, London, Sydney, Toronto, San Francisco.

van de Hulst, H. (1981) *Light Scattering by Small Particles*, Dover Publications, Inc., 1981 edn, New York.

Voigt, S., Orphal, J., Bogumil, K., and Burrows, J. (2001) The temperature dependence (203–293 K) of the absorption cross sections of O_3 in the 230–850 nm region measured by Fourier-transform spectroscopy. *J. Photochem. Photobiol. A*, **143**, 1–9.

Volten, H., Munoz, O., Rol, E., de Haan, J., Vassen, W., Hovenier, J., Muinonen, K., and Nousiainen, T. (2001) Scattering matrices of mineral aerosol particles at 441.6 nm and 632.8 nm. *J. Geophys. Res.*, **106**, 17375–17401.

Wait, J. (1955) Scattering of a plane wave from a circular dielectric cylinder at oblique incidence. *Can. J. Phys.*, **33**, 189–195.

Wait, J. (1963) Electromagnetic scattering from a radially inhomogeneous sphere. *Appl. Sci. Res.*, **10**, 441–450.

Wali, K.S. (1997) *S. Chandrasekhar – the Man Behind the Legend*, Imperial College Press, London.

Warren, S. and Brandt, R. (2008) Optical constants of ice from the ultraviolet to the microwave: a revised compilation. *J. Geophys. Res.*, **113** (D14220), doi:10.1029/2007JD009744.

Waterman, P. (1965) Matrix formulation of electromagnetic scattering. *Proc. Inst. Electr. Electron. Eng., (IEEE)*, **53**, 805–812.

Wendisch, M., Hellmuth, O., Ansmann, A., Heintzenberg, J., Engelmann, R., Althausen, D., Eichler, H., Müller, D., Hu, M., Zhang, Y., and Mao, J. (2008) Radiative and dynamic effects of absorbing aerosol particles over the Pearl River Delta, China. *Atmos. Environ.*, **42**, 6405–6416, doi:10.1016/j.atmosenv.2008.02.033.

Wendisch, M., Mertes, S., Heintzenberg, J., Schell, D., Wobrock, W., Frank, G., Martinsson, B., Fuzzi, S., Orsi, G., Kos, G., and Berner, A. (1998) Drop size distribution and LWC in Po valley fog. *Contrib. Atmos. Phys.*, **71** (4), 87–100.

Wendisch, M., Pilewskie, P., Pommier, J., Howard, S., Yang, P., Heymsfield, A., Schmitt, C., Baumgardner, D., and Mayer, B. (2005) Impact of cirrus crystal shape on solar spectral irradiance: a case study for subtropical cirrus. *J. Geophys. Res.*, **110**, doi:10.1029/2004JD005294.

Wendisch, M., Yang, P., and Pilewskie, P. (2007) Effects of ice crystal habit on thermal infrared radiative properties and forcing of cirrus. *J. Geophys. Res.*, **112** (D03202), doi:10.1029/2006JD007899.

Weng, F. (1992a) A multi-layer discrete-ordinate method for vector radiative transfer in a vertically-inhomogeneous, emitting and scattering atmosphere. Part I: Theory. *J. Quant. Spectrosc. Radiat. Transf.*, **47**, 19–33.

Weng, F. (1992b) A multi-layer discrete-ordinate method for vector radiative transfer in a vertically-inhomogeneous, emitting and scattering atmosphere. Part II: Application. *J. Quant. Spectrosc. Radiat. Transf.*, **47**, 35–42.

Wick, G. (1943) Über ebene Diffusionsprobleme. *Z. Phys.*, **121** (11–12), 702–718.

Wiscombe, W. (1977) The delta-M method: rapid yet accurate radiative flux calculations for strongly asymmetric phase functions. *J. Atmos. Sci.*, **34**, 1408–1421.

Wiscombe, W. (1980) Improved Mie scattering algorithms. *Appl. Opt.*, **19** (9), 1505–1509.

Wriedt, T. (2009) Light scattering theories and computer codes. *J. Quant. Spectrosc. Radiat. Transf.*, **110** (11), 833–843.

Wyatt, P. (1962) Scattering of electromagnetic plane waves from inhomogeneous spherically symmetric objects. *Phys. Rev.*, **127** (5), 1837–1943.

Yang, P. and Fu, Q. (2009) Dependence of ice crystal optical properties on particle aspect ratio. *J. Quant. Spectrosc. Radiat. Transf.*, **110**, 1604–1614.

Yang, P., Gao, B.C., Baum, B., Wiscombe, W., Hu, Y., Nasiri, S., Soulen, P., Heymsfield, A., McFarquhar, G., and Miloshevich, L. (2001) Sensitivity of cirrus bidirectional reflectance to vertical inhomogeneity of ice crystal habits and size distributions for two Moderate-Resolution Imaging Spectrometer (MODIS) bands. *J. Geophys. Res.*, **106** (D15), 17267–17291.

Yang, P., Gao, B.-C. Wiscombe, W., Mishchenko, M., Platnick, S., Huang, H.L., Baum, B., Hu, Y., Winker, D., Tsay, S.C., and Park, S. (2002) Inherent and apparent scattering properties of coated or uncoated spheres embedded in an absorbing medium. *Appl. Opt.*, **41**, 2740–2759.

Yang, P., Kattawar, G., Hong, G., Minnis, P., and Hu, Y. (2008) Uncertainties associated with the surface texture of ice particles in satellite-based retrieval of cirrus clouds: Part I. Single-scattering properties of ice crystals with surface roughness. *Geosci. Remote Sens.*, **46**, 1940–1947.

Yang, P. and Liou, K. (1995) Light scattering by hexagonal ice crystals: comparison of finite-difference time domain and geometric optics models. *J. Opt. Soc. Am. A*, **12** (1), 162–176.

Yang, P. and Liou, K. (1996) Geometric-optics-integral-equation method for light scattering by non-spherical ice crystals. *Appl. Opt.*, **35** (33), 6568–6582.

Yang, P. and Liou, K. (1997) Light scattering by hexagonal ice crystals: solutions by a ray-by-ray integration algorithm. *J. Opt. Soc. Am.*, **A14**, 2278–2289.

Yang, P. and Liou, K. (2006) *Light Scattering and Absorption by Non-spherical Ice Crystals. In: Light Scattering Reviews: Single and Multiple Scattering* (ed. A.A. Kokhanovsky), Springer, Berlin.

Yang, P. and Liou, K. (2009) An "exact" geometric-optics approach for computing the optical properties of large absorbing particles. *J. Quant. Spectrosc. Radiat. Transf.*, **110**, 1162–1177.

Yang, P., Liou, K., Wyser, K., and Mitchell, D. (2000) Parameterization of the scattering and absorption properties of individual ice crystals. *J. Geophys. Res.*, **105** (D4), 4699–4718.

Yang, P., Wendisch, M., L., B., Kattawar, G., Mishchenko, M., and Hue, Y. (2011) Dependence of extinction cross-section on incident polarization state and particle orientation. *J. Quant. Spectrosc. Radiat. Transf.*, **112**, 2035–2039.

Yee, S. (1966) Numerical solution of initial boundary value problems involving Maxwell's equations in isotropic media. *IEEE Trans. Antennas Propag.*, **14**, 302–307.

Yoshino, K., Esmond, J., Cheung, A.C., Freeman, D., and Parkinson, W. (1992) High resolution absorption cross sections in the transmission window region of the Schumann–Runge bands and Herzberg continuum of O_2. *Planet. Space Sci.*, **40**, 185–192.

Yoshino, K., Parkinson, W., Ito, K., and Matsui, T. (2005) Absolute absorption cross-section measurements of Schumann–Runge continuum of O_2 at 90 and 295 K. *J. Molec. Spectrosc.*, **229**, 238–243.

Zdunkowski, W., Trautmann, T., and Bott, A. (2007) *Radiation in the Atmosphere – A Course in Theoretical Meteorology*, Cambridge University Press, New York.

Index

Theory of Atmospheric Radiative Transfer, First Edition. Manfred Wendisch and Ping Yang
© 2012 WILEY-VCH Verlag GmbH & Co. KGaA. Published 2012 by WILEY-VCH Verlag GmbH & Co. KGaA.